W9-AOV-523

Managing the Environmental Crisis

Managing the Environmental Crisis

Incorporating Competing Values

in Natural Resource Administration

Second edition, revised and updated

William R. Mangun and Daniel H. Henning

With new foreword by Lynton K. Caldwell

Duke University Press   Durham and London 1999

© 1999 Duke University Press

All rights reserved

Printed in the United States of America on acid-free paper ⊗

Typeset in Stone Serif by Tseng Information Systems, Inc.

Library of Congress Cataloging-in-Publication Data

appear on the last printed page of this book.

To Jeanie, Kimberly, Shaun,
Emily, and Chris
and to Tundra and Forest
for their love and support

Contents

Foreword

Since 1989, when *Managing the Environmental Crisis* was first published, many of the trends then noted have continued, and new issues have emerged. In a changing world, environmental policy and management present a moving target for authors, and conclusions and generalizations invariably call for qualifications, necessitated by the complex, interactive, and cosmic scope of the environment.

Science now conceptualizes the environment as a multilinear, multi-loop feedback system. Unfortunately, the general public has been slow to comprehend this larger view. For many people "the environment" as yet does not extend very far beyond or above "My Back Yard." There is, more-over, a wide gap between the here-and-now economic and personal rights perspectives of many people and the slow-moving cumulative effects of environmental change.

Managing the Environmental Crisis is focused, as it should be, on specific political issues. These are the issues that general readers and students can most readily comprehend and with which law and management directly deal. But there are two larger trends implicit in the book, both of them relatively long-term developments. They are the globalization and politi-cal polarization of the environmental movement. At present they present paradoxes that ultimately must be resolved by synthesis. Balancing equi-ties may be appropriate for specific short-term policy conflicts but seldom offer viable solutions to major worldwide environmental problems. To understand the true meaning of *Managing the Environmental Crisis,* the reader needs to recognize that it is not the environment per se that is managed. Rather, it is human action relating to the environment that is managed through individual motivation and behavior and institutional governance, public and private.

Today there are serious and cumulating hazards to the integrity of the environment. The environmental crisis we face now is not in external nature, but within ourselves. It is a "crisis of will and rationality," as I declared twenty-eight years ago in a publication of the Scientists Institute for Public Information. Humanity is in a race between learning how to manage its affairs for a sustainable future and the destructive trends in the environment that modern civilization has set in motion.

The two trends, globalization and polarization, are parodoxical and thus complicate national policymaking. The globalization of international environmental policy is occurring contrary to traditional assumptions of national sovereignty. For example, science has found credible evidence of a thinning stratospheric ozone layer, which has global consequences, injurious to both plant and animal life (including human health). Evidence of global climate change induced by human activities, notably the discharge of "greenhouse gases" requires an all-nations adherence to restrictions on uses of energy (e.g., fossil fuels), which could severely disrupt present economies. The polarization of public policy has arisen with the realization that major environmental problems can be ameliorated or overcome only through a reconsideration of property rights in land and unconstrained economic growth. The globalization of trade and unprecedented increases in human population growth has exacerbated conflict in both trends, pitting traditional assumptions and contemporary economic advantage against the integrity and sustainability of the global environment.

With both of these trends, the prognosis seems to lead in contrary directions. International trade and investment have become major factors in world affairs, but there has also been growing concern in both developed and developing countries that an unbridled economism that disregards environmental and cultural values is dominating policy. In the United States, the polarization of environmental policy has tended to become a "liberal-conservative" partisan issue, although the opposing factions have their own differences. There is reason to surmise, however, that the weight of opinion will progressively shift toward environmental protection. The growth of volunteer environmental organizations—notably at the college level—and of environmental organizations and publications worldwide suggest that antienvironmentalism may be enjoying a "last hurrah," but the echoes may reverberate for some time to come.

This foreword sets out the context in which the significance of this book might be best understood. Repeating what I said in my foreword to the first edition, I affirm that this book is admirably suited as a basic introductory text on environmental policy and administration. It is solidly documented and written for easy comprehension, and continues to be a welcome addition to the literature of public affairs.

Lynton K. Caldwell
Bentley Professor of Political Science Emeritus and
Professor Emeritus of Public and Environmental Affairs,
Indiana University

Preface to the Second Edition

If one were to have taken a snapshot of environmental conditions ten years ago and then taken another one more recently, several contrasts would be discerned. Bright images of success would appear in some areas of environmental administration; less pleasant scenes would appear in other areas. Such images could be seen both around the world and regionally in the United States. But what may be more important is something less visible, just below the horizon, that may signal much greater problems to come. All over the world, for example, anomalies that are not easily explained are appearing in animals such as frogs and birds. Environmental administrators must investigate such environmental problems and then develop appropriate management strategies to address the most pressing issues.

In the area of natural resource management a major paradigm change has occurred over the past ten years. There is now a greater appreciation on the part of public land managers of the need to manage holistically: to consider entire ecosystems when initiating management actions, and to plan accordingly. The need for such an approach was one of the themes of the first edition of this book.

In 1993 the Clinton administration convened an interagency task force on ecosystem management as part of its effort to reinvent environmental management. The administration has adopted many of the recommendations of the task force, including a requirement that all federal agencies with environmental or natural resource responsibilities adopt an ecosystems management perspective. There is an emphasis on a cooperative approach that involves many different stakeholders, both public agencies and private citizens. This is important because environmental problems do not respect political jurisdictions. The ecosystems management approach is also multidisciplinary in nature, incorporating biological, social, and economic principles in recognition of the fact that many environmental problems are the consequence of decisions made for other social and economic purposes.

The United States now has a more ecologically oriented, better-educated society that expects a quality environment. However, that society still may not be willing to make the sacrifices necessary to address effectively the multiple, highly complex environmental problems that have evolved along with it. For example, the nation has been somewhat successful in addressing conventional air and water pollutants, but more complex problems remain to be addressed. Toxic substances found in the atmospheric and water resources of the Earth in increasing amounts continue to confound control efforts.

Pressure group influence has changed somewhat in approach and perspective over the past ten or fifteen years. In the 1980s the antienvironmental pressure groups were symbolized by the so-called Sagebrush Rebellion in the western states. Participants in the rebellion pressed Congress to transfer public lands in the West to state and local governments as well as to individuals. In response, the federal government reduced its land holdings from over 700 million acres to about 650 million acres. Today there is the "wise use" movement that presses for greater access to public lands and employs misinformation campaigns to achieve its ends. Industrial organizations and state and local governments expressed major concerns over burdensome regulations and demanded administrative relief. Corporations had obtained some regulatory relief from the Reagan and Bush administrations, though it came in the form of regulatory delays, regulatory moratoria, and lack of enforcement. They gained a more constructive kind of relief through a significant streamlining of the regulatory process effected by the Clinton administration, in its efforts to reinvent environmental management. Beyond increased flexibility in meeting regulations, two new programs, the Common Sense Initiative and Project XL, gave industries opportunities to develop innovative approaches to meet environmental regulations more effectively and at less cost.

State and local governments have obtained regulatory as well as financial relief from Congress in the form of the Unfunded Mandates Reform Act of 1995 and additional funding through legislative initiatives. The 1996 Safe Drinking Water Act, for example, established a state revolving fund to aid state and local government compliance with the new requirements of the act. The Clinton administration also promoted additional cooperation with state and local governments by providing greater regulatory flexibility through the Brownfields and Community-Based Environmental Protection programs.

In the 1990s a "brownlash" developed, as more and more individuals began to question the degree of severity of environmental threats. Respected scientists like Dixy Lee Ray, an ecologist and former governor of Washington, wrote books and articles that confused and misled the public through the partial truths that they exposed. The danger in such misinformation campaigns, as Paul and Anne Ehrlich correctly observe, is that public support for the development of strong environmental protection initiatives will weaken and gains be lost while other environmental problems deteriorate. And in fact, public support for environmental protection programs has fallen. Roper Organization polls indicate that support for the environment peaked in 1991 and has gradually declined since. Membership in environmental and conservation organizations has declined somewhat as well.

Some of the decline in public opinion and group membership may be attributable to successes achieved in environmental protection in the United States. For example, the Environmental Protection Agency (EPA) has reported significant declines in the amount of air pollution for five of the six criteria air pollutants over the past twenty-five years. Some highly visible endangered species have been delisted in recent years, such as the bald eagle and the peregrine falcon. Some substantial improvements in water quality have also occurred, most noticeably with point sources such as industrial and municipal discharges.

The world's energy outlook is a little brighter than it was ten years ago, for several reasons. Energy conservation measures, along with a downturn in the economy, have moderated energy demand. Exploration for coal, natural gas, and petroleum has identified new energy sources that have expanded the world's known energy reserves. Oil, for example, is now projected to be economically viable for more than one hundred years at present consumption rates. Additional energy is being generated from renewable energy sources, such as wind turbines and biogas. And research and development into applied uses of solar and hydrogen-based energy cells has led to greater use of solar and hydrogen-based energy. However, distribution of energy sources in an economically viable manner to less developed countries remains a major problem.

In 1992 Congress provided the basis for development of the first comprehensive national energy plan for the United States with the passage of the National Energy Policy Act. This bipartisan effort required the Department of Energy to develop what is now known as the Comprehensive National Energy Strategy. The strategy has three focal points: to protect national security; to conserve energy, and to promote research that will facilitate the provision of future energy supplies. The Clinton administration has also vigorously promoted greater energy efficiency through its purchasing powers and its Green Lights program.

On the international level there has been a veritable explosion in environmental awareness and cooperation. Two-thirds of all international environmental treaties have been signed in the past twenty-five years. The countries of the world have also entered into hundreds of agreements, declarations, action plans, and treaties concerning environmental issues. Yet worldwide there are increasing water shortages, population pressures continue to mount, increasing numbers of environmental refugees push into cities and across national borders, tropical forests are disappearing, and greenhouse gases threaten to warm the entire planet beyond a tolerable limit.

Yes, there have been many successes in environmental administration over the past ten years. But the real management challenges and opportunities may be just emerging. These challenges will require the efforts of the greatest minds as well as the individual actions of billions of people.

Preface to the First Edition

This book emphasizes comprehensive, conceptual, interdisciplinary approaches to environmental administration. Both national and global issues are examined and characterized. Environmental problems encountered in, between, and among industrialized and less developed countries are explored. Natural resource management is considered in the context of environmental administration. Since the values infused in environmental politics ultimately determine human choices and their impacts on the biosphere, deliberate emphasis is given to values as they relate to environmental policy. The decisionmaking processes in environmental administration generally are rife with value conflicts. Values in competition with each other—in human and organizational struggles to arrive at appropriate decisions—determine the kinds and degrees of environmental action that will be undertaken.

Efforts either to protect or exploit the environment are guided by values. Therefore, in order to characterize the policymaking and administrative processes of government, the dynamics of human values and behaviors must be examined. Analysis of values is particularly needed in situations in which environmental considerations come into contact and conflict with nonenvironmental considerations. Values, articulated through human and institutional interactions, provide the basis for determining the role and scope of government in environmental affairs.

In environmental decisionmaking activities, values serve as a frame of reference, a "philosophy," a general guide to behavior, and a vision of "reality." Having emotive as well as symbolic content, values are individual and collective conceptions of what is good or bad, right or wrong for persons and institutions. Environmental administration, which entails the making of value judgments, is the overall process of directing and managing public policies, programs, and activities in environmental affairs in order to promote "the environmental public interest and the public good." Environmental administration should be comprehensive, interdisciplinary, and holistic in its outlook toward the management of complex interrelationships among persons and organizations. Contrary to the popular view that environmental administrators manage the environment, it should be recognized that environmental administrators manage the relationships between people and organizations on the one hand, and the environment on the other.

In the rush of twentieth-century economic and technological development, environmental considerations frequently are ignored, overlooked, or undervalued. This has resulted in the emergence of serious threats not only to the quality of life everywhere on earth, but also to the survival of the biosphere. As understanding of these threats has grown, new laws, policies, programs, and movements identified with the protection of the environment now are in effect in the United States, other industrialized countries, and the less developed countries. There are also international environmental entities with limited powers. In spite of the existence of these indicators of enhanced environmental awareness, many governments have encountered severe resistance to efforts to implement needed environmental policies and programs. Those engaged in efforts at environmental control have faced political pressure both to limit pollution control and to encourage development "at all costs."

The actual incorporation of environmental values in environmental administration depends on the value perceptions and priorities of a given society and its leadership. The incorporation of environmental values into decisionmaking processes often requires changes in the overall values of a given society. Unfortunately, some of these alterations in human attitude come about through the experience of environmental crises or through pressures exerted on decisionmakers by a public intensely, but superficially, aware of potential environmental degradation. More profound value alterations probably require long-term education and training with emphasis given to interdisciplinary, holistic orientations to environmental administration. Such training should lead to better solutions to environmental problems by addressing the complex interactive origins of these problems.

Although some positive changes in human and societal attitudes toward the environment have occurred, the overall tendency has been to resist those changes most needed for actually incorporating environmental values into environmental administration. Patterns of resistance as well as patterns of positive change and efforts at innovation are explored in this book. One reason for the emphasis on national governmental activities of the United States has been to provide information concerning American environmental issues, efforts, and innovations that would be valuable for other countries and international organizations as well as for the United States. Environmental policy and administration cannot be practiced in a vacuum. Global perspectives must inform individual environmental decisionmaking.

The United States is the greatest exploiter of domestic and imported natural resources in the entire world. It is also one of the greatest exporters of pollutants, pesticides, and hazardous materials. These national habits have tremendous impacts upon the global environment. Still, because of its environmental administrative experience, the United States can offer hope, assistance, and even innovative programs for the protection of the global environment, particularly in the less developed countries. All countries of the world now suffer serious environmental problems. Failure to incorporate environmental values into decisionmaking processes has resulted in adverse consequences to the global environment.

This book is organized to cover the broad spectrum of environmental and natural resource administration. It discusses all major areas of public administration and public policy in environmental affairs. Policy and administration are considered to be interrelated and inseparable. Global aspects of environmental decisionmaking are brought to focus in a special chapter on international environmental policy. A glossary of terms utilized in environmental affairs is included.

This book is intended to appeal to persons having a wide variety of interests in environmental affairs. Individuals from the United States, other industrialized countries, less developed countries, as well as international organizations should find the approaches to decisionmaking useful. The interdisciplinary orientation of the book should make it a valuable textbook for students and professors from several academic and professional fields, including public administration, political science, environmental studies, and natural resource areas; its comprehensive nature should make it a useful reference or training book for government personnel directly involved in environmental administration; it should also be valuable for corporate governmental personnel associated with economic development; and its orientation to values should interest the general reader concerned with protecting the biosphere.

Acknowledgments

Because research for and writing of this edition of the book extended over several years, it is not really possible to individually acknowledge the numerous people who contributed to its completion. Information, perspectives, ideas, and critiques came from a variety of individuals and organizations. Students from our environmental and public administration classes over the years presented us with excellent data and suggestions. Government officials, professors, politicians, environmentalists, and other interested persons provided us with ideas, comments, and references that greatly contributed to our work. UNESCO permitted use of an adaptation of an environmental affairs glossary created for environmental education by Daniel H. Henning.

We consider ourselves exceedingly fortunate to have had Lynton K. Caldwell, a pioneering scholar in environmental policy, as the writer of the foreword and as a major reviewer for this book. He served as advisor to the U.S. Senate Committee on Interior and Insular Affairs in drafting the National Environmental Policy Act. The efforts of Kelly Lenora Rudd, Mary Harris, and Ted Barr are gratefully acknowledged, for they made the second edition possible through their constant attention to detail in typing, editing, and formatting through its multiple revisions. Appreciation is also extended to the East Carolina University Department of Political Science for the support provided during the process of completing this edition. A special acknowledgment is given to Pat Guyette, of the East Carolina University Joyner Library staff, for her remarkable efforts in obtaining many distant and sometimes obscure research documents. The reviews of draft materials and additional insights provided by Jean Mangun, Scott Frisch, Carmine Scavo, and Richard Kearney are most appreciated. Jean Mangun's insightful comments as well as the new material she provided for the natural resources and outdoor recreation sections have contributed immeasurably to the book. Finally, we wish to express a heart-felt thank you to Valerie Milholland, Duke University Press editor, for her kind and persistent efforts to bring the second edition into actuality.

Abbreviations

APA	Administrative Procedure Act
ASCS	Agricultural Stabilization and Conservation Service
AUM	Animal unit per month
BLM	United States Bureau of Land Management
BOR	United States Bureau of Outdoor Recreation
CANDU	Canadian Deuterium Uranium
CAA	Clean Air Act
CBEP	Community-based environmental protection
CEQ	United States Council on Environmental Quality
CERCLA	Comprehensive Emergency Response, Compensation, and Liability Act (Superfund)
CFC	Chloroflurocarbon
CFSA	United States Consolidated Farm Service Agency
CITES	Convention on International Trade in Endangered Species
COG	Councils of government
COP	Conference of Parties
CRP	Conservation Reserve Program
CRS	Congressional Research Service
CWA	Clean Water Act
CZMA	Coastal Zone Management Act
DDT	Dichloro-diphenyl-trichloro-ethane
DOE	United States Department of Energy
EA	Environmental assessment
EEZ	Exclusive economic zone
EIA	Environmental impact assessment
EIS	Environmental impact statement
EO	Executive order
EPA	United States Environmental Protection Agency

EPACT	Energy Policy Act
EQIP	Environmental Quality Incentive Program
EREN	Energy and Renewable Energy Network
ESA	Endangered Species Act
ESP	Experimental stewardship program
FAA	United States Federal Aviation Administration
FACT	Food, Agriculture, Conservation, and Trade Act
FAIR	Federal Agricultural Improvement and Reform Act
FAO	Food and Agriculture Organization
FCCC	Framework Convention on Climate Change
FFDCA	Federal Food, Drug, and Cosmetic Act
FIFRA	Federal Insecticide, Fungicide, and Rodenticide Act
FIP	Forestry Incentives Program
FLPMA	Federal Land Policy and Management Act
FONSI	Finding of no significant impact
FSA	Food Security Act
FWCA	Fish and Wildlife Conservation Act
FWPCA	Federal Water Pollution Control Act
FWS	United States Fish and Wildlife Service
GAO	United States General Accounting Office
GEF	Global Environment Facility
GNP	Gross National Product
GPO	United States Government Printing Office
GPRA	Government Performance Results Act
GSX	Name of a hazardous waste management company in South Carolina
HCRS	Heritage Conservation and Recreation Service
HEW	United States Department of Health, Education, and Welfare
HSWA	Hazardous and Solid Waste Amendments
IFMRC	Interagency Floodplain Management and Review Committee
IIASA	International Institute for Applied Systems Analysis
IJC	International Joint Commission
IPCC	Intergovernmental Panel on Climate Change
IPM	Integrated Pest Management
IRPTC	International Register of Potentially Toxic Chemicals
IRS	United States Internal Revenue Service
IUCN	International Union for the Conservation of Nature and Natural Resources (World Conservation Union)
LDC	Less developed country
LD50	Lethal dose to 50 percent of population
LRTAP	Convention on Long-Range Transboundary Air Pollution

LULU	Locally unwanted land use
LWCF	Land and Water Conservation Fund
MBO	Management by objectives
MMPA	Marine Mammal Protection Act
MMS	United States Minerals Management Service
NAAQS	National Ambient Air Quality Standards
NAFTA	North American Free Trade Agreement
NASIS	National Soil Information System
NEPA	National Environmental Policy Act
NESHAPS	National Emission Standards for Hazardous Air Pollutants
NFMA	National Forest Management Act
NIMBY	Not in my backyard
NIPF	Nonindustrial private forest
NORC	National Opinion Research Center
NOZE	National Ozone Expedition
NPCA	National Parks and Conservation Association
NPDES	National Pollutant Discharge Elimination System
NPL	National Priorities List
NRDC	National Resources Defense Council
NRI	United States Department of Agriculture Natural Resources Inventory
NWF	National Wildlife Federation
NRCS	Natural Resources Conservation Service
NSDAF	National Soils Data Access Facility
NWPA	Nuclear Waste Policy Act
NWRIA	National Wildlife Refuge Improvement Act
OCS	Outer continental shelf
OCSLA	Outer Continental Shelf Lands Act
OECD	Organization for Economic Cooperation and Development
OMB	United States Office of Management and Budget
OPEC	Oil-producing and -exporting Countries
ORRRC	Outdoor Recreation Resources Review Commission
ORV	Off-road vehicles
OSM	United States Office of Surface Mining
PBB	Polybrominated biphenyls
PCAO	President's Commission on Americans Outdoors
PCB	Polychlorinated biphenyls
PIK	Payment in kind
PLLRC	Public Land Law Review Commission
POP	Persistent organic pollutant
PRIA	Public Rangelands Improvements Act

RARE	Roadless Area Review and Evaluation
RCRA	Resource Conservation and Recovery Act
RFF	Resources for the Future
RPA	Resources Planning Act
RRA	Regulatory Reform Act
SAF	Society of American Foresters
SARA	Superfund Amendments and Reauthorization Act
SBREFA	Small Business Regulatory Enforcement Fairness Act
SCS	United States Soil Conservation Service
SDWA	Safe Drinking Water Act
SIP	State implementation plan
SMCRA	Surface Mining Control and Reclamation Act
SWDA	Solid Waste Disposal Act
TCP	Transportation control plans
TVA	Tennessee Valley Authority
UMRA	Unfunded Mandate Reform Act
UNCED	United Nations Conference on Environment and Development
UNEP	United Nations Environment Program
UNESCO	United Nations Education, Social, and Cultural Organization
USDA	United States Department of Agriculture
USGS	United States Geological Survey
VERP	Visitor Experience and Resource Protection
WHO	World Health Organization
WMO	World Meteorological Organization
WRC	Water Resources Council
WRP	Wetland Reserve Program
WWF	World Wide Fund for Nature (formerly World Wildlife Fund)

1 Forces Shaping Environmental and Natural Resource Administration

Since the first edition of this book, several major changes have occurred in environmental administration. First, there is now a greater emphasis on the interconnectedness of elements within ecosystems. Management of environmental problems is taking on an ecosystem perspective in the United States. The leading land management agencies now use an ecosystem management approach that incorporates ecological considerations while acknowledging human needs.[1] Since ecosystems do not recognize political boundaries, management strategies must be employed that involve cooperation across jurisdictional lines, as well as between the private and public sector. For example, a watershed management approach has been adopted by federal agencies for water quality management that emphasizes cooperation across jurisdictions.

A second change is a greater emphasis on economic incentives and flexibility in pollution control. The high cost and inflexibility of the traditional "command and control" regulatory approach have yielded to a growth in regulatory incentives and innovative management approaches that increase flexibility in implementation. For example, trading programs have been established in air pollution control for sulfur dioxide emissions[2] and in water pollution control for nutrient loads.[3] The U.S. Environmental Protection Agency has reinvented environmental management with a greater emphasis on flexibility and cooperation with state and local governments through its Common Sense Initiative, Project XL, Brownfields, and Community-Based Environmental Protection (CBEP) initiatives.[4]

A third major change appears to be a devolution of responsibility from the federal government to the states.[5] In some areas of environmental protection, the states have been more innovative than the federal government. In other areas, the states claim that the inflexibility of federal regulations impedes their creativity. Collectively, the states have assumed greater responsibility for environmental protection and clearly dominate in the areas of waste management, groundwater protection, and coastal zone management.[6] The states acknowledge the federal government's role in establishment of national standards for pollution control but demand greater flexibility in their implementation.

Environmental and Natural Resource Administration

Environmental administration constitutes all activities taken by government (local, national, or international) to protect or enhance the quality of the environment in the public interest. It involves numerous governmental departments, agencies, and independent units with environmental and natural resource programs and concerns at virtually every level of government. In the United States, numerous federal departments and agencies are involved in the implementation of environmental policies. Some are charged with the administration of more than 650 million acres of public land, which is about one-third of the nation's total geographical area.[7] A totally comprehensive public policy for environmental administration does not exist in the United States nor any other country.

The administration of environmental programs, just like any other public program, is a human process where individuals and groups interact and work toward achieving certain collective and organizational objectives or values. Politics and public administration are intertwined in the struggle for power to affect governmental policies. Every policy decision in the public sector is a product of individual or group judgment through the political process. Regardless of their areas of responsibility or clientele, public organizations and personnel must operate in a highly political environment influenced by regionally or nationally dominant cultures.

The public administrator must pay particular attention to those pressures that he or she can best control while minimizing the negative effects of those that cannot be controlled or moderated significantly. The purpose of this book is not to prescribe any specific environmental administration theory. The position taken by Peter Savage—specifically with regard to comparative administration and public administration—in general remains true today: "[P]aradigmatic orthodoxy of any kind is unlikely to take root short of force."[8] Nevertheless, this book provides a guide for natural resource decisionmakers. The theme of the book is that environmental and natural resource managers should act continually on behalf of the environmental public interest. They, more than any other public administrators, must guard against public decisions that support short-term economic gain at the cost of substantial and potentially irreversible environmental losses in the long run. Throughout the book, the role of values in decisionmaking is emphasized. Values enter into all decisions, and the quality of the environment will be determined extensively by the degree to which environmental values are taken into consideration in decisions made throughout the public sector. Among the most important influences affecting environmental and natural resource administrators are (1) ecological complexities, (2) a crisis orientation, (3) the environmental public interest, (4) concern for future generations, (5) the intangible nature of many environmental values, and (6) the values and value judgments of agency personnel.

These considerations influence the feasible decision space of the public administrator. The world of the public official is not the neat, clearly defined world that many people perceive; rather, it lacks any specific shape and tends to be amorphous. The ill-defined world of the public administrator can be modified by particular forces, some of which he or she may be able to influence. Therefore, some influences may have greater or lesser impact than others, depending on the time and place of interaction with the administrator as well as his or her own administrative capabilities.

Ecological complexities

Ecology is the study of the relationships among all living things and the physical environment. The ecological concept of interrelations provides a central theme for environmental administration. Management of a specific natural resource at a given time and place involves a complex of interrelations that has the capacity to influence the environment in the present or the future. For example, the management of water resources cannot occur without affecting other natural resources. Since the environment includes all living and nonliving features, other environmental sciences like hydrology, phycology, geology, and limnology play key roles in environmental administration.

Many resource agencies assert that they use ecological approaches in management activities.[9] In reality, this orientation may be confined to superficial treatment and short-range planning. When serious decisions are to be made through political and administrative processes, traditional values of agencies and their influential clientele groups usually become dominant. This means that the student of ecology must also be aware of how complex social and political values are incorporated into actual decisionmaking.

Crisis orientation

Many leaders express deep concern over potentially irreversible forms of environmental degradation that could engender additional negative consequences. Although this attention has not been maintained at a continually high level, especially during the Reagan, Bush, and Clinton administrations (the last of these has been constrained by forces in the Republican majority in Congress), concern over environmental affairs is still evident. Urgent and powerful pressures for environmental change have been and will continue to be exerted in political and administrative arenas. Since the late 1960s, attention has been directed, however, mainly toward environmental crises. Environmental administration, especially, has a crisis orientation in crucial situations concerning highly sensitive areas or species. Alarmed by this crisis orientation, scientists at the well-respected environmental policy research center Resources for the Future observed:

> The crisis syndrome performs a profound disservice to the nation. . . . The problems mostly are long term and require a sustained effort even to understand what is involved, let alone to forge solutions. A roller coaster of public interest (and government and foundation support) does not provide a context congenial to the kind of long-range research, analysis, and social institution building required. Crises occur, to be sure, but they usually represent a trend or deep-seated problem, and the prominence of the one must not obscure the greater importance of the other. . . . The part is confused with the whole.[10]

Environmental public interest

Governments and their personnel are, in theory, oriented toward the public interest in their policies and decisionmaking. In the case of environmental administration, public interest issues arise from the interface of society with its environment. The public interest, however, is an abstract and symbolic concept that refers to the values and general interests of all citizens. It will often conflict with private interests of specific individuals, groups, and organizations. Environmental public interest may be variously justified and interpreted, but it affirms the value of long-term interests of the public over short-term private interests.

Future generations

In environmental administration, concerns must extend to future as well as present generations. Complex values difficult to identify and predict for future generations must be encompassed in judgments and decisions made in the present. The depletion of nonrenewable resources in the present, for example, can affect the quality of human life and the survival potential of living things and systems in the future. Environmental decisions made in the present can constrict or eliminate future options. An endangered species that is eliminated today may have been the crucial link for medical research tomorrow.

Intangibles

Many values and considerations found in environmental administration are of an intangible nature and consequently are difficult or impossible to define or quantify. It is extremely difficult if not impossible to assign economic values to living resources, due to the complexities and ambiguities associated with their being given proper weight and consideration. Tangible, quantifiable values such as board feet of timber or tons of minerals will be given more emphasis than less tangible, nonmarket values such as those associated with watershed protection, wildlife conservation, or scenic preservation. Intangible values include psychological and indirect benefits associated with aesthetic and other aspects of the natural environment that contribute greatly to the quality of human life.

Values

The values and value judgments of agency personnel play predominant roles in the policymaking and other processes of environmental administration. Values form the basis for the perspectives and actions of personnel in their decisionmaking. Thus, they provide a frame of reference or general guide for many of the actions of environmental administrators. In this sense, values are individual and collective conceptions that have emotional and symbolic components about what is important or desirable. Values also include judgments about what is "right" or "wrong." Value judgments also occur when administrators decide that a particular environmental condition should be given greater emphasis than the values of human species.

As a consequence of the factors that distinguish environmental policy and administration from other decisionmaking activities, environmental administrators are less able than others to adopt compartmentalized approaches to problem solving. Holistic approaches should therefore be employed to direct or manage complex relationships that occur among individuals or groups of individuals and their evolving societal and natural environments.[11] Because values and value judgments are involved in the formulation and implementation of environmental policy, decisionmaking processes should be made within an interdisciplinary context where multiple values are considered.[12]

Natural Resource Characteristics

Since much of environmental administration pertains to natural resource development and allocation, it is useful to consider the characteristics of natural resources. All natural resources are produced through natural processes. Some, the renewable resources, can sustain and perpetuate themselves through reproduction, or their survival can be encouraged through employment of proper management techniques. Human beings have the option of interfering or not interfering with these processes. Under a utilitarian philosophy, a natural resource must be used or have the potential for use by humans. However, uses and needs as well as attitudes toward natural resources may change greatly over time and with location and culture.[13] An isolated mountain peak may or may not be a natural resource, depending upon one's perception of it.

Actual management of natural resources by government represents a small proportion of environmental administration in the United States. The term "resource management" is in fact somewhat misleading. The majority of resource managers are concerned basically with management of people's behavior relative "to" natural resources rather than "with" natural resources.

Resource management implies some manipulation of natural resources and their environment, whereas the main function of resource managers is to interpret or decide upon the various uses of resources or degrees of use to be granted to individuals and nongovernmental organizations. For example, resource managers decide how much grazing or logging would be acceptable on public lands. Their decisions on use depend on the self-sustaining needs of the renewable resources. Thus, resource managers actually manage the interaction of humans with the environment rather than the environment itself.

Natural resource administrators usually recognize that the great preponderance of their time is spent on problems affecting the population at large. Their decisions concerning human behavior in relation to the environment involve the value determinations and perceptions described above that are further complicated by the changing complexities of science, technology, and society.

Cultural differences, for example, can affect management perspectives. Forestry management practices in the United States are minimal in comparison with Germany, where intensive control and manipulation of forest ecosystems are undertaken by government foresters.

Fluctuating resource availability also affects decisionmaking. In their decisions environmental resource administrators must make value judgments within a political environment of competing interests and powers. These judgments are complicated by the decisionmaker's limited perspective of the world, which is further influenced by his or her experience or field of specialization. Agency ideology also influences the decisionmaker. The entire decision process hinges, again, upon value perceptions.

Although management implies control, complex ecosystems are not readily controlled, even through the use of the most advanced technology. Ecologists caution us about the philosophy of the noosphere, which proposes not a biospherical world but one dominated by the human mind. Most ecologists also agree that people are not wise enough to understand the results of their actions.[14] The view that humans, through management by science and technology, will be able to produce the type of controlled environment or world that will do nothing but serve their needs and wishes competes with the more ecological view of human beings as responsible creatures who respect and value themselves, future generations, and other forms of life.[15]

The ability of humans to grasp the consequences of their actions is complicated further by the efforts of those scientists who participate in efforts to provide misinformation to citizens on environmental problems. In their book *Betrayal of Science and Reason* Paul R. Ehrlich and Anne H. Ehrlich provide alarming insight on the potentially negative effects of misinformation strategies. They describe the backlash against environmental controls that began in the 1980s and continued into the 1990s. Advocates of this "brownlash" challenge beliefs that environmental problems continue to be the threats that they were once perceived to be because human ingenuity has substantially addressed the problems. The Ehrlichs provide compelling arguments that such antienvironmental rhetoric threatens our future by potentially weakening public understanding and support for needed environmental controls.[16]

Natural resources vary greatly in physical and geographical distribution. They may be widely dispersed. Often, as in the case of a watershed, their distribution exceeds political and agency boundaries. Thus, a major task of environmental administration is to relate the administrative process to natural resource distribution and need. Natural resources, moreover, differ in utility and accessibility. Thus, a *potential* natural resource may not be an *actual* one. Many developing nations may be "rich" in natural resources but social, economic, or locational constraints make their value unredeemable.

Qualitative and quantitative definitions of natural resources and environment also may change in response to varying needs and different time periods. For example, growing population bases, changes in prevailing value systems, increases in levels of ecological awareness, and changing degrees of technical and economic development all affect these definitions. Although environmental processes continue to operate according to physical laws, people determine the quality and quantity of natural resource use through administrative decisions. For instance, a timber company might choose to log lower-quality trees at lower elevations rather than higher-quality trees located on a slope because of the higher energy costs required in logging the latter.

Although natural resources may be highly elastic and flexible, they are also destructible and exhaustible. Renewable natural resources, as forms of life, supposedly sustain and perpetuate themselves through natural processes, but these can be aided or hampered by people. Resource management based on strong economic and political factors to the exclusion of ecological ones can, for example, deplete a natural resource, to the point where it will not be able to recover to its former quality or quantity. Fortunately, such short-sighted management can be controlled. The Clinton administration initiated ecosystem management approaches in the Bureau of Land Management, U.S. Fish and Wildlife Service, USDA Forest Service, and National Park Service, because of concerns over the state of natural resources on federal lands.[17]

Some people believe natural resources are replaceable by other natural resources or synthetic products. Through modern science and technology, it is certainly possible to improve upon natural resources, as is illustrated by forest genetics. However, certain natural resources are irreplaceable or have very limited replacement potential in their essential life-sustaining characteristics. A prominent example is water. Interdependencies also exist that preclude the administration or manipulation of a single natural resource without influencing and affecting other related natural resources. For example, diversion or drainage of water from marshlands may adversely affect plants and wildlife. The professional literature and the news media have abundantly illustrated the catastrophic consequences of resource diversions that result in serious problems for humans as well as for other forms of life. The U.S. Congress is so concerned about the diversion of resources worldwide that it has asked the World Bank to make safeguarding the environment a priority in funding development projects.[18]

The values of a more urbanized and educated citizenry ("urban values") may include appreciative, nonutilitarian attitudes toward some natural resources. Such attitudes contrast with the utilitarian perspective of many rural communities, which depend upon occupations requiring exploitation of natural resources.[19] Many environmental agencies, such as the USDA Forest Service and the Natural Resources Conservation Service, are also identified with utilitarian concerns. They still manifest an attitudinal time lag concerned with the utilitarian aspects of resource management that has failed to keep pace with the urbanization of environmental attitudes in the United States.

Utilitarian value structures treat natural resources in terms of use or potential use by humans, but there are other, appreciative or nonutilitarian value structures that use criteria of a qualitative and intangible nature. Such value structures appear to exert a powerful influence on current environmental administration. These qualitative definitions and attendant demands have always been present in some form throughout history. But they are now much more prevalent in the environmental concerns of the nation. This shift has resulted in the recognition that a natural resource should be administered for its own sake. It has also resulted in a wider acceptance of nonutilitarian benefits for people. The popularity of nonconsumptive use of wildlife for recreational purposes in the United States illustrates this point. In 1996 more than 62.9 million Americans aged sixteen and over actively participated in observing, feeding, and photographing wildlife and spent more than $29 billion in pursuit of this activity.[20]

The natural resource characteristics described in this section imply a need for better management of human use of natural resources. First, management must orchestrate appropriate levels of utilization. Second, it must incorporate conservation processes that perpetuate or renew living natural resources by (1) maintaining ecological processes and life support systems, (2) preserving the genetic diversity or range of species, and (3) ensuring the sustainable utilization of species and ecosystems, as stated in the *Caring for the Earth: A Strategy for Sustainable Living* proposed by the World Conservation Union, in conjunction with the United Nations Environment Programme (UNEP) and World Wide Fund for Nature (WWF).[21] The implementation of such measures is deeply influenced by the forces and values of the cultural setting in which environmental administration occurs, factors that are described in the next section.

Cultural Factors

Environmental administration does not occur in a cultural vacuum. Environmental policy and administration are best understood when consideration is given to those forces, values, and other factors of the culture in which these activities take place. Although many of these considerations apply to the cultures of various countries, emphasis is given in this book to cultural factors in the United States.

Democracy
Various aspects of democracy have important positive and negative implications for environmental administration. The concepts of majority rule, representation, public participation, and individual dignity require envi-

ronmental administration to be responsive to public opinion and public interest. Under majority rule, the growing force of public awareness of environmental issues clearly contributes positive pressures for responsive governmental action. A totalitarian form of government, by contrast, is not obliged to recognize diverse interests when determining its policies for the environment. On the other hand, democracy may interfere with long-range policies and planning for present and future generations. Further, a democracy may not be able to make changes in environmental policy and administration rapidly enough to address adequately environmental problems. Charles E. Lindblom observes, "Democracies change their policies almost entirely through incremental adjustments. Policy does not (generally) move in leaps and bounds."[22]

Science and technology

Within the last fifty years science and technology have had dynamic and profound influences on lifestyles and economies worldwide. These influences defy assessment or prediction concerning their effect on the environment. Governments, moreover, have strong commitments to scientific research and technological development representing billions of dollars in annual expenditures. The federal government of the United States, for example, invests twice as much in research and development as the private sector. In spite of the benefits that often accrue from these commitments, uncontrolled change, rapid consumption, and ecological impacts continue to be important limiting considerations for science and technology. And some problems simply have no scientific or technical solution, despite the tendency to seek such solutions aggressively, without consideration of human values or morality.[23] It should be further recognized that, in this era, science and technology can no longer claim value neutrality for environmental administration or any other area.[24]

Thus, moral considerations apply to the management of science and technology. In efforts to resolve the complex problems and issues of the world, scientists must consider the implications of their innovations for the environment. Government should limit scientific and technological activities at least to the extent that the direction and emphasis of these activities are influenced by moral considerations. Science and technology are not ends but means. As Mustafa Tolba, former executive director of the United Nations Environment Programme, has said: "Governments have faced the reality that man is both creature and molder of his environment and that the power afforded him by scientific and technological advances has given him a new capacity to alter his planet's life-support system in significant and even irreversible ways. The same power carries with it the concomitant responsibility to act with prudent regard for environmental consequences."[25]

Progress and materialism

In many subcultures the value of progress in any form is unquestioned. This premise includes a short-range orientation that is principally concerned with quantity—economic benefits, newness, and expediency—and not with long-range quality or environmental considerations. These developmental criteria are applied to natural resources as well as to other areas. Unfortunately, considerations of ecological quality, aesthetics, and other intangibles may be sacrificed to the desire for economic progress. An important aspect of economic progress is its emphasis on goods and services that results in greater demands and pressures upon environmental administration. Some natural resource agencies are administered with a developmental perspective. Such orientations have evolved from the frontier mentality of unlimited opportunities and materialism that emerged as an essential element of the American dream. Often, environmental administrators are pressured to elevate short-term material standard of living rather than maintain or improve the long-term quality of life and environment. Those agencies managed under multiple use and sustained yield principles are most heavily affected by such pressures.

Pragmatism

Pragmatism appears as a major philosophical theme of American culture. In this framework great emphasis is placed upon the practicality of environmental administration. Practical solutions and results may appear to be realistic and concrete for some resource problems on a short-term basis, but in long-range terms they may yield fragmented and inadequate programs that do not take into consideration values essential to an integrated approach. Also, with the concern for practical and concrete results, many long-range considerations that do not lend themselves to immediate or quantitative evaluation are not included in decision analysis. This places further emphasis on short-range pragmatic solutions within the political framework of environmental administration.

Gabriel Almond notes a tendency for Americans to be optimistic toward attaining ends—"things can be done"—but they fall back on narrow-focus improvising when it comes to means—"know-how."[26] Americans, despite the advantages of new technical approaches to problems, fail to reap the net advantages of thoughtful policy planning, achievable through greater intellectual discipline. It has become increasingly apparent that the pragmatic approach may avoid consideration of long-range and causal problems and considerations of the environment while dealing with short-term and symptomatic ones in a practical and expedient framework.

The middle class

Although American culture may not be considered class-conscious, the majority of Americans identify strongly with the middle class. Major factors in this identification are competition, materialism, and conformity. Positive emphasis is given to education, accomplishment, and in maintaining parity with one's neighbors. This orientation heavily influences environmental administration. In contrast, many cultures provide various forms of social and institutional support that reward the individual for not committing himself to materialism.[27] A cultural paradox is present here, in that much of the support for conservation of the environment comes from the American upper-middle class, as is the case in many other nations.[28] The middle class also supplies much of the personnel for both private and public sectors in economic and environmental affairs.

Urbanization

In the United States about 79.8 percent of the population is concentrated in metropolitan areas and 20.2 percent in nonmetropolitan areas.[29] Population distributions in other countries are rapidly changing as people migrate in large numbers from rural to urban areas, particularly in developing countries. This process is referred to as "inmigration." Because of inmigration and population increases in general, there are now more people around the world living in urban than rural areas. The spread of urbanization results in an increase in urban values, with an associated decline in rural cultures. Today's urban values and needs place different pressures upon environmental administration than did those of the agrarian economy of earlier days. While some city dwellers view the urban environment as the only "environment," without recognizing its limited, dependent, and artificial nature, urban environmental attitudes tend to be appreciative of the environment, nonutilitarian, and recreational, in contrast with utilitarian attitudes reflective of rural communities.[30]

From a sociological viewpoint, the urban situation produces social group relationships that are characterized by impersonality and manipulation. This in turn fosters a tendency toward centralized government, in contrast to the grassroots representation of rural communities upon which many resource agencies place a high value. Moreover, under the "one-person, one-vote principle," the increase in urban population has caused a shift in political power from the rural base. This power shift allows today's urban population to exert a stronger influence upon environmental administration than traditional rural constituencies.

Mixed economy

The economy of the United States is a blend of conditions and policies with free enterprise, service, and welfare elements directed toward growth and profit. Governmental influence upon and intervention in the economy is taken for granted. Limited attention, however, is paid to the social and environmental costs or externalities of such manipulation. Much attention is paid to increasing the Gross National Product and the standard of living, both in real terms. As natural resource limitations and environmental pressures grow, this attitudinal circumstance may undergo a reorientation toward quality and survival. This is particularly true with the scarcity concept in operation, where several economic institutions compete for the allocation of environmental *segments*. Ecological interdependencies and the environmental costs of negative influences may then complicate the economic-environmental system. For example, forest and agricultural productivity are threatened by emissions from air pollution sources both locally and hundreds of miles away via air currents.[31] The sustainable growth movement is directly concerned with such interdependencies.[32]

The economic policies of the 1980s heavily influenced the mixed economy of the United States. Besides massive cuts in domestic government programs and declines in certain private sectors of the economy, "Reaganomics" stressed a reduction and decentralization of the role of the national government in economic, social, and environmental affairs. Budget reductions and other economic policies of the Reagan administration greatly changed and reduced the scope and effectiveness of the agencies and personnel responsible for environmental administration.[33] The National Performance Review of the Clinton administration led to further substantial reductions in agency personnel in the 1990s, under the claim of making federal agencies more efficient.

The first environmental quality report (1981) of the Council on Environmental Quality (CEQ) issued under President Reagan clearly indicated two important principles of Reaganomics: (1) balancing costs and benefits with the recognition that *internal* costs to private sectors and the national government should be considered in evaluating different approaches to environmental protection; and (2) allowing the marketplace to work while leaving the achievement of environmental goals and protection of environmental standards to free-market mechanisms and private initiative wherever possible.[34] Such principles clearly placed private interests above the environmental public interest. The Clinton-Gore initiative for better environmental results has by contrast promoted the use of innovation and flexibility in order to produce a federal government approach that worked better and cost less.

Communications revolution

The relatively recent communications revolution now exposes practically everyone to a variety of media influences.[35] The impact of this revolution appears to be a paradox for environmental administration. On the one hand, the mass media create increased demand for material goods among all segments of the public, resulting in subsequent depletion of natural resources and environmental deterioration. On the other hand, the various media also present information on environmental issues and problems to citizens. The internet provides citizens with an overwhelming amount of information of a timely and critical nature on topics ranging from environmental quality to detailed analysis of government programs. Armed with appropriate information, citizens can organize and attempt to influence and modify environmental policy decisions effectively through expression of concern for environmental quality and survival. Brownlashers also use the internet to present their misinformation. Evaluations of various public opinion surveys show that environmental protection continues to have the strong support of a majority of Americans. In an attempt to explain the upturn in public opinion for environmental quality concerns, one researcher, Riley Dunlap, concluded that

> because of the efforts of scientists, environmentalists, and the media, the public has come to view environmental problems . . . as increasingly serious threats to human health and well-being. Although there seem to be no hard data for documentation, it is reasonable to assume that since the late 1970s the public has heard more and more about problems such as toxic wastes, acid rain, ozone depletion, and the greenhouse effect. These problems typically pose more serious threats to humans than did the major environmental issues of the 1960s and early 1970s.[36]

But in the late 1980s and 1990s, there has been a considerable, vocal backlash to the growth in environmental laws and regulations, as noted by Paul Ehrlich and Anne Ehrlich, two of the pioneer thinkers in the environmental movement.[37]

Social and psychological aspects

Through the social sciences we are becoming aware of various factors and forces that influence modern humanity and thus environmental administration.[38] As a result of high mobility, urbanization, and other complexities, many Americans now feel a sense of alienation and anomie, lacking a realistic identity with governmental and environmental concerns. The consequences are apathy and irresponsibility.

Rapid change and ambivalence are normal for contemporary American life. Yet, as Stanley Milgram observes, this can lead to overload: "Overload refers to a system's inability to process inputs from the environment because there are too many inputs for the system to cope with or because successive inputs come so fast that input A cannot be processed when input B is presented."[39] The end result is usually confusion of values for long-range governmental and environmental planning.

Traditional values, attitudes, and priorities for most Americans are crumbling, with no clear replacements in mind. Surveys reveal a departure from optimism and a decline of public confidence in government.[40] A gap between attitudes and values, on the one hand, and behavior, on the other, exists for most Americans, with regard to materialism and environment. A popular lifestyle for several million Americans, however, is associated with values centering on material simplicity, human scale, self determination, ecological awareness, and personal growth.

In addition to being affected by the atmosphere of constant change, human behavior falls into a pattern of reaction to crisis situations. Public attention and governmental actions engender large inputs over the short term rather than continuous inputs on a given problem or issue over the long range. Some individuals have observed that this may be the typical response pattern where environmental issues are concerned.

Finally, U.S. citizens often band together to provide mutual support and accomplish common ends. According to Alexis de Tocqueville, America is a nation of organization joiners. He also observed that a principal thrust of Americans is to subdue nature with an image of themselves.[41] His generalizations still hold true. Pressure groups and organizations exert strong influences on the government and the environment.

Religion

Although America may be a conglomeration of religions, it is still strongly influenced by Christianity and certain theologically based misconceptions. Lynn White Jr. states, "Especially in its Western form, Christianity is the most anthropocentric (human-centered) religion the world has seen." Christian assumptions have been used to justify technological exploitation of the earth and its living things without adequate recognition of responsibility toward other forms of life.[42] Obviously, exploitative values and actions are sometimes based on such misconceptions and exert considerable pressure on environmental administration. However, a growing movement in Christianity has begun to refute these misconceptions and to advocate environmental stewardship. It draws on ethics and the actions of working toward harmony between people and nature, without exploitation, and with controlled materialism based on spiritual values.[43] Joseph Sittler, a noted theologian, observes that the Bible itself contains numerous references to ethics and actions with respect to the recognition of principles of order, limitations, and boundaries. Yet he also observes that Christians and non-Christians alike in America have difficulty accepting any ethics that call for limitations on their actions when pursuing progress. Constraining human behavior in order to protect the environment is unpopular.[44] The environmental movement also has assumed an ecumenical character through the World Wide Fund for Nature International Network on Conservation and Religion, headquartered in Switzerland. This network, embracing all major world religions, was initiated at the Interfaith Ceremony held on September 29, 1986, in Assisi, Italy. The network publishes a bulletin entitled *The New Road.* More recently, the legal scholar Chuck Barlow has attempted to convince the Christian Right that they should employ political action to protect the environment because such action would be an ethically moral thing to do and would be in accordance with stewardship principles found in the Bible.[45]

The West

Because the vast majority of public lands and natural resources in the United States are located west of the Mississippi River, general characteristics of the American western subculture as identified by both Donnally and Jonas are pertinent: (1) the emphasis on the development of natural resources, particularly water; (2) strong natural-resource pressure groups, especially in water, power, ranching, and mining; (3) an acute awareness of the presence of the federal government in the West and the strong influence of westerners on federal natural resource agencies; and (4) an emphasis on individual independence and personality.[46]

Opportunity for independent economic exploitation of natural resources is highly valued in western states. Attitudes and values in much of the West center on the traditional frontier notion of exploitation and give little recognition to environmental problems and considerations. Many westerners tend to view federal public lands as provincially owned, belonging only to local or state residents. The federal government in fact owns the majority of land in many western states. Federal administrators of public lands in the West, under the concept of multiple use, are subject to heavy pressure from local communities for economic development.[47]

Frontier tradition

In terms of progress and other previously discussed considerations, the frontier tradition still operates in America, regardless of lags in time and values. Frederick J. Turner in his classic work, *The Frontier in American History*, considered American democracy to be the outcome of experiences of people in dealing with the frontier. He judged frontier interaction to have profoundly influenced American values. America had the richest frontier in the history of the world, with its large rivers and fertile soil. Although, as Turner documents, the American frontier had disappeared by 1890, symbolic frontier concepts and values still permeate American life, as they do in some other nations that have frontiers containing large areas of undeveloped wildlands.[48] Some of these frontier exploitative and negative attitudes still remain in both public and private sector environmental administration.

Population

Growing populations and increasing per capita consumption has accelerated resource use, resulting in increasing environmental impacts that possibly have engendered irreversible damage to segments of the biosphere. The human population of the world was expected to pass 6 billion in 1998.[49] It is estimated that it took about 123 years for the world's population to increase from 1 to 2 billion, but each additional increment of a billion took thirty-three years, fourteen years, and thirteen years respectively. The transition from 5 to 6 billion is estimated to occur over eleven years. At current rates the world's population is likely to be 12.5 billion people by 2050. However, the United Nations believes that through the implementation of substantial plans of action, including population controls, the population could stabilize at 7.27 billion.[50] Population growth is at the root of most domestic and international environmental problems. Yet most of this population growth—95 percent over the next 35 years— will occur in less developed countries, in contrast to developed or industrialized countries that will generally achieve zero population growth.[51] An American child, as a member of the largest consumer nation of the world, consumes many times more resources than does a child from a developing country. Paul Ehrlich argues that an American child should therefore be considered a significant part of the population explosion, in terms of his or her resource consumption.[52] The population of the United States now exceeds 262 million people. Current projections suggest a U.S. population of 392 million by 2050.[53] The United Nations Environment Programme (UNEP) estimates that the United States with 6 percent of the world's population consumes 25 percent of the fossil fuels, while the developing nations, with about 75 percent of the world's population consume only 20 percent.[54]

Increasing world population and attendant stresses upon the Earth's resources could exceed the ability of the planet to sustain itself. When the carrying capacity of a given area, region, or nation is exceeded through the stresses imposed by overpopulation, severe environmental repercussions naturally result. Moreover, environmental degradation in one part of the world can cause negative environmental effects elsewhere. These effects are not limited to the physical environment. From a governmental point of view decreases in human freedom, personal dignity, quality of life, and individual human welfare can attend population growth and increased resource use. Population pressures are already creating ecological victims and refugees. More centralized bureaucratic control may also be imposed. Government may not move to solve but rather try to eliminate human problems through repatriation and forced starvation of affected populations. Without adequate population control, a proliferation of authoritarian and totalitarian forms of government is fairly certain to emerge. Under such circumstances environmental administration, as part of the totalitarian governmental process, would move to restrict and control individuals, although whether or not for their collective survival is an open question.

Global interrelationships

The political and economic security and stability of all nations are affected by global resources and environmental and population factors. Such global interrelationships intrude into the domestic affairs of all nations. Typical concerns include (1) availability and prices of renewable natural resources, (2) conservation of the domestic renewable resource base, (3) migration from areas of resource impoverishment, (4) disputes over international waters and other resources, (5) energy availability, (6) international trade, (7) conservation of genetic resources, and (8) climate change and pollution of the global atmosphere.[55] Many serious environmental problems do not respect national boundaries, are basically global in nature, and demonstrate the need for global cooperation. Control of the manufacture and use of the chemicals contributing to the depletion of the ozone layer that surrounds the earth and protects the earth from ultraviolet radiation provides a good example.

There are six critical environmental issues that affect U.S. interests in the global environment. They are: the quality of the atmosphere, depletion of fresh water, loss of soil productivity, loss of genetic diversity, tropical deforestation, and toxic contamination/hazardous materials.

These pressing issues will demand an inordinate amount of time, money, and effort in most countries of the world. The noted scholar Lynton K. Caldwell has said that these issues are critical because "unless reversible in the very near future, [they] may result in irretrievable damage to planetary life-support systems . . . [and] because . . . remedial means are presently either not available . . . or would require complex socioeconomic changes that even willing governments could not bring about. . . . Yet none of these trends are today beyond remedy if the will to reverse them can be mobilized."[56] The development of comprehensive, highly integrated environmental policies and administration is a key element for future success. However, timely intervention and severity of threat often opt against such comprehensive approaches and, ad hoc, less well-conceived measures are necessary. Some environmental problems may require isolated efforts to ensure a timely and adequate response. If agencies could develop strategic planning that provides direction for such ad hoc approaches toward an integrative perspective, greater long-term positive effects might result.

The material in this chapter provides a firm basis upon which to assess and evaluate the implementation of environmental and natural resource policies described in the remaining chapters. The six unique considerations of environmental administration—ecological complexities, crisis orientation, environmental public interest, future generations, intangibles, and values—interact to shape the policies and management decisions used to protect and enhance the environment and its store of natural resources. Another important source of influence is the struggle between the economic and environmental pressure groups, which affect policy through the legislation and court decisions that they promote through their actions.

Public policy reflects the culture in which it is formulated and conducted. In American political culture the pragmatic and pluralistic characteristics are dominant in general policy areas. This orientation results in the absence of a comprehensive, unified policy for the environment. However, at least one common ideological basis exists for all environmental programs—the involvement of government with social goals and problems relating to the environment and natural resources.[1] Without a broader general ideological base, however, policies adopted at every level show different values. The end result is a collection of fragmented, short-term, and often conflicting policies throughout the range of environmental policy. Some of the recent decline in environmental concern may be attributed to the failure to develop a deep commitment to an ideology of environmental values throughout society.[2]

Modern public administration appears to operate on a crisis basis, oriented toward immediate and critical problem solving. Today's political climate and changing conditions do not appear to allow much stability in policy. Given the thousands of legislative acts pertaining to the environment, and given the numerous environmental agencies with their respective policies in flux, both macro- and micropolicies have a tendency to reflect a quest for short-term, expedient solutions. Regarding this approach, Norman Wengert notes: "Pragmatism is pluralistic and eclectic, focusing on problems and performance rather than principles, upon action rather than upon ideas. The pragmatic test of 'will it work' or 'how does it work' de-emphasizes ideology in the sense of a developed synthetic system of beliefs and values to govern actions."[3]

Amid the pragmatic and pluralistic approaches to the creation of environmental policies, the one unifying theme is that of unity or wholeness. The National Environmental Policy Act of 1969 (NEPA) approaches a legal formulation of this theme. The intent of NEPA is "[t]o declare a national policy which will encourage productive and enjoyable harmony between man and his environment; to promote efforts which will prevent or eliminate damage to the environment and biosphere and stimulate the health and welfare of man; to enrich the understanding of the ecological systems and natural resources important to the Nation; and to establish a Council on Environmental Quality."[4]

The act stipulates that systematic, interdisciplinary, and interagency approaches to achieving or maintaining environmental quality must be followed by all federal agencies. These and other requirements of the act are considered supplemental to each agency's originating legislation and policy. It also stipulates that each agency is to report any inconsistencies and deficiencies in legislation, policies, or procedures that interfere with compliance with the purposes and provisions of the act.

The great majority of federal agencies are single-purpose organizations whose congressional mandates have created policies and procedures, as well as clientele, that limit the agencies' responsiveness to broad environmental needs. In that sense the limitations and definitions of policy responsibility have served to insulate agencies from various complexities and demands. A greater awareness of environmental problems in the general public, however, and the increased political strength of the environmental movement have combined to produce new legislation that forces agencies to give greater consideration to the overall environment. Often, where these new mandates call for institutional changes in established agencies, there is resistance and conflict. Such conflict is unnecessary, since the primary missions and influences generally are not altered by the new environmental initiatives.

Because there are indications of an increasing tendency for environmental policy to be formulated and implemented by public executives, the individual and collective value systems of these executives are crucial. Policy is a statement of principles and objectives based on values that serve as guides for the operation and limitation of an agency. Although policy is largely based on legislation, it usually involves broad, idealistic guidelines that are interpreted subjectively by the individual administrator. Given the dynamic nature of the policy process, ambiguity naturally enters the picture with changes in time, individuals, problems, and institutions. This is particularly evident in crisis-oriented administrative policymaking, where policy often fluctuates as new demands arise. Policy is an expression of values that compete for dominance in conflict situations.

Much of the role of leadership of the national government in formulating and implementing environmental policy was established in the 1970s. Legislation in this most significant and productive period included the National Environmental Policy Act (1969); the Clean Air Act (1970), which created national ambient air quality standards (NAAQS); the Toxic Substances Control Act (1976), regulating the manufacture of toxic substances; the Federal Water Pollution Control Act Amendments (1972), establishing a national pollution discharge elimination system (NPDES); the Safe Drinking Water Act (1974), protecting water supplies; the Resource Recovery and Reclamation Act (1976), establishing hazardous waste management guidelines; and the Surface Mining Control and Reclamation Act (1977), establishing surface mining regulations. As Walter Rosenbaum indicates, these acts' "collective purpose was to compel through public policy a reversal of two centuries' environmental negligence and to force a new public responsibility for future environmental protection."[5] (For a comprehensive list of environmental laws, see table 1 at the end of this chapter.)

In contrast, environmental policy in the 1980s was characterized by a preference for economic development over environmental constraint, as well as a desire to keep government out of environmental affairs. Lacking sufficient strength on Capitol Hill to push through the rollback of environmental legislation that it desired, the Republican Party relied on administrative discretion rather than legislation as the preferred instrument of change. In the 1990s, the position of power shifted to the legislative branch when the Republicans assumed control of both the House of Representatives and the Senate, but lost the presidency to Bill Clinton. Former senator Gaylord Nelson described the 104th Congress (1995–1996) as the most antienvironmental Congress in history.[6] Republicans used two basic strategies to alter environmental policies. First, they proposed numerous legislative proposals to substantially change existing environmental legislation. Second, they attempted to reduce funding for environmental agencies to weaken or halt implementation of existing environmental regulations. At one point the Republicans, under the leadership of Newt Gingrich, attempted to deny the EPA necessary appropriations to implement all environmental laws within its jurisdiction.

Through procedural means such as Executive Order 12291 (46FR13193) of February 17, 1981, the Reagan administration effectively stalled the environmental regulatory process. Virtually every environmental regulation was subjected to intense economic scrutiny under EO 12291 and, hence, delay. In his book *Environmental Policy under Reagan's Executive Order,* V. Kerry Smith pointed out what he perceived to be the fallacy of such an approach, as well as why it proved so successful in causing delay at the time:

> [T]he order assumes that it is possible to prepare benefit-cost analyses for fairly detailed environmental regulations. It assumes a general acceptance of methods for benefit and cost estimation, a level of tech-

nical information, and a clear understanding of the effectiveness of each proposed regulation. In most cases none of these components exists in ideal terms. Often substantial judgment is required to piece together even a small fraction of the information that would ideally be desired for a complete benefit-cost analysis. Indeed, this issue lies at the heart of many of the areas of controversy with EPA's regulatory activities—the state of available information and scientific understanding of the problems involved may be inadequate. Thus there may be ample scope for reasonable people to disagree over the impacts of specific levels of pollution.[7]

Paul Portney points out that, due to the failure of the Republicans to create statutory changes, Reagan's environmental policies could be relatively easily reversed by future administrations.[8] However, the Bush administration continued the policy of using economic impact analysis of proposed regulations, continuing to slow down the promulgation of rules to implement new environmental laws. Bush even imposed an occasional moratorium on rule promulgation. In response to the Bush administration's efforts to slow down environmental regulations, the Natural Resources Defense Council relied on citizen suit provisions in the Clean Air Act to sue the EPA when numerous rules for implementation of the 1990 Clean Air Act Amendments had not been promulgated by the deadlines established by Congress.

The Republican-dominated 104th and 105th Congresses, in the 1990s, addressed the long-term weakness in executive orders and legislatively formalized economic impact requirements. The Unfunded Mandate Reform Act (UMRA) of 1995 (Public Law 104-4) requires federal agencies to use cost-benefit and cost-effectiveness analysis in the development of major regulations. Congress established additional analysis requirements designed to reduce the impact of regulations on small businesses. The Small Business Regulatory Enforcement Fairness Act (SBREFA) of 1996 (Public Law 104-121) allows judicial review of agency regulatory analysis. It also establishes a new procedural requirement. Congress must receive a copy of a new regulation with its regulatory analysis sixty days before it is to go into effect. Congress can then vote to block the regulation through a joint resolution of disapproval.[9]

Conservation and Environmental Movements

The conservation and environmental movements can be compared and contrasted as major influences in environmental policy and administration. They are social movements, defined by Joseph Gusfield as "socially shared demands for change in some aspect of the social order," that encourage growth and recognition, and tend to generate public controversy.[10] For practical purposes, the environmental movement may be viewed as an extension and elaboration of the conservation movement. Yet the environmental movement, which emerged essentially as a mass ideology in the late 1960s, is a much more powerful and broader force; it is also much more politically, ecologically, and legally oriented.[11]

The conservation movement emphasizes the "wise" use of natural resources for present and future generations, but at times the term may simply mean preservation of the status quo. As a philosophy and process, conservation was and is highly valued in America. Few politicians or public administrators would publicly oppose this ideology. Although not specifically referred to as "conservation," various measures and regulations were carried out to promote wise use and protection of natural resources in America as early as the colonial period. Examples of early conservation activities are regulating how many deer one could take, reserving forest areas for Royal Navy timber, and reserving land for town parks.

Officially the conservation movement in America may be traced to the 1908 White House Conference on Conservation that was composed of state governors and conservation leaders who were called together by President Theodore Roosevelt. Gifford Pinchot, the chief of the USDA Forest Service from 1899 to 1908, was a moving force behind early conservation efforts as well as this conference. In 1907 Pinchot chose the term "conservation" to describe the movement; it was based on "conservancies," a name conferred on government forests in India. Much of the emphasis given to early conservation in the United States in fact centered on forestry and forest problems.[12]

Pinchot is credited with proposing the direction for early conservation efforts—the use and management of natural resources for the greatest good of the greatest number of people over the longest time. This goal, however, was borrowed from the utilitarian ideal of the English philosopher Jeremy Bentham. Given the broad concept of wise use (not to be confused with the antienvironmental group of the 1990s), numerous different interpretations and approaches were introduced into the conservation movement by different individuals and organizations.

In the conservation movement of the twentieth century, a chasm exists between individuals promoting conservation for wise use and development and those promoting conservation for purposes of preservation and appreciation. The latter is considered to be the province of the more educated and upper-middle-class segments of society.[13] This chasm in the conservation movement has had considerable impact on natural resource agencies that are often concerned with wise use of a single resource. Those who believed in conservation management principles operated within comfortable definitions and limits of their responsibilities relative to specific natural resources until quite recently. As environmental problems became more severe, however, the conservation movement began to broaden its horizons and accept a more complete ecological perspective. This can be seen in the ecological perspective of the National Wildlife Federation (NWF), an organization that developed through the conservation movement. Currently, NWF is involved in a wide variety of environmental issues with a broader ecological emphasis. In this sense, aspects of the conservation movement blend into the emergent environmental movement, making distinctions between the two difficult.

While the conservation movement is by its nature *related* to ecology, the environmental movement is *based* on ecology and an understanding of the total environmental picture, including population problems. Many leaders of the conservation movement were government scientist-administrators such as Gifford Pinchot and Major J. W. Powell, director of the U.S. Geological Survey, as well as individuals from other professions such as John Muir, naturalist and founder of the Sierra Club. When the environmental movement emerged, the spokespersons and leaders were professional ecologists like Eugene Odum, Barry Commoner, and Paul Ehrlich. Today, a wider spectrum of ecologists and others who "think ecologically" alert the public about imminent environmental problems that stem from negative human interaction with the physical environment. They take active roles in attempts to shape and alter government policies that affect the quality of the environment.

At one time the conservation movement contained relatively small segments of the total population on an organized basis, but it could count on some support from the general public. Through their dramatic and abundant coverage of environmental issues and topics since the late 1960s, the news media have greatly helped the environmental movement become a mass ideology with strong and active support as well as interest. A certain amount of this response might be attributed to the success of the movement to educate the public about environmental problems. A substantial proportion of the American population now recognizes that they depend upon and relate to the complex environment in which they live. Many people realize that they have responsibilities to both present and future generations, human and otherwise. People are surrounded by examples and cumulative effects of environmental deterioration almost on a daily basis. Consequently, the general public has become more aware of and responsive to ecological messages and environmental problems.

Public opinion provides a valid indicator of the extent of public concern and support for the environmental movement. In a summary of several polls taken in the 1970s and early 1980s Robert Hamrin states: "In the midst of high inflation and general economic difficulties, one might expect the public to be abandoning strong environmental values. This has not occurred." Hamrin also states that polls taken in this period generally reveal the findings that most Americans strongly favor environmental protection, despite its large costs; moreover, they want policymakers to keep environmental hazards to an absolute minimum. Much of the environmental movement is the result of changing values and priorities of the American public. The Market Opinion Research Corporation said in its report to management, in February 1977, "It would be foolish for anyone to conclude that the public is less than adamant about environmental quality." [14]

In analyses of public opinion polls Riley Dunlap reports strong support for environmental quality issues, despite the strength of antienvironmental forces during this time period.[15] CBS News/*New York Times* polls conducted during the period 1981 to 1986 indicate that the general public increasingly supported environmental protection, regardless of the cost. The public was asked, "Do you agree or disagree with the following statement: Protecting the environment is so important that requirements and standards cannot be too high, and continuing environmental improvements must be made regardless of cost." Positive support on this question grew from 45 percent in 1981, to 58 percent in 1983, and eventually to 66 percent in 1986.[16] Public opinion indicated that the Reagan administration's regulatory efforts related to the environment were also not satisfactory to the public. Cambridge Reports, Incorporated, found that 59 percent of the public thought there was "too little" rather than "too much" governmental regulation for environmental protection (only 7 percent indicated the latter). Dunlap reports that "[t]hese results, along with those for the NORC [National Opinion Research Center] spending item, indicate that . . . the public was nearly 10 times as likely to see a need for increased governmental efforts and spending on behalf of environmental protection as to take the opposite view." [17]

In a 1993 "Health of the Planet Survey," 65 percent of Americans polled indicated that they would be willing to pay more for environmental quality.[18] The 1996 American National Election Study (NES), conducted by the University of Michigan's Institute for Social Research, reported that 40.1 percent of the American public believed that more should be spent on the environment, 50.4 percent felt that it should remain the same, and only 8.8 percent believed that environmental spending should be decreased.

The 1996 NES also indicated that more than twice as many Americans believed that the environment was more important than jobs.[19] Furthermore, an article in the *New York Times* on January 26, 1996, reported that a pollster hired by the Republican party found that only 35 percent of the voting public would vote to reelect members of the House of Representatives who voted to cut EPA spending.[20]

The environmental movement is a powerful unifying force in the United States and elsewhere. This force continues, despite differing degrees of emphasis on values, priorities, and problems that often result in internal and external discord as well as conflicts between and within the environmental and conservation movements. In the United States there is a multitude of national organizations that have numerous state and local chapters or groups; more than 7 million Americans are members of them.[21] And this only reflects the tip of the iceberg of public concern in the environmental arena. Numerous ad hoc environmental groups and committees are formed to deal with specific problems and issues at all levels.

The conservation movement has always been concerned primarily with the "wise" use of natural resources, whereas the central message of the environmental movement has been and remains the need to preserve fully functioning ecosystems. The former movement is centered in humans, the latter in the biosphere. Arne Naess, a Norwegian ecophilosopher, argues for a "deep ecology (or environmental) movement." He considers "deep ecology" a value priority philosophy that advocates environmental harmony and systemic equilibrium over other values. Combating pollution and resource depletion would be but a small part of the deep ecologist's concern. Deep ecology views humans as just one species among many others. Humans have no special right to dominate or destroy the environment, including other life-forms.[22] Some of the impact of the deep ecology movement is demonstrated in the volume edited by George Sessions, *Deep Ecology for the Twenty-First Century*. Sessions's book contains thirty-nine essays that describe how the concept of deep ecology has helped to shape the environmental debate between activists and policymakers.[23]

Naess argues for bringing holistic deep ecology and its associated values, along with its scientific aspects, into policy and prescriptive areas for change. Given the nature and degree of ecological interpretations, much of the environmental movement points toward social, economic, and political changes that will have to take place to meet growing environmental problems and crises. Although few conservationists advocate drastic changes to forestall a doomsday or apocalyptic event, the foundation for the changes recommended by some ecologists and environmentalists is the prediction of imminent environmental disasters. And, they point out, environmental change will impact society, too. Barry Commoner states, "Every one of the ecological changes needed for the sake of preserving our environment is going to place added stresses within the social [and governmental] structure. We really can't solve the environmental crisis without solving the resulting social crisis."[24] Developing this line of thought, William Bryan remarks, "Environmentalists are, therefore, social activists, with varying degrees of commitment toward effecting change consistent with such [spaceship earth] ecological and social goals."[25]

The environmental movement relies on the news media to alert the public of existing and emerging environmental threats. Newspapers (and increasingly television), specifically, play an effective and continuing role in activating public awareness. The American orator Wendell Phillips observed that "we live under a government of men and morning newspapers."[26] This statement is well illustrated in the area of federal health regulations. According to Edward Burger, the activities of regulatory agencies are influenced heavily by public perceptions of risks of exposure to dangerous food additives, carcinogenic substances in water supplies, and a variety of hazardous materials in the workplace. Public awareness of and response to health risks can help dictate regulatory responses. How regulatory agencies carry out their legal mandates and how zealously they pursue specific issues are strongly affected by the public's view of what regulatory roles should be played and which items are perceived as risks to health and safety.[27]

Media-induced public awareness of the dangers of hazardous wastes was a major factor in the passage of the Superfund Amendments and Reauthorization Act over the threat of veto by President Reagan. Media attention to such inflammatory issues as the controversies in Times Beach, Missouri, and the Stringfellow Pits, in California, produced widespread support for passage of the Superfund amendments. At Times Beach, the federal government was forced to close down and purchase the entire community, due to dioxin contamination from a federal government source. In the Stringfellow Pits, illegal hazardous waste dumping led to threats to groundwater supplies for a heavily populated area. With such media exposure, only those members of Congress with the most secure positions could afford to vote against this $8.5 billion bill. Public concern for health and safety was a major factor in the passage of the Superfund legislation.[28]

The connection to public health is particularly significant in view of the complexity of the hazardous waste issue and the difficulty of assessing health risks in this area. As Vladimir Bencko et al. state, "The extent of environmental pollution and resulting human exposure to hazardous toxic chemicals is difficult to assess . . . [while] the assessment of the degree of pollution risk is increasingly a major health concern. The total extent of environmental pollution is difficult to assess, both quantitatively and qualitatively."[29] Public perception or interpretation of the degree of risk, particularly with respect to human health and safety, can provide strong political support for governmental action to deal with known and unknown aspects of environmental problems. In response to the growing public interest in risk and how it is taken into consideration in decision-making, the EPA issued a report entitled *Risk Assessment and Management.*[30] However, the public's ranking of environmental risks often differs from that of EPA experts. For example, while the public often ranks hazardous waste sites and industrial pollution of waterways as high risks, the EPA experts rank these as medium to low risks.[31]

Public perception of risks operates within the framework of complex and dynamic factors, not all of which result in positive problem-solving efforts. Noting the emergence of antienvironmental forces in the early 1980s, Stephen Duggan states: "There is, indeed, a movement growing which would, if successful, turn back many if not most of the environmental gains of the past 10 years. . . . The real problem, as I see it, is not purposeful obstructionism but the complexity of the issues. Environmental issues are often quite complicated."[32] Such complexity occurs from the interaction of diverse forces, such as increasing energy shortages, a declining economy, and fear of government regulation of business.

In the 1990s, antienvironmental rhetoric reached new heights. Among the more prominent books that mislead or misinform the public about environmental issues is Gregg Easterbrook's *A Moment on the Earth.*[33] Dixy Lee Ray has two books that challenge prevailing scientific opinion on environmental issues: *Trashing the Planet* and *Environmental Overkill: Whatever Happened to Common Sense.*[34] Her background as a marine biologist, former Chairman of the Atomic Energy Commission, and governor of the state of Washington provide Dr. Lee a high level of credibility. It is this credibility that provides further ammunition for antienvironmentalists who also serve as talk show hosts, such as Rush Limbaugh. Other backlash books that appeared in the 1990s include Ronald Bailey's *Eco-Scam: The False Prophets of Ecological Apocalypse* and his collection of essays written by prominent scientists, *The True State of the Environment,* as well as Bast, Hill, and Rue's *Ecosanity: A Common-Sense Guide to Environmentalism.*[35] The primary danger of such backlash or brownlash literature, according to Paul and Anne Ehrlich, is the potential erosion of support for the formulation and implementation of environmental laws sufficient to deal with environmental problems.[36]

Many heralded the election of Ronald Reagan to the presidency as an opportunity to reduce the negative impact of environmental regulations on business. Certainly many of Reagan's appointments—James Watt as secretary of the interior and Anne Gorsuch as administrator of the EPA, specifically—signified a sharp break from environmental policies of the immediate past administration. Long after these officials had left office, the impact of the policies that they initiated were and are still being felt.[37] A case in point is Secretary Watt's massive leasing of public lands for energy development in environmentally sensitive areas. Some of the leases run for fifty to one hundred years. Watt's successors in Republican administrations continued his policies. In 1987, for example, Secretary Donald Hodel announced intentions of the Department of the Interior to open the Arctic National Wildlife Refuge to oil leasing in spite of potentially serious environmental ramifications.[38]

Ironically, James Watt's dramatic and radical antienvironmental course assisted in revitalizing the environmental movement. Leaders of major environmental organizations such as the NWF, Sierra Club, and Audubon Society acknowledged large membership increases in response to Secretary Watt's activities. Donald Rheem reports increases in the top ten U.S. environmental organizations from 1980 to 1986.[39] See table 2 at the end of this chapter for a statistical assessment of the strength of the environmental movement. A petition to replace Secretary Watt was signed by over 1 million people from virtually all national environmental groups, representing about 6 million members. Watt resigned in October 1983. But his resignation was submitted only after an embarrassing public statement having nothing to do with his antienvironmental activities.[40]

Along with the successes of the environmental movement, there has been a resurgence of the conservation movement. The less radical approach of the conservation movement emphasizing wise use and development of natural resources is more appealing to some persons than the more restrictive or prohibitive posture of the environmental movement. Thus, many public officials and individuals from the private sector feel more comfortable with the term "conservation," which gives the impression of being more compromising and politically acceptable. Nevertheless, many individuals and organizations in both the conservation and environmental movements use both terms interchangeably.

As an extension of the conservation movement, the environmental movement both influences and expands narrow agency ideologies and technical and scientific approaches. Many of these influences bring needed ecological and societal value considerations into environmental policy formulation and implementation while generating a greater public awareness. Wise stewardship of the environment for the public interest calls for such value recognition and change on a sound ecological and societal basis for the establishment of effective environmental policies.[41]

Pressure Groups

In all aspects of environmental policy, pressure groups exert powerful influence through policy formulation and implementation. Wengert emphasizes that a major characteristic of environmental policies is extensive pressure group activity, especially in natural resources. This activity tends to be highly diverse and polarized.[42] Thousands of pressure groups are involved in the environmental policymaking process. Through the media and other means pressure groups try to persuade the public to support their positions in order to further pressure public agencies at all levels of government. The general concern of pressure groups is to gain access to and influence over the governmental decisionmaking process whenever and wherever policy affecting their special interests is being formulated or carried out.

Commenting on the activities of pressure groups, Wengert observes the struggle to be for

> power to influence and control factors of major, even vital, importance to the group and to its members. The political struggle is more often for something than against other groups . . . [that is,] for advantage and position rather than a fight with specific adversaries. It may also involve the protection of status quo and a desire to veto change. At the same time, the success of one group may mean the failure of another, so conflict and hostility among groups may exist, although it is not often an immediate goal and end in itself. . . . [T]he struggle is based upon winning friends, alliances, and alignments as a means of influencing the course of governmental decisions.[43]

On a very broad basis, organized pressure groups operating in environmental policy areas may be divided into two general categories: (1) economic or business groups that are affiliated with private interests, such as the National Association of Manufacturers, the National Cattlemen's Association, and the National Association of Reclamation, seeking economic gain; and (2) nonprofit citizen groups, such as the Sierra Club, the Audubon Society, and the NWF, which are associated with environmental and broad public interests. Generally speaking, groups related to economic interests are registered lobbyist organizations, while ideologically motivated groups are often considered educational, although they also seek to influence various types of legislation and policy.

This classification is based on a formal definition of organized pressure groups with a connotation of specialized interest. A more realistic and comprehensive concept of these groups encompasses all groups or organizations that influence, or are influenced by, the complexities of environmental policy. In the private and public spheres large organizations make a majority of the decisions that influence environmental policy at the governmental level. It is also increasingly difficult to identify various informal and complex aspects of the decisionmaking processes for large governmental agencies and private organizations that have a stake in the environment.

Governmental agencies informally operate as environmental policy pressure groups. With the broad legislative directives regarding the establishment and missions of the large number of federal agencies involved in environmental affairs, many environmental policy decisions are products of pressure exerted by these units. Legislative liaison sections and personnel of federal agencies contribute a large percentage of congressional legislation. Agency personnel influence environmental policies and programs at various levels through informal relations.

Most federal agencies are clientele-oriented organizations that serve the special interests of "customers" through their missions. Symbiotic relationships are often maintained between an agency and its clientele by means of informal, behind-the-scenes political operations leading to support for specific policies, programs, and budgets. One example is the relationship between the Bureau of Reclamation and the National Reclamation Association. Many agency-clientele relationships are economically oriented, and this produces little attention to second-order or ecological considerations that are more in the public interest. The complex informal associations formed between personnel of the agency and its clientele have a pragmatic effect on policymaking. Specialized and compatible dominant interests in both government agencies and private organizations form the basis for these symbiotic associations.

With the rise in public concern for the environment and with increased media coverage, a very rapid increase has taken place in the number of environmental and conservation pressure groups. In the United States there are now more than 40,000 environmental organizations that overall have more than 20 million members.[44] In 1998 the three largest organizations, the NWF, the World Wide Fund for Nature, and the Nature Conservancy, had memberships of about 4.4 million, 1.2 million, and 900,000 respectively. In the 1998 fiscal year, the top three organizations had budgets of $96 million, $60 million, and $131 million respectively.[45]

Increasing membership in mainstream organizations was only one type of reaction to the decline in environmental concern by the political establishment during the Reagan years. A small but significant segment of the environmental movement became radicalized. Symptomatic of this process is the emergence of Earth First! (the exclamation mark is part of its name), a small but intensely idealistic group dedicated to environmental preservation. This group uses political tactics previously associated with civil rights activists.

Some of the conservation/environmental pressure groups, such as the NWF and the Wilderness Society, identify with specific issues, as their titles would indicate. Within the environmental movement, however, even groups with specialized orientations are beginning to become involved in a variety of environmental issues; thus indicating a greater awareness of the relationships of their specialized orientation to the overall environment and its problems.

Environmental organizations display diverse values, interests, and internal and external conflicts, as well as pluralism. This results in a willingness to adopt a variety of political tactics. Within the environmental movement there is a tendency for groups to cooperate and form coalitions for political purposes, particularly when these groups are facing severe opposition. There has also been a countertendency toward radicalization, a pattern for removal from the older coalition-building groups.

Nevertheless, the more conservative environmental organizations share in common with the extremist groups a growing tendency to become involved in political affairs. Alarmed by what they perceive as serious threats to the environment posed by the policies and appointments of officials unsupportive of environmental interests, even old-line conservation groups are breaking tradition and undertaking major political efforts. This new thrust includes organizing members around the nation to be a political force in protecting environmental gains made in the past. A part of this new effort is directed toward maximizing grassroots power through partnership with other organizations at the local level. Environmental nongovernmental organizations are growing worldwide, promoting activism and "world civic politics."[46]

Although these environmental pressure organizations are based on voluntary memberships and contributions by citizens, an executive director and a professional staff usually handle daily operations. In the larger organizations, the professional staff may contain legal, scientific, communication, and other experts. An elected president and other officers or a council may provide overall policy and direction. Organizational magazines, newsletters, and other communications from the professional staff bring issues to the attention of the general membership for political action. Typical action includes letters to legislators and public officials, petitions, public hearing testimony, campaign activities, contributions, projects, and other types of citizen involvement. Direct-action tactics borrowed from the civil rights movement have also been utilized on occasion, such as Greenpeace's attacks on foreign national fishing ships engaged in illegal fishing activities.

Much of the political action is intended to secure support for environmental policies and activities in line with the values, interests, and positions represented by the environmental groups. The officers, staff, and general membership carry on complex interactions with various agency and legislative decisionmakers. They influence policy formally through public testimony and participation in budget hearings. Informally, they engage in behind-the-scenes brokerage politics and constantly monitor agency actions on behalf of the environmental public interest.

With the growing political involvement of environmental organizations in the United States, complexities have arisen under the Federal Regulation of Lobbying Act of 1946. The law prescribes that tax-deductible donations cannot be used for lobbying or for other attempts, such as advertising, to influence legislation. Consequently, nonprofit organizations, including those active in environmental affairs, are severely hampered and threatened by the extent of their lobbying. If they lobby too much, they may undercut the basis for their financial support. In contrast, profit-oriented groups can deduct any money spent for lobbying from their taxable income. These and political costs may be passed on to consumers in the form of higher prices.[47]

Most environmental pressure groups are citizen-oriented, nonprofit organizations with meager budgets based on volunteer contributions. They often come under the same tax classification as church and educational pressure groups that also engage in political activities. Although many conservation organizations were formed to advance educational objectives (including the education of politicians and officials), some organizations, such as Friends of the Earth and Zero Population Growth, have waived their tax-exemption status so that they may engage directly in political and lobbying activities, including contributing to political candidates. This exemplifies a growing recognition that pressure groups considerably influence the formulation of environmental policy. Yet severe interpretation and enforcement of the lobbying act by the Internal Revenue Service (IRS) or its implied threat of enforcement could result in decline of financial support for most environmental pressure groups.

The position of economic pressure groups under the lobbying act of 1946 is in sharp contrast with the ambiguous and threatened position of environmental pressure groups. Economic pressure groups are usually registered as lobbyists under the act. The American Mining Congress, the National Association of Manufacturers, the National Cattlemen's Association, and the Chamber of Commerce are prime examples. Although a single industrial organization may have its own pressure group or lobbying staff, many economic pressure groups are representatives of several organizations or various segments of a specific industry. Business and industry provide these groups with large amounts of money, however, to conduct their affairs. The full amount of money is seldom fully declared, as the lobbying act requires. Economic pressure groups are able to employ outstanding professional staffs, including top legal aides and specialized services. They engage in a full spectrum of lobbying and public relations activities (sometimes contracted to public relations firms) in efforts to influence the policy process.

Economic pressure groups seek policies favoring growth and development while resisting legislative and administrative action designed to control or regulate industry. Although a large number of new jobs were created in recent years through environmental protection activities like pollution control and land reclamation, these activities have also increased costs for some segments of industry. Prior to enactment of environmental control measures these costs were "externalized" to the general public. The declining profits experienced by some segments of industry are linked directly to mandated internalization of formerly externalized costs. Environmental groups argue that industry should have been paying these costs all along. Economic pressure groups are generally concerned with opposing all efforts to establish such mitigation measures that detract from or threaten their industrial interests. They are also concerned with securing from government special advantages, such as cheap leasing of public lands, subsidies, and contracts for public works projects. Growing economic problems, high unemployment levels, along with the general support of economic interests by the Reagan administration, placed economic pressure groups in more powerful positions in the 1980s, while environmental pressure groups were forced into a defensive posture. Although the economic health of the United States improved measurably in the 1990s, economic interest groups continue to press the Republican majority in Congress to introduce numerous bills to roll back environmental legislation in the 104th and 105th Congresses. The Republican-dominated Congress was so responsive to the economic pressure groups that Earth Day founder and former senator, Gaylord Nelson, characterized the 104th Congress as the "worst environmental Congress."

The politicization of environmental policy has placed economic and environmental pressure groups into intense, complicated positions. Ideological polarization between them has led to legislative struggles. First-order considerations—materialistic gratification, profit taking, or unrestricted free enterprise—are pitted against second-order considerations—ecosystem preservation, internalization of formerly externalized costs, or maintenance of quality of life. The values of other technologically oriented organizations in both private and public sectors further complicate the picture. The power struggles between formal and informal pressure groups produce constantly changing and multifaceted environmental policies.

An emerging pattern involving environmental pressure groups is occurring, particularly at the local level. These pressure groups are challenging and appealing micropolicy decisions made by governmental agencies at all levels. Some pressure groups are local, ad hoc organizations that have developed in response to a given environmental issue or problem, while others are grassroots organizations with state and national affiliations. These groups are interested in modifying and influencing policy decisions through expression of their values and opinions.

Environmental administrators in a democratic society must be sensitive and responsive to such an informed citizenry. This new activism may be ascribed, in part, to the "communications revolution" that exposes almost everyone to environmental problems through various media.[48]

The National Forest Management Act encourages local participation in the development of management plans. Local news media inform citizens of controversial use strategies and arouse citizen action groups to congregate at USDA Forest Service public hearings and demand restriction of environmentally damaging activities within the national forests. These groups often draw upon the expertise of local university professors' knowledge about the forests through their personal research. The sum effect is extensive citizen involvement in what used to be fairly routine management decisions ranging from clear-cutting of timber to opening up areas for more intensive recreational use.

In a pluralistic society far greater participation and influence comes through the representatives of pressure groups. Many environmental pressure groups are urban-based and must often act from a considerable distance on local policy decisions. By emphasizing local input and recognition, environmental and economic pressure groups can have an even greater role in local environmental policy. Regardless of who is involved, much of any pressure group's effectiveness depends upon the value system of the local government agency's local administrators and on his or her perception of and responsiveness to values and interests of given pressure groups.

Most environmental agencies are staffed by local administrators who have various technical and scientific specializations and backgrounds. But, as Henry Ehrmann warns,

> A merely technical education, however excellent, does not prepare one sufficiently for the task of developing long-range policies. Pressure of time, always particularly great in these administrations, is added to the pressure of special interests. The shorter the perspective and the narrower the field, the more the civil servant will be inclined to be swayed by the persuasive, pragmatic [argument] preferred by his counterpart speaking for the private sector of the economy.[49]

Environmental Law

Pressure groups, governmental agencies, special legal organizations, and individuals use the law as a means of attaining particular environmental objectives. By merely implying that a given policy will be subjected to legal proceedings, an environmental group can exert sufficient pressure on a particular agency or industry to correct the violation, thus avoiding costly and lengthy court action. Nevertheless, environmental pressure groups have been forced increasingly to turn to the courts, due to greater resistance or apathy to their objectives from the federal government and industry. As a result, public land managers are often involved in several legal suits at the same time. Opponents to clear-cutting in the Shawnee National Forest, in southern Illinois, for example, filed one lawsuit after another to stop the cutting of nonnative pine trees under the premise that the pine trees provided habitat for pine warblers. The irony in this action was that the USDA Forest Service was attempting to restore the ecosystem to the oak-hickory community that was present in the 1700s prior to European settlement.

Through recourse to the law, environmentalists can influence environmental policy on an interactional basis. Court and administrative hearing opinions related to the environment forge much of environmental policy. With the thousands of federal, state, and local laws pertaining to the environment as well as the abundance of administrative laws and regulations and agency procedures, the fields of environmental policy and law are complex and interwoven. Since environmental policy and law often are articulated in ambiguous and idealistic terms, both are also subject to interpretation through legal and administrative processes.

The burden of proof for determining the extent of damages normally falls upon the environmental organization filing suit against an agency for failure to implement the law. Thus, the group advocating maintenance or improvement of environmental quality has the *very* heavy burden of showing that the status quo of environmental quality should be preserved or the status quo of poor environmental quality should be altered.

Much environmental law represents the public interest on an equity basis. The law of equity—sometimes known as the law of conscience—is a well-recognized part of the legal system that calls for the public interest to be paramount over private interests. In this sense, NEPA, requiring government and industry to be responsible to the environmental public interest, should be considered an equity law. However, court decisions interpreting NEPA hold that it creates a mere procedural requirement and does not force federal agencies to do anything in particular. In contrast, a number of state environmental policy acts have retained their equitable force. In any case, until institutional changes occur, it will be up to environmental groups, through their choices of situations, to litigate to determine how environmental policy decisions will be made.

Probably the most severe problem confronting environmental groups anticipating litigation is a shortage of funds. In contrast, industry and government can readily afford legal assistance to engage in judicial and administrative proceedings. They can afford to retain scientific experts whose services alone can be extremely expensive. The Reagan administration—well aware of the financial constraints of environmental and other public interest groups—pushed strongly to deny the award of legal fees in suits against the federal government.

In addition to environmental groups, there exist public interest law firms that can more readily engage in environmental litigation because they are supported wholly or in part by foundation grants. Among such groups are the Environmental Defense Fund and the Natural Resources Defense Council (NRDC). Even these environmental advocates operate in gray areas, since only a portion of the tax-deductible contributions supporting them may be used for lobbying. An interesting exploitation of the concept of public interest is illustrated by the activities of the Mountain States Legal Foundation, an anti–environmental law foundation underwritten by conservative western business people. While he served as its first president, James Watt initiated legal actions against federal strip mine regulation, a specific wilderness plan in the state of Wyoming, a National Park Service ban on motorized rafting in the Grand Canyon, and various other environmental laws, regulations, and policies.[50]

Numerous laws advocating environmental protection and quality are already in existence. There is also a vast body of common or precedent law regarding nuisance and trespass that can apply to pollution and other environmental controls on the national, state, and local levels. Regardless of legal justification for enforcement of environmental laws, political and agency considerations inhibit many government lawyers from taking aggressive positions on behalf of the environmental public interest. Such considerations also encourage these lawyers to represent the economic and vested interests associated with government. Much of the work of environmental advocates and pressure groups therefore seeks to enforce and adequately interpret existing legislation—that is, to get government to enforce its own laws. They also perform watchdog activities to assure that the agencies follow proper procedures under administrative law. They take legal actions against governmental agencies who are not complying with or enforcing environmental law.

In considering the role of the Natural Resources Defense Council and other organizations oriented to environmental law, Stephen Duggan, a former NRDC board chairman, states:

> We do not pass laws; we merely try to see that the existing environmental laws are properly enforced. We do not dictate public policy; but we do try to illuminate the potential consequences of policies which would harm the environment . . . but enforcement of these laws [Clean Air and Water Acts, etc.] has largely depended, as former EPA administrator Russell Train put it, on NRDC and other groups "holding bureaucratic feet to the fire."[51]

As an example, the NRDC and five other environmental groups filed suit in a federal court against the Department of the Interior to stop its secretary from eliminating wilderness protection for 800,000 acres of western public lands. The NRDC also sued the EPA under the Bush administration for failure to meet congressionally established deadlines for the promulgation of rules under the 1990 Clean Air Act Amendments.

A complicated question for environmental law is that of "standing to sue." In the early 1970s the courts generally allowed standing to persons speaking on behalf of the public interest relative to the environment, even though they themselves were not injured. A plaintiff has to establish "the existence of a chain of causation between the allegedly illegal government action and an injury to some portion of the environment used by the plaintiff. It is not necessary to allege any physical injury to the plaintiff or any economic damage; aesthetic injury is enough."[52] By 1975, however, the Supreme Court and other courts returned completely to more traditional interpretations of standing to sue. This left the situation muddled and ambiguous.[53] Yet it is recognized that provisions in the National Environmental Policy Act, the Federal Water Pollution Control Act, and the Clean Air Act provide citizens with the right of "standing to sue" in environmental litigation.[54] A unique and thought-provoking case for "standing to sue" is made by Christopher Stone in *Should Trees Have Standing? Toward Legal Rights for Natural Objects*. Stone argues that natural objects (trees, mountains, rivers, lakes) should have legal rights just as corporations do, and consequently should be entitled to public-interest legal representation in environmental protection cases.[55]

Much environmental law centers on efforts to give environmental values an equal weight with developmental and other values. A. Dan Tarlock indicates that most federal agencies (despite the Fish and Wildlife Coordination Act of 1958) neglected consideration of environmental values before 1970 because they were not consistent with carrying out their primary missions as defined by Congress. Thus, congressional legislation like NEPA and other environmental laws sought to provide "equal dignity" for environmental values and required federal agencies to take them into consideration. Much "public interest" environmental law work concentrates on legal efforts for defining and securing this authority and protecting environmental values.[56] Court trends in the environmental field are toward conservative and economic interpretations of the law. They are also toward transferring discretion and original jurisdiction from the courts to federal agencies that often may be more responsive to and aligned with economic and private interests. This situation is further complicated by a wide and diverse range of opinions by scientific and other experts on environmental issues and law.[57]

Ideology can also play a role in court rulings made by judges. The United States Circuit Court of Appeals for the District of Columbia has a pivotal role in environmental regulatory decisions. This occurs because the D.C. Circuit Court hears all appeals from the EPA administrative court. Judges are supposed to make decisions on the basis of the law and, in the case of a court of appeals, whether the lower court abided by the appropriate procedures. In a detailed study of the decisions of the D.C. Circuit Court in the 1980s and 1990s, Richard Revesz revealed several interesting patterns. During this period of time a disproportionate number of the Circuit Court judges were appointed by Republican presidents. Revesz examined both industrial and environmental challenges to regulations. The reversal rates for industrial challenges showed no significant patterns; the judges were essentially neutral. In contrast, for environmental challenges, the reversal rates of Democratic judges were consistently higher than for Republican judges. It appears that Republican-appointed judges tend to be less supportive of the perspectives of environmentalists and are less willing to tolerate environmental challenges to regulations created by Republican administrations. Revesz suggests that since a disproportionate number of D.C. Circuit Court judges have been nominated to the Supreme Court this may affect the way judges vote. They may want to consistently please their party.[58]

In a reflection on the first twenty years of environmental law, Sheldon Novick summarizes the implications of its evolution:

> What we see looking back is that pollution control is a new subject for federal law. Everything is new—the institutions, the ideas, the rules of law. All have been freshly invented. A large and very complex system has been built over the past twenty years, its separate parts built separately and sometimes without much awareness of the whole structure.
>
> First, and most commonly, existing sources of pollution—the operating factories and mines whose effluents prompted these laws at first—are treated with care. . . . Flexible schedules . . . in the pollution laws [however] are complicated—they drive action toward goals, but they also put off difficult or expensive problems. . . . [t]he standards themselves are realistic. . . . Pollution control law, in short, is a largely completed structure. Like the tax laws, the structure is so big and complex, and it embraces so much of national life, that big changes would be disruptive, and would require a major political effort for which no support appears likely, after the brief effort at change early in the Reagan administration. The pollution law we have is therefore the law we are likely to keep for a long time. . . . It is probably time to make the structure permanent . . . [i]t might be [also] time to review the pollution control laws and remove the pointless inconsistencies. . . . Still, the federal laws express our values and our purposes, and whatever failings they are pretty solidly founded. Talk about fundamental change is probably pointless.[59]

Republicans in Congress, in the 1990s, obviously did not agree with Novick's conclusion. The Republican attempts to reduce funding for implementation of environmental laws, the riders that they put on unrelated legislation to change environmental regulations, and their overt attempts to change the regulatory process, most specifically with regard to environmental laws, attest to their belief that environmental law can be changed.

The above material on environmental law emphasizes an American context. It is also important to think, however, in terms of a concerted international approach to environmental law. Pollution does not respect abstract political boundaries. Air pollution generated along the Ohio River has negative implications for trees and fish in Canada, likewise pollution from Czechoslovakia affects Germany—witness the ailing, world-renowned Black Forest.[60] The Rhine River gathers pollution from all the countries it runs through and then passes it on to the Netherlands. Haz-

Table 1   Major Federal Environmental Legislation

Date	Title	Important Provisions
Water Pollution Control		
1899	Rivers and Harbors Act (Refuse Act), 30 Stat. 1152	Required permit from chief of engineers for discharge of refuse into navigable waters.
1948	Federal Water Pollution Control Act P.L. 80-845, 62 Stat. 1155	Gave the federal government authority for investigations, research, and surveys. Left primary responsibility for pollution control with the states.
1956	Federal Water Pollution Control Act Amendments, P.L. 84-660, 70 Stat. 498	Established federal pollution policy for 1956–70 period. Provided (1) federal grants for construction of municipal water treatment plants, (2) complex procedure for federal enforcement actions against individual dischargers.
1961	Federal Water Pollution Control Act, Amendments, P.L. 87-88, 75 Stat. 204	Strengthened federal enforcement procedures.
1965	Water Quality Act, P.L. 89-234, 79 Stat. 203, 79 Stat. 903	Created Federal Water Pollution Control Administration.
1966	Clean Water Restoration Act, P.L. 89-753, 80 Stat. 1246	Increased grant authorizations.
1970	Water Quality Improvement Act, P.L. 91-224, 84 Stat. 91	Established liability for owners of vessels that spill oil and created new rules regarding thermal pollution.
1972	Federal Water Pollution Control Act Amendments, P.L. 92-500, 86 Stat. 816	Set policy under which the federal government now operates. Provided (1) federal establishment of effluent limits for individual sources of pollution, (2) issuance of discharge permits, (3) large increase in authorized grant funds for municipal waste treatment plants.
1974	Safe Drinking Water Act, P.L. 93-523 (as amended by P.L. 96-502), 88 Stat. 1660	Directed EPA to set standards (as amended) applicable to all public water systems, to protect human health from organic, inorganic, and microbiological contaminants and for turbidity in drinking water.
1986	Safe Drinking Water Act Amendments, P.L. 99-339, 100 Stat. 642	Established primary (enforceable) and secondary (advisory) national drinking water regulations based on maximum contaminant levels of specific pollutants. Prohibited

Table 1   Continued

Date	Title	Important Provisions

Water Pollution Control

		use of lead in pipes or solder for public water systems.
1977	Federal Water Pollution Control Act Amendments, P.L. 95-217, 91 Stat. 1566–1609	Relaxation of some standards under 1972 amendments. Relaxed existing industrial antipollution standards on suspended solids, fecal bacteria, and oxygen demand of discharge if it can be shown that the cost of equipment exceeds benefits.
1987	Federal Water Pollution Control Act Amendments, P.L. 100-4, 101 Stat. 7–78	Required states to identify nonpoint sources and develop plans for a comprehensive water pollution control program including nonpoint sources.
1996	Safe Drinking Water Act, P.L. 104-182, 110 Stat. 1613	Established a state revolving fund to provide financial assistance to small communities to help them meet federal drinking water standards. Required citizen notification of water quality, safety, and violation of standards.

Air Pollution Control

1955	Air Pollution Control Act, P.L. 84-159, 69 Stat. 322	Authorized a federal program for research, training, and demonstrations relating to air pollution control (extended for four more years in 1959).
1963	Clean Air Act, P.L. 88-206, 77 Stat. 392	Gave the federal government enforcement powers through enforcement conferences similar to the 1956 approach to water pollution.
1965	Motor Vehicle Air Pollution Control Act, P.L. 89-272, 79 Stat. 992	Added new authority to 1963 act, giving U.S. Department of Health, Education, and Welfare (HEW) power to prescribe emission standards for automobiles as soon as practicable.
1967	Air Quality Act, P.L. 90-148, 81 Stat. 485	(1) Authorized HEW to oversee state ambient air quality and state implementation plans, (2) set national standards for auto emissions.
1970	Clean Air Act Amendments, P.L. 91-604, 84 Stat. 1676–	Greatly expanded the federal role in setting and enforcing standards for ambient

Table 1   Continued

Date	Title	Important Provisions
Air Pollution Control		
	1713	air quality and established stringent new emission standards for automobiles.
1974	Clean Air Act Amendments, P.L. 93-319, 88 Stat. 248–259, 261, 265	Technical amendments. Some relaxation of standards.
1977	Clean Air Act Amendments, P.L. 95-95, 91 Stat. 685–796	Required states with air quality nonattainment areas to adopt plans for full compliance by 1982. Deferred further reductions in automobile toxic fumes until 1981.
1990	Clean Air Act Amendments, P.L. 101-549, 104 Stat. 2339	Established acid rain controls, air toxics program, additional provisions for reduction of ozone problems in nonattainment areas.
1993	Federal Employees Clean Air Incentives Act, P.L. 103-172, 107 Stat. 1995	Provided incentives to federal employees to use mass transit and carpool.
Solid Waste and Resource Recovery		
1965	Solid Waste Disposal Act (Title II of P.L. 89-272) as amended by the Resource Recovery Act of 1970 (P.L. 92-518 and P.L. 93-14, 973), 79 Stat. 997	(1) Promoted the demonstration, construction, and application of solid waste management and resource recovery systems, (2) provided technical and financial assistance to states, local governments, and interstate agencies in the planning and development of resource recovery and solid waste disposal programs, (3) provided for national research for improved management techniques, (4) provided for federal guidelines.
1976	Resource Conservation and Recovery Act, P.L. 94-580, 90 Stat. 2795–2839	Provided technical assistance for the development of management plans and facilities for recovery of energy and other resources from discarded materials. Regulated management of hazardous materials. Regulated management of hazardous wastes.
1984	Hazardous and Solid Waste Amendments Act, P.L. 98-616, 98 Stat. 3222–3277	Requirements added: to regulate small-quantity generators (less than 1,000 kg/month, more than 100 kg/month), to ban land disposal of hazardous wastes to

Table 1   Continued

Date	Title	Important Provisions

Solid Waste and Resource Recovery

		extent required to protect human health and the environment.
1980	Comprehensive Environmental Response, Compensation and Liability Act, P.L. 96-510, 94 Stat. 2767	Established Superfund from taxes on petroleum and chemical derivatives to clean up abandoned hazardous waste sites. Mandated the establishment of a National Hazardous Substance Response Plan. Established a system of generating revenues for the cleanup funds (1) by imposing taxes upon the manufacture, production, or importation of petroleum and certain chemicals, (2) by imposing user fees upon owners and/or operators of waste disposal facilities, and (3) by appropriation of general revenues.
1986	Superfund Amendments and Reauthorization Act, P.L. 99-499, 100 Stat. 1613	Provided $8.5 billion for development of permanent solutions for hazardous waste sites that reduce the amount of hazardous waste. Title III established the Emergency Planning and Community Right-to-Know Act (P.L. 99-499) that requires states to develop emergency response plans for toxic releases and to assure that communities are alerted to the presence of dangerous chemicals.

Noise Pollution Control

1968	Air Act to Require Aircraft Noise Abatement Regulation, P.L. 90-411, 82 Stat. 395	Amended Federal Aviation Act of 1958 to require aircraft noise abatement regulation.
1972	Noise Control Act, P.L. 92-574, 86 Stat. 1234	(1) Provided for coordination of federal research and activities in noise control, (2) authorized establishment of federal noise emission standards for products distributed in commerce.
1978	Quiet Communities Act, P.L. 95-609, 92 Stat. 3079–3081	Provided for federal technical assistance to communities to develop noise control programs.

Table 1   Continued

Date	Title	Important Provisions

Chemicals

1947	Federal Insecticide, Fungicide, and Rodenticide Act, P.L. 80-102, 61 Stat. 163	Authorized federal regulation of pesticides and related chemicals.
1972	Federal Environmental Pesticide Control Act, P.L. 92-586 (as amended by P.L. 94-51, P.L. 94-140, P.L. 95-396, P.L. 96-359, and P.L. 98-620)	Authorized federal regulation of pesticides and related chemicals including banning, manufacture, commercial sale, and use.
1976	Toxic Substance Control Act, P.L. 94-469	Required testing and necessary use restriction on certain chemical substances.
1978	Federal Insecticide, Fungicide, and Rodenticide Act (FIFRA) Amendments, P.L. 95-396, 92 Stat. 819–838	Clarified state authority to regulate the sale or use of pesticides and gave approved states primary responsibility for pesticide violations.
1996	Food Quality Protection Act, P.L. 104-170, 110 Stat. 3290	Amended FIFRA and the Federal Food, Drug, and Cosmetic Act to permit some amount of pesticide in food, modifying the impact of the Delaney clause that totally banned the presence of pesticides in food. Required periodic review of pesticide registrations.

Water Resources and Land Use

1964	Water Resources Research Act, P.L. 88-379, 78 Stat. 329	Established water resource research centers to promote a more adequate program of water research.
1965	Water Resources Planning Act, P.L. 89-80 (as amended by P.L. 94-112 and P.L. 95-404), 79 Stat. 244	Provided for the "optimum" development of the nation's natural resources through coordinated planning of water and related land resources. Established a water resources council and river basin commission.
1968	Wild and Scenic Rivers Act, P.L. 90-542 (as amended by P.L. 96-487 and P.L. 98-444), 82 Stat. 906	Provided for designation and restricted use of wild and restricted rivers.
1972	Coastal Zone Management Act, P.L. 92-583 (as amended by P.L. 94-370. P.L. 94-464,	(1) Provided for assistance to states to develop and implement management programs for use of land and water resources

Table 1 Continued

Date	Title	Important Provisions
Water Resources and Land Use		
	and P.L. 99-272), 86 Stat. 1280	of the coastal zone areas, (2) encouraged participation and cooperation among the public, federal, state, local, and regional authorities in development of coastal zone management programs.
1976	Federal Land Policy and Management Act, P.L. 94-579, 90 Stat.	Provided an organic act for the Bureau of Land Management (Department of the Interior). Directed that unless otherwise specified, the management of public lands be on a multiple use and sustained yield basis.
1976	National Forest Management Act, P.L. 94-588, 90 Stat. 2949	Required comprehensive assessment of present and anticipated uses, demands for and supply of renewable resources from the nation's public and private forests and rangelands.
1972	Marine Mammal Protection Act, P.L. 92-522 (as amended by P.L. 95-136 and P.L. 97-58), 86 Stat. 1027	Regulated the taking of marine mammals and replenishing species or population stock that has diminished.
1973	Endangered Species Act, P.L. 93-205 (as amended by P.L. 95-632, P.L. 97-304, and P.L. 98-327), 87 Stat. 884	Regulated the taking and use of threatened and endangered species. Created mechanisms for identification and protection of critical habitat for these species leading to land use restrictions.
1937	Federal Aid in Wildlife Restoration Act (as amended by P.L. 91-503, P.L. 92-558, P.L. 98-502, and P.L. 10-233), 50 Stat. 917	Authorized the imposition of excise taxes on hunting equipment to create a federal aid fund for the states to increase wildlife populations.
1950	Federal Aid in Sport Fish Restoration Act (as amended by P.L. 91-503, P.L. 98-369 and P.L. 100-17), 64 Stat. 430	Authorized the imposition of excise taxes on fishing equipment to create a federal aid fund for the states to restore sport fish populations.
1980	Fish and Wildlife Conservation Act, P.L. 96-366, 94 Stat. 1322	Authorized the development of nongame programs and funding mechanisms.
1990	Coastal Zone Act Reauthorization Act Amendments, P.L. 101-508, 104 Stat. 1388	Required coastal states to develop nonpoint source water pollution controls

Table 1   Continued

Date	Title	Important Provisions
Comprehensive Environmental Acts		
1969	The National Environmental Policy Act, P.L. 91-190 (as amended by P.L. 94-52 and P.L. 94-83), 83 Stat. 852	Established a national environmental policy; required information on and co-ordination of federal projects and programs impacting upon the environmental; established the Council on Environmental Quality.
1970	Environmental Quality Improvement Act (title II of P.L. 94-224), 84 Stat. 114	Required each federal department and agency conducting or supporting public works activities that affect the environment to implement the policies established under existing law.

Date	Title
Health, Safety, and Consumer Acts	
1960	Federal Hazardous Substances Labeling Act, P.L. 86-613, 74 Stat. 372
1965	Service Control Act P.L. 89-286, 79 Stat. 1034
1966	National Traffic and Motor Vehicle Safety Act, P.L. 89-563, 80 Stat. 718
1968	Flammable Fabrics Act, P.L. 90-189, 81 Stat. 568–573
1968	National Gas Pipeline Safety Act, P.L. 90-481, 86 Stat. 616
1969	Federal Coal Mine Safety and Health Act, P.L. 91-173, 83 Stat. 742
1969	Child Protection and Toy Safety Act, P.L. 91-113, 83 Stat. 187
1970	Federal Railroad Safety Act, P.L. 91-458, 84 Stat. 971
1970	Hazardous Materials Transportation Control Act, P.L. 91-548, 84 Stat. 977
1970	Occupational Safety and Health Act, P.L. 91-596, 84 Stat. 1590
1971	Federal Boat Safety Act, P.L. 92-75, 85 Stat. 213
1972	Ports and Waterway Safety Act, P.L. 92-340, 86 Stat. 424
1972	Consumer Products Safety Act, P.L. 92-573, 86 Stat. 1207
1974	National Manufactured Housing Construction and Safety Standards Act, P.L. 93-383, 88 Stat. 700
1974	Motor Vehicle and School Bus Safety Amendments, P.L. 93-492, 88 Stat. 1470
1974	Hazardous Materials Transportation Act (Hazmat Act), P.L. 93-633, 88 Stat. 2156
1976	Energy Policy and Conservation Act, P.L. 94-163, 89 Stat. 371
1980	Low-level Radioactive Waste Policy Act, P.L. 96-573 94 Stat. 3347
1981	Environmental Research, Development, and Demonstration Authorization Act, P.L. 94-475, 90 Stat. 2071
1984	Nuclear Waste Policy Act, P.L. 97-425, 96 Stat. 2201

Table 1   Continued

Date	Title

Health, Safety, and Consumer Acts

1984   Asbestos School Hazard Abatement Act, P.L. 98-377, 98 Stat. 1287

1985   Food Security Act, P.L. 99-198, 99 Stat. 1354

1990   Pollution Prevention Act, P.L. 101-508 (Title VI)

1990   Hazardous Materials Transportation Uniform Safety Act, P.L. 93-633, 88 Stat. 2156

1990   Oil Pollution Act, P.L. 101-380, 104 Stat. 484

1991   Intermodal Surface Transportation Efficiency Act, P.L. 102-240, 105 Stat. 1914

1992   Energy Policy Act, P.L. 102-486, 89 Stat. 371

1995   Alaska Power Administration Asset Sale and Termination, P.L. 104-58, 109 Stat. 557

1996   Land Disposal Program Flexibility Act PL 104-119, 110 Stat. 830

1996   Federal Agricultural Improvement and Reform Act, P.L. 104-127, 110 Stat. 888

Adapted and updated from Advisory Commission on Intergovernmental Relations, *The Federal Role in the Federal System: The Dynamics of Growth: Politics, Pollution, and Federal Policy* (Washington, D.C.: GPO, 1981) and U.S. Code.

ardous waste is transported illegally from one country to another for disposal or other economic reasons, as the infamous Seveso accident attests.[61] So much hazardous waste was being shipped to other countries, especially less-developed countries, that the Basel Convention of 1989 was created. Beginning in 1998, exportation of all forms of hazardous waste from the twenty-nine Organization for Economic Cooperation and Development (OECD) countries to non-OECD countries is prohibited.[62]

There are four basic reasons why an international legal approach is necessary:

1. The environment knows no frontier; air, water—both rivers and oceans —and wildlife cannot be contained by national boundaries.
2. It makes economic sense because costly environmental damages can be avoided.
3. Natural resources located outside jurisdiction of any state (the so-called global commons) need to be conserved.
4. Mankind as a whole has responsibility for the conservation and/or management of acceptable natural areas or resources for scientific, cultural, ecological, or aesthetic reasons.[63]

Table 2   Top Twelve U.S. Environmental Organizations

Organization	Staff	Budget (in $ millions)	Headquarters	Membership
National Wildlife Federation	400	96	Virginia	4,400,000
World Wild Fund for Nature	N/A	60	Washington, D.C.	1,200,000
The Nature Conservancy	2000	131	Virginia	900,000
Ducks Unlimited	300	71	Tennessee	553,000
Greenpeace USA	300	37	Washington, D.C.	500,000
National Audubon Society	300	44	New York	600,000
Sierra Club	294	43	California	550,000
Natural Resources Defense Council	165	27	Washington, D.C.	350,000
National Parks and Conservation Association	50	14	Washington, D.C.	350,000
Environmental Defense Fund	160	24	New York	300,000
The Wilderness Society	98	14.7	Washington, D.C.	255,000
Defenders of Wildlife	60	7	Washington, D.C.	180,000

Source: Adapted from data in Christine Maurer and Tara E. Sheets, eds., *Encyclopedia of Associations: An Associations Unlimited Reference,* 34th ed., vol. 1, part 1 (Detroit, Mich.: Gale Research, 1999).

One of the main problems in international environmental law is the lack of enforcement. Either treaties are not ratified or economic and social realities prevent their implementation at levels to protect the environment. This reality hinders the development and the implementation of adequate problem resolution strategies globally.[64]

3 Environmental Administration in a Decisionmaking Context

Since the first edition of this book was published, several events have occurred that have shaped the environmental decisionmaking context in the United States. The first event was a change in emphasis from discrete land management decisions to a broader, all-encompassing ecosystem management approach. This approach requires decisionmakers to consider many more variables when making a decision. Ecosystem management requires the decisionmaker to identify who the stakeholders are in the decision process and then to involve them in development of any action plans. Such a cooperative endeavor fosters a more positive relationship between levels of government, as well as between the federal government and the private sector. It also provides greater opportunities for problem resolution through the pooling of resources and the reduction of hostilities.

The multidisciplinary nature of ecosystem management helps to break down communication barriers between individuals involved in the decision process who have been trained in totally different fields. Even within environmental management, professionals trained in hydrology, range management, and wildlife management, for example, have considerably different perspectives. Fields outside environmental management often don't even train their professionals how to address environmental variables. The ecosystem management approach developed by the United States government now requires managers to address multiple perspectives when making land management decisions.

When this book first appeared, the regulatory "command and control" approach to decisionmaking was still strongly in place. Speculative approaches based on technological standards that forced industries to develop new equipment and processes to reduce pollution were fairly well accepted. Since then, economists have convinced many government policymakers that a regulatory approach based on economic incentives would be superior. The general belief is that industry will be encouraged to develop more cost-efficient pollution control approaches in order to appreciate economic gains from economic incentives. Less efficient companies should be able to purchase emission rights from more innovative companies that have already reduced the amount of pollution that they generate in excess of legal requirements. It is believed that economic subsidies will stimulate recalcitrant corporations to install pollution control equipment that they would otherwise not be inclined to purchase.

The long-term effects of an economic incentives approach, some economists suggest, will be a reduction in regulatory enforcement costs for agencies. Industries will also be able to reduce costs while meeting pollution reduction goals more flexibly and more efficiently.

Another management approach that has developed involves an integration of command-and-control regulations and economic incentives. Just as the command-and-control regulatory approach has been criticized by many economists, many environmentalists are concerned that the economic incentives approach is merely granting corporations an inherent "right to pollute." An integration of the two approaches would involve the establishment of "ecological ceilings" beyond which the environment would suffer irrevocably. Corporations could use all of the various economic means at their disposal until their emissions or discharges approached the "ecological ceilings" at which point prohibitions would be enforced.

Another major development has been the increased devolution of environmental administration responsibilities to state and local governments from the federal government. This devolution in authority began with acceptance by the states of the primary responsibility to implement federal environmental laws such as the Clean Air Act, Clean Water Act, and Safe Drinking Water Act. The federal government developed nationally uniform standards and the states, in turn, created laws equal to or better than the federal law. However, over time the states became more and more concerned that federal regulations were becoming too inflexible and too costly to implement. Through laws like the Unfunded Mandate Reform Act (UMRA), the Government Performance Results Act (GPRA), and the Regulatory Reform Act (RFA), Congress provided substantial relief for state and local governments. Partially in response to these acts the EPA has developed a series of innovative environmental approaches that have streamlined the regulatory process and have given state and local governments greater flexibility in meeting federal standards. At the same time, the EPA has significantly increased the level of technical support for state and local governments to help them more effectively meet federal standards through such efforts as the Community-Based Environmental Protection and Brownfields programs. The EPA also has initiated cross-jurisdictional approaches to the management of pollution such as the watershed management approach, which incorporates consideration of both point and nonpoint sources across multiple jurisdictions within a single watershed. All these factors have had a profound impact on the environmental decisionmaking process.

Decisionmaking in Environmental Administration

The environmental administrative process consists of the ways and degrees that policy is both generated and implemented at various stages and levels. It is a continuous process that involves many changes and value interpretations. As Lynton K. Caldwell observes, "If administration may be defined as 'the art of getting things done,' then its scope and content will necessarily be as flexible as the methods available to the administrator, the objectives [values] he seeks, and the milieu in which he operates."[1] Much of the actual operating policy and its effects in environmental affairs depends upon the administrative process as it takes form through agencies and personnel. In this process, environmental values and considerations often may not be incorporated into actual operating policy and administrative operations, despite legislative and other policy mandates.

Agency Aspects

The consideration of environmental values in management decisions is often resisted by governmental agencies. Agencies are social and political institutions with their own sets of values and vested interests. Bureaucratic ideology, moreover, is concerned with survival and expansion of an organization. Consequently, the degree of administrative acceptance of or resistance to environmental values depends upon bureaucratic ideology or the collective value system of the agency. A comprehensive environmental approach is seldom attained in the administrative process because of pressures from diverse clientele groups as well as an agency's own security and expansion interests. As a bureaucratic institution, an agency is concerned with its own welfare first. Other interests are secondary considerations. Environmental legislation requiring consideration of environmental values in agency deliberations necessarily falls into the latter category.

In the administrative process of an organization, according to Talcott Parsons, "goals are those services or products that an organization sets out to produce. An effective organization is an organization whose outputs coincide with those stated goals."[2] However, many organizations do not follow stated goals. In order to survive, organizations develop pseudo goals. These pseudo goals take precedence over the stated goals and become a normal part of the organization's processes. Dependence on these substitutes is further complicated by other values demanding a place in the primary goal structure.[3] In the case of a governmental organization, political and social values may become more important considerations than primary goals, while means and procedures may become more significant than ends. For example, members of the Senate Armed Services Committee have felt that there was too much emphasis on "technical, managerial, and bureaucratic skills" in the Department of Defense, to the detriment of "defense mission objectives" and "leadership skills in wartime."[4]

The goal selection process has various complexities. Goals or values are seldom specified and change over time. Some of the complexities involve identification and analysis of who defines the goals, by what processes, and how priorities are determined. Complex problems such as those pertaining to environmental quality are often presented to an organization in value terms relative to that organization's unspecified goals. This ambiguity creates a vacuum that permits the organization, consciously or unconsciously, to substitute secondary goals more oriented toward its immediate interests than toward the interests of society or the environment.

The preoccupation of agencies with secondary objectives or pseudo goals through goal displacement can result in a lack of responsiveness toward environmental considerations. A budget can become an end rather than a means for achieving real organizational goals, and it can serve to severely limit expenditures for environmental programs. For example, the Reagan administration used the budgetary process to severely restrict the effectiveness of the EPA in its implementation of environmental regulations that affect business.[5]

Similarly, a resource management plan can become a dogma for economic purposes rather than a flexible guide for meeting changing needs and addressing environmental concerns. When James Watt became the secretary of the interior, he established a management-by-objectives (MBO) system. One of his major objectives was the promotion of energy development activities on public lands. Since the value of coal to be extracted from public lands is far more readily determined than the value of wildlife habitat or scenic vistas on undisturbed lands, the latter were given limited consideration.

A governmental organization may have its own sense of rationality and morality that can differ sharply from an outsider's views. Resistance to environmental value considerations, therefore, can be a natural product of the organizational process. According to Douglas Price, "Dissatisfaction must build up before there is a serious search for alternative programs."[6] Sufficient dissatisfaction will cause the organization to search for alternatives and to change its policies. In the environmental administrative process much dissatisfaction is produced through crisis situations and through information presented in the communications media, expression of public sentiment, and political awareness inside and outside organizational confines of such situations.

Sufficient dissatisfaction existed with the regulatory processes of the EPA that the Clinton administration required it to reinvent environmental regulations. Under guidance of Vice President Gore's National Performance Review, the EPA improved environmental regulations in several ways. The EPA consolidated similar requirements for certain industries, eliminated requirements that were duplicative or unnecessary, and rewrote regulations in language more readily comprehended by the average citizen. For those industries willing to invest in innovative technologies that would improve environmental protection, the EPA extended compliance schedules. Overall, the EPA granted industries more operational flexibility for greater public accountability.[7]

Decisionmaking and Values

In considering decisionmaking in agencies, it is appropriate to recognize the value basis that underlies the various issues and choices. Decisionmaking processes ordinarily include debates and conflict when goals are formulated in relation to problems to be solved and when alternative ways of achieving these objectives are laid out for decisions.[8] Choices are often not made directly between values but between options that differ to the extent that they embody particular values (or neglect them), or in the emphasis that some values possess in relation to others. Many executives are strongly attached and devoted to their beliefs or values.[9] These emotional convictions in turn shape the pattern of decisions or organizational strategies for their organizations.

Values are important throughout the decisionmaking process. The way problems are shaped, defined, and perceived—including the urgency and importance of various problems—is basically a function of values. When people recognize that something should be done about a problem, they are making what is called a "value assertion." The values of individuals who implement the decisions are important. This is so because they can have a powerful effect on the shape of the ultimate policy after the normative tasks of setting objectives and ranking options have been completed at higher levels.[10] In this sense it is difficult to separate the various roles of generalist and specialist personnel relative to their impacts and influences on decisionmaking processes.

The environmental administrative process, therefore, needs to deal with the range and complexity of values as the underlying basis of decisionmaking.[11] Basically, values are individual or collective conceptions about what is important or desirable with emotional, judgmental, and symbolic components. Values are formed by groups of attitudes (a state of mind or feeling) that represent a behavioral predisposition toward a given environmental object or factor. Attitudes in turn are produced by groups of beliefs that collectively cluster around a given environmental object or factor. Values, however, can produce behavior in contrast to attitudes that represent a behavioral predisposition. In the final analysis, every decision involves some form of value judgment with some values being sacrificed or reduced for the sake of others. For the purposes of this text, values encompass goals, beliefs, attitudes, and traditions that have significant influence on human interactions and the exercise of power. This definition also encompasses value systems of individuals and organizations.

Values operate in the following framework:

1. Values: including identification and analysis of involved values and their conflicts, as well as their related alternatives and options, with attention to consequences and projections of values.
2. Human interaction: relating and correlating the above to individuals, groups, and organizations that are directly or indirectly involved with the values and interests associated with the problems or issues on a formal and informal basis.
3. Power or authority: relating the above to formal and informal authority and/or power in terms of influence and/or spheres of influence, as based on values and human interaction.
4. Decision or policy: analysis and evaluation of the authoritative allocation of values for the final stage or determination, with recognition that much depends on how the selected values will affect other values associated with the implementation of the decision or policy.

Within this framework, decisionmaking determines a governmental policy that will shape an environmental or developmental action, nonaction, or degrees of any of these. Emphasis is given to competing and often conflicting values that operate through formal and informal human interactions in the struggle for power and authority, that is, the authoritative allocation of values. In a problem or issue, the values involved may represent a variety of interests with positive or negative results for the environmental public interest. Serious environmental problems have resulted from decisions that failed to consider and incorporate environmental and societal values properly.

The decisionmaking process must take into account the situational, incremental, and often tentative aspects regarding values. Further considerations involve time constraints, uncertainties, and set organizational response patterns; all three place severe limits on value analysis and content. Usually, decisionmaking in environmental affairs is an ongoing, dynamic process and few decisions are of a final nature. The process often includes successive decisions based on new and interrelated problems or consequences over time. In this process, values become clouded and unclear, and the optimum values and alternatives may be excluded or neglected in actual decisionmaking considerations.

The overriding requirement for decisionmaking in environmental affairs is the exercise of value judgments. Such judgments may be biased both internally and externally, whether clearly or indistinctly, and each individual will express his or her value orientation. Budget and time constraints, as well as group priorities, will also affect value judgments. For example, in order to be reelected, a politician may seek to accomplish short-term goals that are at odds with the longer-term goals of environmental scientists. The public, often driven by sensational press coverage, may demand immediate redress of environmental problems. These biases manifest themselves in ambiguity or confusion.

Ambiguity and confusion of values characterize the latter part of the twentieth century. Determination and application of values to environmental problems are further complicated by the fact that values are constantly changing. Social goals are in a constant state of change. Thus, scientific and technological efforts to meet social goals are often ill timed.[12] Such efforts may no longer be relevant to the social goals identified at any given time. Furthermore, by their very nature values are difficult to describe and analyze relative to the way power is distributed and decisionmaking takes place in society. Although a great many individuals and groups may wish to see their environmental agendas enacted, only a few will successfully compete for the limited power of decisionmaking.

Values are an essential part of decisionmaking because they strongly influence the objectives that individuals pursue and the means they select for achieving these objectives. It would appear obvious that in most "real life" situations some values will have to be sacrificed for the sake of others. Equally obvious is that no one can assume a value-free role in environmental problem solving. In the decisionmaking setting, old values may be changed or strengthened while new ones are learned. There will be constant organizational pressure to support some values while rejecting others. Decisionmaking occurs against a background of conflicting, but intermixed, personal, and organizational pressures.[13]

Much of decisionmaking involves uncertainties and conflicts that interfere with group agreement on values and alternatives. This is particularly true with regard to the numerous values and complexities related to the environment. Total agreement by all the decisionmakers on a problem is rare; decisionmakers tend to use an incremental approach that allows for the incorporation of different values and views in a problem. Due to the uncertainties and time constraints associated with obtaining all the needed facts, incrementalism allows for inputs of different views or values on a problem by a specific group of involved decisionmakers. The incremental approach involves: (1) searching for an alternative that satisfies each of the decisionmakers, although no one believes it is the ideal solution; (2) accepting the first proposal to which no one strongly objects; (3) avoiding pressing the search for basic values so far as to threaten cohesion and alliances.[14] Incrementalism assumes that personnel can embrace the same proposal for different reasons, so that exposing the ultimate goals or values is likely to get in the way of agreeing on what to do. Further incremental decisionmaking is directed toward marginal changes and the status quo. Thus it generally ignores long-term values while focusing on immediate problem solving on a short-term, practical, and piecemeal basis. The latter often excludes important environmental values and considerations while being attuned to economic and quantitative factors.

It seems clear that incrementalism in decisionmaking tends to deemphasize underlying values and distracts group members and leaders from considering them. Compromise on lowest-common-denominator solutions may neglect or disregard the value clarification and change needed to incorporate important environmental values and considerations into decisionmaking. If the assumptions and views of personnel on issues are not clearly identified and analyzed along value lines, then the decisionmaking processes through incrementalism can only deal with values on an indirect and partial basis. Personnel need to be more aware of the value nature of their views and assumptions, and to consider a broader range of environmental perspectives in order to improve comprehensive decisionmaking. When value intentions or assumptions are incorporated into proposed or actual alternatives, value change and growth may occur, particularly in getting at the long-term consequences of an action. For example, the environmental and societal impacts and values associated with some alternatives may be inconsistent or negative.

Herbert Simon pointed out many years ago that decisions are shaped and channeled long before they are officially made.[15] When decisions or recommendations reach the higher levels of agencies and departments the problem has been defined, the alternative solutions (and their value aspects) have been narrowed, and factual data have been gathered to support the recommendations. Important values, including those pertaining to the environment and society, may have been excluded from consideration already. Higher-level administrators cannot easily reclaim values and alternatives discarded earlier in the decisionmaking process, nor can they ask the right questions at the proper time in view of the specialized nature of the problem.[16]

Higher-level administrators usually have a vast and complex range of conflicting decisions and policies to make in their policymaking roles. These include conflict resolutions between various competing values and interests on a much larger scale than those of a given agency or of a specific unit within an agency. Therefore, it is paramount that important values be incorporated into agency and unit programs at the initial phases of the decisionmaking process.

A series of incremental decisions incorporating selected and limited values of personnel at various levels eventually accumulates into major policy changes. Throughout the decisionmaking process, personnel need to seek out and articulate a wide range of values that are important and relevant to the issues and problems under analysis. Robert Behn and James Vaupel argue that far greater emphasis should be given to thinking about the problem before analysis begins.[17] It is in this thinking stage that values should be clearly identified and, then, incorporated as the analytical process proceeds. Throughout the various operational stages of decisionmaking—such as problem definition, identification of alternatives, analysis of alternatives, and selection of acceptable courses of action—this value orientation should be incorporated.[18] This, in turn, requires more attention to values by higher-level personnel who are engaged in decisionmaking. Decisions in the environmental public interest are more likely to occur as a result of such a process.

Decentralization

Power to make decisions is widely dispersed throughout environmental and natural resource agencies. Because decisions affecting the environment must be made at all levels of administration, decentralization of power must occur. Decentralization does not entail a lack of control. As Henry Mintzberg has said, "Control over the making of choices (for decentralization)—as opposed to control over the whole decision process—does not necessarily constitute centralization."[19] Decentralization in decision-making consists of subordinates communicating to superiors inferences they draw from a body of evidence. The decision is then based on the subordinate's analysis. Any uncertainty associated with an answer or recommendation is filtered out by the subordinate while the superior "makes" the decision. This process is like a book review in which the reviewer draws inferences and then communicates his or her conclusions to the reader who does not read the book.[20]

Decentralization implies that decisions are made in the field rather than in centrally located offices. Decentralization also implies greater responsiveness to and better perception of environmental concerns with close proximity of decisionmaking to the problems. But such intimacy also results in greater opportunity for local political and economic interests to influence and manipulate decisions toward nonenvironmental values and away from the environmental public interest.

A carefully orchestrated decisionmaking process that stipulates consideration of important values at appropriate times can minimize this potentiality. Herbert Kaufman aptly describes such a process in his study of the forest ranger. In noting that events and conditions in the field are anticipated and described in terms of courses of action, Kaufman observes, "The field officer then need determine only into what category a particular circumstance falls. Once this determination is made, he or she then simply follows the series of steps available to that category. Within each category, therefore, the decisions are pre-formed."[21] In describing the USDA Forest Service administrative manual (with its similarities to other agency manuals), Kaufman further states: "The provisions describe what is to be done, who is to do it, how (and how well) it should be performed, when (or in what sequence) each step should be taken, where the action should take place, and even explain the 'why' of the policies—the reasons for their adoption, the objectives they are expected to attain."[22] Thus, many apparent ranger-level or decentralized decisions are in fact made for the rangers by centralized sources of the agency through established procedures. However, as Knott and Miller note, "No organization can program or predefine in the rules *all* the individual behaviors that are necessary for organizational success."[23] Agency personnel who simply follow a set of written rules may fail miserably in certain situations.

Stipulations of the National Environmental Policy Act of 1969 (NEPA) require agencies to review and change various policies and procedures that conflict with standards and requirements of the act. Yet it is recognized that procedures, manuals, and informal ways of doing things are slow to change, particularly at the field level. Change, as well as decentralized decisionmaking, is further complicated by ineffective communication among levels of organizations. Dysfunctional activities and elaborate defenses reduce the probability that accurate information will flow through an organization.

Because of lack of trust and a desire not to make waves, important aspects of information and truth fail to be communicated among upper, middle, and lower levels of an organizational hierarchy.[24] Indeed, it can be argued that one purpose for creating various organizational levels is to prevent communication from taking place. The end result of this may be superficial treatment of environmental problems in decentralized decisions; the whole truth will not be told because of possible negative connotations. Consequently, communications between central headquarters and field tend to emphasize the conservative, technical, and superficial on a positive basis without covering real problems and issues.[25] Ben Heirs and Gordon Pehrson observe that "in some organizations, it is a simple rule of survival that one should only present proposals or views of the future that support or embellish 'safe' ideas already well-accepted in the decisionmaking process."[26]

The field person's individual mental picture is definitely important in the decentralization process because his or her values will naturally be reflected in the decisions made. With professional and organizational socialization that indoctrinates them to be responsive and sympathetic to local economic interests, personnel in the field may often incorporate local community values into their official actions. The location of administrative offices in small towns or villages places a field person in continuous professional and personal contact with individuals and organizations strongly committed to economic uses of natural resources and to development.

In contrast, field personnel tend to have more transitory contacts with recreationists and environmentalists concerned with noneconomic uses. Frequent rotations in job assignments are often necessary to reduce the influence of local political and economic interests. In the case of forestry (as well as other resource professions) local interests still may have a definite influence on the field person, even at a considerable distance from the resources over which he or she has responsibility. A given individual, on the other hand, may have a value bias toward economic considerations, regardless of location.

The Environmental Public Interest

The administrative process both legally and ethically must serve the environmental public interest of a region, nation, or the planet as a whole.

Administrative personnel are constantly requested to make decisions that favor economic over environmental values and considerations. This is especially the case with regard to public lands in the United States, which are located primarily in the West and away from urban centers. On a professional and social basis, much of the daily contact of agency personnel is with individuals and organizations in small communities who are concerned with economic and commercial use of public lands and natural resources. Many resource agencies, moreover, have policies and procedures that require responsiveness and special attention to local economic interests. Role perceptions, outside pressures, organizational structures, perceived costs, personality characteristics, identification with certain outside reference groups, and other influences on personnel and agencies all affect and shape their values and decisions.[27]

An understanding of what the "environment" entails, like the symbolic concept of the public interest, is subject to various interpretations based on the values and perceptions of the individual or organization. Conflicts naturally occur when differing values and perceptions of the environment collide in the decisionmaking process. Although the concept of "public interest" has been used for some time, its environmental dimensions imply unique considerations that include ecological complexities, future generations of all life-forms, and intangibles under a holistic or integrated view of environment and society.

In the context of environmental administration, agencies and personnel usually claim that their decisions are in the public interest. However, this may or may not involve the above-mentioned unique environmental considerations. The public interest is often associated with assumed public benefits or needs when private interests may often be the real beneficiaries of a decision. Like any other concept, public interest is subject to individual and group opinions. Supposedly, public opinion is reflected in the public interest in the process of governmental decisions. But this must occur through subjective value interpretations and judgments of agency decisionmakers operating under broad legal frameworks and discretion. A comprehensive public opinion is seldom fully expressed or understood as a determining force within any time period.[28] Further, consensus on public opinion may simply mean that the majority are not visibly speaking out at a given time on a given, often obscure, issue.

Often much of the administrative process is devoted to determining the public interest in a specific environmental decision. Richard W. Behan, in his article "The Myth of the Omnipotent Forester," stresses the tendency of professional foresters to assume that they know what is best for the land and that they can tell the public how lands should be managed. It may be noted that foresters are hardly unique in this regard. Their arrogance is shared by architects, planners, and other professionals accustomed to making land use recommendations and decisions. About the forester, Behan adds, "It is when the professional forester arbitrarily determines those ends, or even clumsily tries to, that he most seriously violates our classless society and democratic politics." Behan also believes that environmental pressure groups are properly most hostile and challenging when foresters involve themselves and their agencies in problems by attempting to determine the social ends of natural resources. He says that "it is when we invoke this rationale by judging, in its terms, 'goodness' and 'badness,' that pressure groups properly challenge our leadership. 'Goodness' and 'badness' in our society are collective value judgments, and land expertise is no better a qualification than many others for making them."[29]

Administrators play an important part in the process of determining the social ends of an abstract public interest. With an awareness of this fact, pressure groups attempt to influence administrative policy and decisions through the bargaining and compromise of brokerage politics in the highly political process of resolving resource use conflicts. Walter Rosenbaum notes that "governmental officials operate on the premise that major organized groups affected by a public policy should have an important voice in shaping and administering it."[30]

At various stages and levels of policy- and decisionmaking, many agencies have advisory committees and consultants from the private sector to assist them on complex environmental problems. Experts or specialists of this nature may be well qualified in the resource or environmental field under study. However, they are not necessarily qualified to determine what constitutes the public interest. But it is reasonable that they should develop and make recommendations for the public interest in an environmental problem or plan. The Clinton administration has responded to such concerns by expanding the number of individuals that participate in the decisionmaking process in both regulatory arena of pollution control and decisions involving natural resources. The Common Sense Initiative and Project XL allow corporations to develop and propose innovative strategies for regulatory compliance.[31] The EPA has entered into performance partnerships with the states that give the states greater flexibility to target priority areas while reducing its oversight in those areas of strong performance.[32] Under federal ecosystem management initiatives the land management agencies are directed to enlist the cooperation of state, local, and private stakeholders in the development of cross-jurisdictional ecosystem management strategies.[33]

A valid consideration in the selection of expertise from the private sector might be the individual's demonstrated competence in making recommendations in the environmental public interest as well as in the specific problem area. In this regard, the use of nontechnical experts and individuals as advisers on the relationships of society to a resource or environmental problem is worth exploring. Individuals with vision, sensitivity, and broad knowledge, coupled with familiarity with environmental considerations, could reflect the spirit of the public interest by making contributions to the value bases for decisionmaking.

Many agencies use advisory committees made up of representatives of various groups that have special or clientele interests in their programs. Because of such relationships, the recommendations of advisory committees often cannot truly be equated with or viewed as the public interest. Even within these committees power struggles often occur, reflecting the values and biases of groups represented and the dominant interests of the more powerful groups. Thus, distorted views of the environmental public interest and disproportionate degrees of influence may often be demonstrated by advisory committees.[34] Little weight consequently can be given to the study and articulation of the public interest in environmental affairs and problems when representatives are pressing for the values and interests of their own groups. Furthermore, whole segments of the population may be excluded from inclusion on advisory committees. When the Reagan administration took over the reins of government the newly appointed EPA administrator, Anne Gorsuch, dismissed the existing members of the EPA's Science Advisory Board and replaced them with scientists more favorable to the administration's policies. It should also be noted that federal agency personnel are supposed to follow the guidelines of the Federal Advisory Committee Act when they are in contact with members of such committees.

Public Participation

Because of various laws and changes in governmental and public attitudes, public participation now plays a more vital and visible role in the decisionmaking processes of environmental administration. It adds unique and new dimensions, especially in value considerations. Public participation in environmental assessments, planning, and decisionmaking has become an important factor for agencies and personnel at various levels to consider when making judgments and determinations. The solicitation and proper utilization of public participation inputs are therefore integral parts of the administrative process. In fact it is often only through public participation that certain types of information, evaluation, and public support can be obtained for environmental problem solving and decision-making. Public participation would facilitate consideration of such public values as lifestyles, quality issues, and other complex areas of the interface between society and environment.

The statutory basis for public participation in the United States is the Administrative Procedure Act of 1946 (APA). By the 1970s, citizen intervention in administrative proceedings of federal agencies was common. Under the APA agency decisions are subject to judicial review and citizens are given the right to participate in the formulation of federal rules and regulations affecting their lives, to the extent that these activities do not interfere in the performance of an individual agency's daily business. Essentially, this means that citizens have the right to present evidence and testimony that is to be considered in formal rule making.

Public participation is generally defined as that part of the decision-making process which provides opportunity and encouragement for the public to express its view. It assures that proper attention will be given to public concerns and preferences when decisions are made. Such participation includes involvement or consultation in planning, decisionmaking, and management activities dealing with environmental affairs. The public actively shares in the decisions that government makes in environmental affairs by having individual and group values taken into account. Effective public participation requires the availability of adequate nontechnical information, public encouragement, and opportunities to use that information.

The EPA has responded to the rising public demand for environmental information. The agency has initiated a major information-sharing program to facilitate citizen participation through the Internet.[35] There is so much information available that it is overwhelming at times. Statistical information is available from the Center for Environmental Information and Statistics that allows citizens to obtain detailed information about the air or water quality, the production of hazardous wastes, and release of toxic substances within their counties or watersheds. Users can search for information by facility name, geographic area (e.g., zip code, state, county), or chemical substance to find the type of environmental data they want. On-line up-to-date information about current air quality conditions in cities is available as well as historical emission levels and compliance for nine thousand point sources. Information about current regulatory analyses is available on-line as well. New management initiatives are presented for citizen review and education. All of this information is available on the internet at http://www.epa.gov.

Public participation inputs include suggestions, information, questions, views, and critiques expressed by members of the general public in efforts to influence decisionmaking in environmental affairs. The inputs may be made through both formal and informal participatory processes and may be solicited or unsolicited. Such processes usually involve a search for the public interest. The public interest is often subject to value interpretations and justifications through public participation and the administrative process. Peter Navarro warns that special interests and ideologues are capturing the policy process in the guise of serving the public interest. He says that "conceptually it [the process] allows for the possibility that private interest motives can indeed determine 'public policy.'"[36] Nevertheless, it is the responsibility of agency personnel to determine the overall interest of the general public in their decisions and to assure adequate public participation in the decisionmaking process.

The EPA suggests that public participation should

1. promote the public's understanding and involvement in planning and implementing programs and proposed actions with emphasis on the nontechnical aspects
2. keep the public informed about significant issues, problems, and changes in programs or proposed subjects, including associated values and alternatives
3. make sure that government and personnel understand public concerns and values and that they are responsive to them, including public identification of issues and alternatives
4. demonstrate that the agency, formally and informally, consults with interested or affected segments of the public and takes public viewpoints and values into consideration when decisions are made
5. foster public involvement and activities that focus on identifying problems, laying out and exploring all of the alternatives to resolving the problems, and setting forth a preferred alternative
6. foster a spirit of mutual trust, support, and openness between government, personnel, and the public through a variety of informal and formal contacts for public participation.[37]

Five basic functions are needed to ensure effective public participation— identification, outreach, dialogue, assimilation, and feedback. With respect to *identification,* it is important and necessary to identify groups or members of the public who may be interested in or affected by a forthcoming action. This can be done by developing mailing lists, requesting additional names from those already included, using questionnaires or surveys to discover levels of awareness, and establishing informal contacts by other means. Besides a general list, a specific contact list can also be developed for a particular program or project.

The public can contribute effectively only when they have accurate and timely information. Through *outreach* efforts an agency can provide such information on pertinent issues and related decisions. Information should be presented in a general, nontechnical manner for ease of understanding. Information that is too technical will usually discourage public participation. The agency must ensure that appropriate information is made available to potentially concerned citizens through the news media, Internet, or other public service announcements and personal communications. The content should include background information, a timetable of proposed actions, summaries of lengthy documents and technical material where relevant, a delineation of issues, and specific encouragement to stimulate active participation of interested parties. Wherever possible social, economic, and environmental consequences of proposed actions should be stated clearly in the outreach information.

Dialogue should be carried on between personnel responsible for the forthcoming action or decision and the interested and affected members of the public. This includes an exchange of views and open exploration of issues, alternative courses of action, potential consequences, and value considerations. Dialogue may occur through meetings, workshops, hearings, or personal interaction and may include the establishment of special groups such as advisory committees and task forces.

The *assimilation* of public viewpoints and preferences into final conclusions consists of putting together the results of the "outreach" and the "dialogue" phases. In its decisions and actions an agency must demonstrate that it has understood and fully considered public concerns. Assimilation involves two elements: documentation in which the agency briefly and clearly presents the public's view, and a responsiveness summary in which the agency identifies and describes the types of public participation activities undertaken. Participants from the public should be identified. Summaries should highlight important comments obtained through the participation process and the agency's responses. Evaluations of public participation should contain both quantitative and qualitative aspects and should be directed to the basic issues involved.

An agency should provide *feedback* information to participants and interested parties concerning the outcome of any public involvement. Feedback may be in the form of personal letters or phone calls if the number of participants is small enough. When numerous participants take part, an agency may mail a response summary to those on the list, place it on the Internet, or may publish it. The feedback should include a statement of the action that was taken and the effect of public comment on that action.

A value emphasis throughout the public participation process can provide an overall basis and central theme for soliciting and incorporating public inputs into decisionmaking. This approach includes the relationships of values to alternatives and issues of planning and management for an agency. Problems often occur when there is failure to define or relate issues to significant public interests or concerns relative to values. The general orientation of the greater majority of the public is toward values that form the basis for their nontechnical concerns and interests. In this sense the general public can act effectively to educate decisionmakers about environmental values and concerns. Decisionmakers in turn can influence public opinion in environmental affairs. Further, a major concern in the public participation process is that all parties be aware of the alternatives or choices and that these choices be clarified through an emphasis on values.[38]

Value orientations in public participation also provide the advantage of early identification of public inputs and the recognition that values underlie complex issues and problems. William Whalen, former director of the U.S. National Park Service, has noted: "It is essential that we identify and surface, at the earliest stage of public review, the difficult problems and thorny issues that have the greatest potential for causing public concern and reaction. . . . [B]y allowing the public the fullest and earliest possible involvement in developing alternative solutions we can build open lines of communication and trust between us and the many segments of concerned citizenry."[39] Whalen's remarks suggest that solicitation of public input can be used not only to identify the environmental public interest but also to pinpoint potential areas of controversy that may then be addressed and defused prior to becoming political problems for agencies. The establishment of "lines of communication and trust" can also have the effect of protecting agency priorities, rather than identifying and acting upon the basis of the public interest.

Agencies often carry out extensive public relations programs in order to build public support and gain political influence. These programs describe and justify the policies, missions, and activities of the agencies and make the agencies look good.[40] Complications arise, however, when an agency, through such programs, dictates social ends to the public and implements them through public relations techniques. Such programs can then be used as tools for promoting agency ideology and the vested interests of specific clientele. An administrator can, for example, articulate a value decision and then sell it to the public.

A biased premise with selected supporting data does little to enlighten public opinion and encourage meaningful participation. In fact, it may be repugnant to those who like to consider the alternatives rather than be "sold" a position. Many agencies neglect their responsibilities to inform people objectively about the facts and alternative solutions to environmental problems and offer instead a closed system. For instance, much of the public relations literature from governmental agencies is confined to platitudes and niceties that are presented in an almost insulting, simplistic manner. The environmental and conservation movements attempt to counter such efforts with realistic coverage of complex ecological interdependencies ranging from local matters to global concerns. As a result, a better informed citizenry has developed. A better informed public can more effectively influence policy decisions through expressions of values and opinions. Obviously, environmental administrators must be sensitive and responsive to such a trend.

Barry Commoner points out that the duty of scientists and administrators in environmental affairs is to furnish information that will enable their fellow human beings to use judgment in the human use of science and technology. Without this, Commoner states, "we will have deprived humanity of the right to sit in judgment of its own fate."[41]

Public participation in the administrative process may contribute in important ways by (1) forcing consideration of all different interests for making well-balanced and comprehensive decisions; (2) exercising healthy pressures on agencies to be fair and not to take sides; (3) encouraging agency decisions that are more acceptable to the public because the public was involved; (4) protecting agency personnel from undue pressures by special-interest groups; (5) urging decisions in the public interest; and (6) guaranteeing more adequate consideration of health and environmental values and factors.[42] Public participation permits a wide variety of values to be articulated and analyzed, including people's perceptions and responses to possible changes in the status quo in terms of lifestyles and the environment.

The influence of public perceptions in administrative processes is illustrated by the following example. A national meeting of USDA Forest Service supervisors was held in Snowbird, Utah. The meeting was convened because of an increase in the number of legal appeals filed by the public against the majority of national forest use plans. Forest supervisors informed the chief of the USDA Forest Service that the public was very unhappy about the overemphasis given to logging and other exploitative uses in national forest plans. After this meeting, agency officials were directed to give greater emphasis to public interest considerations in their plans. Even if little changed, at least, USDA Forest Service planning was made to appear more like what the public expected it to be.[43]

The degree of public involvement depends upon both the general attitudes of the government and the interests of the public. The central issue is the degree of trust and confidence that the public has in the management agency and observes that the participation process tends to increase that trust and build public confidence in the agency. In commenting on the role of public participation in national park management, Harold Eidsvik refers to a comment by Edwin Winge: "Public involvement does offer long-range benefits, the most pragmatic of which is that it results in better decisions. Park Service managers have discovered through experience that when they are willing to modify their professional judgments by considering ideas and opinions (values) of concerned citizens, the final decision that results is not only more acceptable to the public, it also is more satisfying to the Service."[44]

However, many citizens do not share this viewpoint. They doubt whether public participation does indeed affect the decisionmaking process. Derrick Sewell and Susan Phillips evaluated a number of public participation programs and found that

> while most agency representatives would claim that increased citizen involvement has led to increased inputs by the public into the decisionmaking process, citizens and citizen groups remain skeptical that this has in fact occurred. Even when increased input is acknowledged, such individuals or groups are suspicious that inputs of other actors (such as bureaucrats, politicians, or developers) are given much more weight in the final decisionmaking. In most instances the public is given no indication of *whether* its views were considered, and even if they were, *how* such views influenced the final outcome.[45]

Consequently, public decisionmakers must provide feedback to citizens who do participate in the policy formulation process to demonstrate to them that their views do matter.

Analysis of the impact of citizen participation on USDA Forest Service decisions concerning RARE II (Roadless Area Review and Evaluation) provides additional insight on the effect of citizen input into agency decisionmaking. Paul Mohai attempted to verify which of two perspectives most accurately describes the role of citizen participation.[46] One belief is that decisionmaking in a natural resource agency like the USDA Forest Service is strongly molded by its professional ideology. As a result of their professional training, according to Ben W. Twight, agency personnel tend to operate in a closed organizational structure somewhat insulated from public concerns.[47] Consequently, they are likely to make administrative decisions based on their training, regardless of the political consequences on the USDA Forest Service in the form of citizen protests. In contrast, Paul Culhane, while acknowledging the role of professionalism in natural resource agencies, believes that the USDA Forest Service and Bureau of Land Management (BLM) are responsive to public input.[48] He argues that citizen participation requirements in post-1970s environmental legislation and the activism of environmental groups ensure agency responsiveness to citizen concerns. Mohai concludes that

> one could take the position that the RARE II decisions did not represent a true compromise or that the Forest Service did not respond to all its publics equitably. Such judgments are difficult to prove or disprove and depend very much on one's own value system. What is important to recognize is that the Forest Service is apparently influenced by public input, whether it is equitable or not. The agency is also influenced by its professional ideology, whether that is rational or not or whether that serves the public interest.[49]

Environmental Impact Statements

Under the provisions of NEPA, agencies are required to conduct environmental impact assessments (EIAS) for all proposed actions and programs that may significantly affect the environment. The purpose of these assessments is to minimize potential negative impacts and irreversible commitments of resources. If no significant impact is determined, an agency is required by federal regulations to issue a Finding of No Significant Impact (FONSI). If the agency determines that a significant impact is likely to occur, an environmental impact statement (EIS) must be written. The EIS provides information on ecological inventories, potential impacts, program or project alternatives and a recommendation as to the preferred alternative. Once drafted, the EIS is subjected to interagency review by all federal agencies that may have expertise pursuant to any aspect of the EIS. Finally, the EPA is required to review and comment on all EISs. Provisions also call for public participation throughout the above process.[50] Lynton Caldwell states: "The purpose of the procedure—environmental impact analysis—was to force federal officials to consider the possible consequences of decisions having major implications for the quality of the human environment."[51]

Among other requirements for federal agencies concerning EISs and related activities, NEPA stipulates that they shall

> (Sec. 102 [B]) identify and develop methods and procedures . . . which will insure that presently unquantified environmental amenities and values may be given appropriate consideration in decisionmaking along with economic and technical considerations; (C) include in

every recommendation report on proposals for legislation and other major/federal actions significantly affecting the quality of the human environment, a detailed statement by the responsible official on—

(i) the environmental impact of the proposed action,

(ii) any adverse environmental effects which cannot be avoided should the proposal be implemented,

(iii) alternatives to the proposed action,

(iv) the relationship between local short-term uses of man's environment and the maintenance and enhancement of long-term productivity, and

(v) any irreversible and irretrievable commitments of resources which would be involved in the proposed action should it be implemented.[52]

These "action forcing" provisions of NEPA prompt federal agencies to incorporate environmental values and components into their policy and planning operations. NEPA is regarded as a form of statutory intervention into the regular administrative procedures of federal agencies. "But the National Environmental Policy Act leaves the missions and structures of the federal agencies unchanged. There is no power directly authorized by the Act to prevent or modify any environment-affecting action by any agency of government. Court orders restraining federal projects have been based on agency failure to conform fully to the procedural requirements of the Act."[53] Courts (and most other governmental institutions) will generally not intervene in agency judgments and decisions, regardless of their value positions and nature of their bias. A review of NEPA-related court decisions indicates that the Supreme Court has only upheld decisions that support the procedural aspects of NEPA and not substantive considerations.[54]

Nevertheless, preparation of environmental impact statements does bring facts and scientific perspectives into the administrative process. Where significant environmental impacts are likely to occur, proposed public actions are subject to serious environmental review. Furthermore, the requirements of NEPA have led to modest but real restructuring of administrative procedures in agencies.[55] Analysis of environmental effects and relationships had to be incorporated into agency planning and decisionmaking. Interdisciplinary approaches had to be incorporated into agency planning in order that complex environmental issues could be addressed. The act made practical the utilization of the theoretical unity of science itself. It forced administrative changes in resource-oriented agencies and science provided the tools for making those changes meaningful. "Science . . . provided the substantive element in redirecting national policy for the environment through procedural reform. The critical procedure—the environmental impact statement—became the vector, carrying integrated interdisciplinary sciences into the shaping of public policy."[56]

An environmental impact statement is a document prepared on the possible negative and positive effects and influences of a proposed project or development that would significantly impact the environment and society. It provides information to decisionmakers and the public on the suggested undertaking and lists alternatives to the proposed action, including taking no action. A federal EIS is usually based on an environmental assessment (EA), which is a preliminary assessment that determines the need for the more thorough and formal EIS. An EA uses a systematic and interdisciplinary approach to assess the interaction of physical, natural, social, and economic factors. Probable effects and consequences of the proposed action on these systems are identified. It supplies information of this nature for decisionmaking and the EIS. Like the EIS, an EA basically consists of asking and answering appropriate questions.

An EIS supplies an early warning and information system to the agency preparing it as well as to other agencies and the public. Many different public and private constituencies are made aware of proposed actions that might have significant environmental impacts. This is particularly important when many proposals are of an irreversible nature and would permanently foreclose other options and environmental values. Thus, environmental impacts should be a prime consideration in the earliest stage of project planning.

Although numerous questions can be raised about the effectiveness of EISs in altering federal agency actions, they are now a routine and integral part of agency operations for planning and decisionmaking. Recent agency attitudes point toward the preparation of analytical summary EISs, contrasting sharply with past EISs. At the outset of the implementation of NEPA requirements, agencies prepared voluminous EISs that often precluded adequate review. Agencies preparing those long and detailed EISs feared having to redo and expand their environmental assessments.

The nature and extent of "environmental consideration" depends on the extent of administrative discretion that an agency is given, the quality of its personnel, and its level of public support. Many environmental abuses and problems could be prevented or greatly reduced with greater emphasis on environmental values by agencies. Often adequate consideration is hampered by insufficient and uncertain knowledge. The absence of such information complicates the prediction of complex ecological and other consequences of proposed developments and programs. Decisions are further complicated by differences in scientific opinions supplied by various experts—the "right" opinion often being the one that reflects the hierarchical power, dominant ideology, and vested interests of an agency. Even with the requirements of EISs for proposed actions, the strong influence of the agency and its vested interests (including clientele) is felt, despite interagency and public review. An agency need only *consider* environmental aspects in its determinations and discretion, despite the severity of the environmental problems and negative consequences that are projected to occur. Once the procedural requirements have been met, pressure groups have extreme difficulty in challenging agency decisions even in the courts.

Due to his role in the development of the EIS requirements in NEPA, Lynton Caldwell is repeatedly asked to comment on NEPA and assess its role in the federal decisionmaking process. He observes that:

> There are at least three ways in which NEPA cuts across the grain of traditional management theory and practice. First, it contradicts long-standing command-control assumptions. Second, it qualifies and complicates narrowly defined mission assumptions and commitments. Third, it holds the risk of embarrassing career advancement in the public service. Thus it follows that the beliefs of public officials regarding their roles, responsibilities, and opportunities in public management influence their attitudes toward the implementation of NEPA.[57]

The environmental impact statement process has had profound effects on federal agency policies. Serge Taylor identifies two modes of influence that EISs have had on decisionmaking in the administrative process:

> The first is internal—the analysis provided by the EIS analysts to the decisionmaker. The second is external—the increased political resources (in the form of information, authority, and legal resources) provided environmental interest groups and environmental agencies to challenge a development agency's technical premises, present competing alternatives, and have their preferences count more in the final balancing.[58]

The EIS process, in some form or another, has been adopted by many state and local governments and by other national and international agencies. Thus, this section is relevant to more than federal environmental administrators.

Regulatory Aspects

Most of the environmental legislation passed over the past twenty-five years is regulatory in nature. As a consequence, natural resource agencies have greatly expanded their regulatory powers and controls. Agencies like the EPA and the United States Office of Surface Mining (OSM) are considered regulatory in nature. Some agencies like the USDA Forest Service and the BLM have regulatory functions as parts of their overall operations. And a traditional land management agency like the U.S. Fish and Wildlife Service has acquired significant regulatory powers through such acts as the Endangered Species Act and the Fish and Wildlife Coordination Act.[59] The latter is like a national environmental protection act for fish and wildlife, whereby the impact of proposed federal projects on fish and wildlife habitat is ascertained prior to the granting of development permits. Generally, administrative regulations of natural resource agencies have few punitive powers and do not have the strong controls that are given to agencies with police powers and recourse to the courts. In this sense the strong regulatory controls of the EPA contrast with the much weaker advise-and-consent powers of most natural resource agencies. Theodore Lowi points out, "Most regulated conduct is subject to less serious restraints. Moreover, the restraints are intended not to eliminate the conduct but to influence it toward more appropriate channels or locations or qualities of service."[60] In this manner regulatory efforts of natural resource agencies are directed primarily toward the minimization of the negative consequences of federal and private activities or developments.

The primary task of agency personnel in the regulatory process is to translate law into operating policies. Their ability to develop satisfactory policies is related to the amount and type of administrative discretion they are granted to decide when, where, and how to interpret the law. Often the degree of discretion is correlated with the ambiguity or indefinite nature of the law and situation.

As a result of the complexities involved in the development of pollution control laws, Congress gives broad discretion to the administrator of the EPA. Within the general guidelines established by Congress, EPA officials promulgate technical regulations for the implementation of the laws. This pattern worked effectively until the 1980s, when, in an effort to reduce the regulatory burden on business, Reagan appointees started to delay the issuance of rules required by the Resource Conservation and Recovery Act (RCRA) related to toxic substances. This prompted environmental groups to file a number of lawsuits successfully challenging the EPA for the intentional delaying tactics.[61] Office of Management and Budget (OMB) officials further delayed environmental regulations through review powers obtained from Executive Order (EO) 12291. However, District Court Judge Thomas Flannery declared "that OMB has no authority to use its regulatory review under EO 12291 to delay promulgation of EPA regulations arising from the 1984 Amendments of the RCRA beyond the date of a statutory deadline."[62] The judge acknowledged that this might be an intrusion into the flexibility of executive agencies, but ruled that the court must uphold the law as passed by Congress.

The less precise Congress is in its laws, the more discretion agencies have in policy implementation. Thus, vagueness in wording of policy objectives, ambiguity over standards among experts, flexible compliance deadlines, and discretionary enforcement are some of the major factors that give agencies considerable administrative discretion. These factors provide opportunities for political bargaining as well as value judgments and conflicts. Without authoritative and definite rules, competing interests can exert influence in this discretionary vacuum and, consequently, on administrative judgments and policies.[63]

Ira Sharkansky points out that industries being regulated are the "most assiduous and the most successful in affecting agency rules and decisions. . . . [C]ozy relationships and outright illegalities tilt decisions in the direction of industries."[64] The quality of regulation in environmental protection varies from one setting to another, and regulation appears to "depend on the nature of legal mandates, the resources provided, and simply good or bad luck in the severity of the problems faced."[65] Alan Stone defines regulation as "a state-imposed [governmental] limitation on the discretion that may be exercised by individuals or organizations, which is supported by the threat of sanction. The term *regulation* is pertinent when decisionmaking in a branch of activity is apportioned between what may be termed the private and public spheres."[66] Thus, regulatory decisionmaking involves discretion and limitation in both the public and private sectors.

In commenting on the success of the regulatory decisionmaking process from the agency perspective, William Drayton suggests that "regulatory law enforcement, from the time a violation is detected onward, is a mess. . . . If jawboning fails to induce compliance, regulators must either give up or litigate, and litigation is uncertain, slow, and costly. . . . As a result, massive delays occur, public and private resources are wasted, scofflaws are rewarded, and voluntary compliance is undermined."[67] To avoid this sort of wasteful confusion, Drayton recommends using "recapture" standards that charge violators an amount sufficient to make compliance as economically attractive as possible. This permits agencies to adopt a host of economic remedies and options in between ineffective jawboning and legal action in the courts.

Much of the resistance and criticism associated with regulatory activities is economic in nature. Some years ago Frank T. Cary, chairman of the Business Roundtable Task Force on Government Regulation, remarked:

"The regulatory pendulum has swung too far the other way [and] has imposed on business excessive costs which often exceed their benefits. And since 1973 these costs may have been cutting our rate of productivity increase by nearly half a percentage point every year. In the imposition of regulations, I believe our government has been inflexible on methods of compliance and insensitive on costs."[68] The Republican assaults on environmental regulations in the 104th and 105th Congress (1995–96, 1997–98) attest to the continuing perception that the EPA continues to be unresponsive to business concerns.

There is also the danger of "agency capture" of the regulators by those who are regulated. Marver Bernstein theorizes that regulatory agencies evolve through life cycles. At the outset the regulatory agencies attack the industries vigorously in order to eliminate abuses. After a period of time the industries respond, and the situation improves or the agency becomes frustrated with its inability to force the industry to comply. Eventually, the regulatory agency views the industry as an important constituency necessary for its long-term survival and becomes responsive to the needs and desires of the industry that it was created to regulate initially.[69]

Thomas Murphy argues in a 1980 article that is still relevant that regulatory reform is needed and that it should emphasize economic analysis. He believes that this would force agencies to evaluate the costs of proposed regulations and consider less costly alternatives for getting the desired benefits. Murphy recommends putting regulatory "teeth" in substantive regulations where numerous remedies may be included to gain compliance. The "teeth" range from withdrawal of all government contracts to fines and imprisonment. For example, the EPA's two thousand pages of regulations on hazardous waste disposal contain provisions for violators to be fined $25,000 a day and, upon a second violation, to spend up to two years in prison. Murphy concludes that regulations are basically laws delegating power from Congress and the courts to regulators. Regulatory matters are considered to be too complex, detailed, or specialized for either Congress or the courts. Hence the regulators make and interpret the law while also enforcing and judging it.[70]

Where burdens and costs to the private sector are concerned, there are other factors that need to be taken into consideration. A Resources for the Future (RFF) study on this topic judged the effects of environmental regulations on the U.S. economy to be adverse but minor. Other factors such as the energy crisis and changes in the labor force may be far more responsible for economic difficulties. Further, most economic indicators show the costs but ignore benefits that result from environmental controls in terms of overall effects and social well-being.[71] In an analysis of the social costs of environmental quality regulations, Michael Hazilla and Raymond Kopp find that regulatory impact analyses tend to ignore less easily measured social costs and intertemporal regulatory impacts on households and firm decisionmaking. Instead, regulations impact in the form of engineering costs associated with installation of pollution control equipment and the related operating and maintenance costs.[72]

The direct effects of pollution control expenditures have been relatively small. Analysis indicates that the direct costs of environmental regulation probably are only 8 to 12 percent of the decline in growth rates. In arguing for efficiency and economic incentives to offset the effects of indirect costs, Kent Price suggests: "Perhaps more important than the direct costs, however, are the adverse effects on the economy that often result from the poor implementation and administration of regulations. It is impossible to account fully for costs associated with regulatory delay, for example, or from increased paperwork burdens, but the costs are nonetheless real." A very important cost is the uncertainty that is inherent in the regulatory process. Private interests do not know to what degree which regulations will be implemented, let alone what standards will change in the future. These uncertainties affect investment and economic production.[73]

Paul Weaver argues that regulatory policy is not economic policy but social policy, which transcends the economy and marketplace and does not make sense economically. As social policy, it is government intervention to advance the public interest apart from or opposed to the outcomes of the marketplace. Thus, regulatory policy includes class politics with different sets of values or "worldviews" for society that government, in turn, asserts in its regulatory activities. Weaver considers a cardinal principle (or value) of environmental regulations to be that of "internalizing the externalities" with industry and consumers paying the social (and environmental) costs of goods and services.[74]

It would be a mistake, however, to presume that internalization of externalities is exclusively a political process. Science and technology have thoroughly permeated our society to the extent that scientific and technological factors influence the process.[75] Environmental regulation has a high scientific and technological content that complicates the implementation of effective decisions. Data gathered out of scientific and legal necessity serves multiple interests. While decisionmaking must incorporate political values, there is still a general consensus that the assembly and interpretation of relevant and scientifically valid information is essential to environmental policymaking. Beyond that, there is no general agreement as to what roles highly competent scientists should play in the decisionmaking process.

Political considerations can override scientific and technical ones. In contrast to previous presidential administrations, the Reagan administration, for example, developed and implemented dramatic reversals through its deregulation policy, drastically affecting the quality and quantity of environmental regulation. This policy involved "sharp cutbacks [funding, personnel, programs, and emphasis] in the enforcement of virtually every kind of environmental regulation."[76]

Craig Reese notes that budget and staff cuts at the EPA, the Council on Environmental Quality (CEQ), and Office of Surface Mining resulted in a significant reduction in environmental regulation enforcement efforts.[77] Also the reorganization of both the EPA and OSM resulted in a deemphasis of enforcement activities. The apparent assumption was that business and industry would voluntarily comply especially if environmental regulations were simplified. Under the Reagan administration, CEQ's guiding principles changed toward (1) an emphasis on regulatory reform including extensive use of cost-benefit analysis to determine the value of regulations; (2) reliance on the free market for resource allocation; and, (3) a shift of responsibilities for environmental protection to state and local governments when feasible. This reduced the role of the federal government in controlling the social costs associated with resource development and pollution.[78] Furthermore, the shifting of environmental regulatory responsibility to the states reduced the capacity of government to deal with environmental problems such as the degradation of air and water quality, which are not confined within state and other political boundaries.

Serious questions can be raised about the nature of regulatory measures and activities of various agencies. Regulatory agencies and personnel are often captured by the very industries that they are trying to regulate. Although public opinion surveys show that most Americans still want strict enforcement of environmental laws, even if it requires economic sacrifice, political and governmental support for environmental protection in the 1980s weakened substantially.

The degree of support can be questioned further in terms of the actual incorporation (and depth) of environmental values and perceptions into the operations of the regulatory bodies themselves. Low internalization of such values is reflected in weak enforcement of regulations. For example, the EPA is considered the major American regulatory agency for executing federal laws for protecting the environment. Yet the agency is often involved in serious problems and controversies over regulatory enforcement as its responsibilities and the complexities of its tasks increase. The EPA is attacked by environmentalists, industry, and federal development agencies alike. Formed in 1970 to consolidate in one agency much of the federal authority and expertise for the control and abatement of pollution and other environmental problems, the EPA is unique in regulatory politics. Noting its strong congressional mandates and adversarial image, Gregory Daneke argues that the EPA was designed to "(1) avoid capture by the industries it was to regulate, and (2) function in such a way as to limit its own discretionary powers and increase legislative responsibility."[79]

The EPA and other regulatory agencies face a number of serious problems. Among them are (1) trying to fulfill missions and meet deadlines that are not scientifically or technologically feasible; (2) trying to balance the high front-end costs of control technologies against intangible environmental and public health benefits; (3) acting effectively to regulate new energy technologies during a period of declining fossil fuel reserves; and (4) promoting nonproductive control technologies to a society already experiencing the belt-tightening effects of a declining economy.[80] Under these conditions, regulatory activities tend to become increasingly relaxed and ineffective.

As noted previously, the EPA was particularly hard hit in the 1980s, suffering personnel, budget, and program cuts, all of which reduced the agency's ability to conduct its regulatory functions. Richard Andrews observes: "It is unfortunate that much of the legitimate conservative agenda that Reagan's administration might have achieved thus appears to have been lost to bad judgments that easily could have been avoided and to continuing ideological rigidity. It is also tragic that so much damage was done in the process both to environmental protection and to the cause of regulatory reform."[81]

Assuming the continuance of prevailing trends toward deregulation during a period of political conservatism, regulatory efforts should be directed toward negotiation and planning for effective public and private controls and innovations. Cooperative environmental planning by the public and private sectors would facilitate the achievement of social goals.[82] Gail Bingham observes that proponents of regulatory negotiation consider current regulatory procedures too adversarial for resolving disputes and that this legalistic framework encourages "costly rounds of administrative appeal and litigation" that postpone "the achievement of desired goals, and may prevent exploration of mutually acceptable alternatives."[83]

Various innovations are obviously needed in order for regulatory reforms to produce better and more sound rules. Over the past several years requirements for rule making have made it a more open process. Those affected by the rules have to be heard. Rule making by the EPA now requires procedures that have definite steps, which include defining alternative approaches and explaining why the recommended approach is preferable. This involves routinely evaluating a vast array of potential environmental, energy, and social impacts that may result from the proposed actions under consideration.[84]

Regulations are an increasingly important part of natural resource agency administrative operations. In reviewing the numerous environmental laws and regulations that apply to the BLM, for example, agency personnel consider future trends and developments to entail

> more laws and regulations concerned with environmental protection. As resources become inevitably more scarce and concurrent demand becomes more intense, the conflicts will also increase. The necessity to resolve these conflicts will put greater pressure on government at all levels to become the mediators of these conflicts. Government's normal approach to this mediation is to enact new statutes and promulgate new regulations.[85]

Regardless of the merits of the law and regulations, much depends on how they are administered and enforced by the agencies and their personnel. The environmental public interest may be negatively affected through administrative discretion that gives too much to the regulated industry by placing private interests above public ones. A good example is the manner in which western cattle and sheep owners continue to get excessive grazing privileges on public lands at unrealistically cheap prices. As early as 1981 Sabine Kremp noted, in her assessment of the BLM grazing policy, that "the grazing fee underestimates the full value of the additional grazing by a sizable amount (50 percent or more)."[86] Problems also occur because of too much or too little regulation or control in a given situation.

The Reagan administration broke away from the theme of discretion for environmental administration and regulations. Howard McCurdy concludes that the three "new" themes of the Reagan administration were (1) use of economic criteria with economic incentives and real prices attached to environmental decisions and the expansion of market influence; (2) "new federalism" with a glorification of the states and a cutback of federal involvement in pollution control and land management programs and regulations; and (3) a new kind of administration with less government (including policy and administration) and less regulation, in contrast to traditional public administration considerations.[87] A similar approach was maintained by the Bush and Clinton administrations. However, constrained by a Republican majority in both houses of Congress, the Clinton administration has tended more toward a policy of compromise with regard to regulations.

Regulations are the primary means for society through government to have some control over human relationships and behavior toward the environment. Regulations provide controls for mitigating or preventing negative and often irreversible consequences for the environment and society. Regardless of the administrative approach, regulations involve the interface of values between public and private sectors and between organizations and individuals in terms of the environment. This interface brings out both the rational and irrational aspects of regulations affecting the degree to which they will be enforced. Thus, the strength and depth of underlying values and their power base determine much of the regulatory process and its results.

The regulatory process also involves interagency relations with various forms of conflict and cooperation. For example, several agencies may have a common jurisdiction or "symbolic territory" while at the same time possessing their own mandates and individual sets of regulations. Agencies may have mandates that conflict with one another or require clearance from other regulatory bodies. One agency may have regulatory functions over another that may offer resistance. For example, the EPA has had considerable difficulty getting the Tennessee Valley Authority (TVA) to comply with air pollution requirements. The EPA eventually had to take court action against and enter into negotiations with TVA to gain compliance. Federal, state, and local agencies implement federal laws and regulations. Interrelationships between different levels of government and agencies are fluid and the courts generally protect state sovereignty. Successful cooperation in regulatory arrangements often involves informal activities composed of networks of individuals from various agencies formulating and negotiating policies concerning what is acceptable to the system.[88] The regulatory process is one important aspect of interagency relations.

Interagency Relations

The National Environmental Policy Act contains numerous provisions calling for interagency coordination and review of environmental quality. For example, with respect to environmental impact statements NEPA specifies: "Prior to making any detailed statement, the responsible federal official shall consult with and obtain the comments of any federal agency which has jurisdiction by law or special expertise with respect to any environmental impact involved. Copies of such statements and comments and views of the appropriate federal, state, and local agencies, which are authorized to develop and enforce environmental standards, shall be made available."[89]

The environment thus provides a new focus and potential for meaningful and effective interagency relations. There are dozens of federal departments or agencies that have responsibilities in the area of environmental affairs. With the addition of numerous state and local government agencies, interagency relations obviously require a focus for unity of action. The development of an environmental focus will not, however, end perennial political conflicts between agencies. An environmental context will only subject their incompatible objectives to fuller consideration in terms of issues and values. As Norman Wengert stresses,

> The environment is a unity, but resources are discrete—hence a problem of coordination arises when an effort is made to reconcile resource-centered programs and agencies with comprehensive environmental goals. . . . Coordination of public programs dealing with or affecting many different facets of the environment superficially, at least, would seem to leave much to be desired. In many cases little or no coordination is apparent, separate programs going their separate ways. In other cases coordination is only *pro forma* and superficial. Given that those planning and conducting particular programs with obvious interrelationships have not consulted together, and in many cases programs dealing with similar or related problems may be seeking different and even conflicting goals or objectives, the pluralism of our society has been projected to government activities.[90]

Resistance to interagency coordination also takes the form of an agency's strict adherence to lines of responsibility in specific environmental situations. Under restrictive guidelines personnel are often frustrated and kept from effective interagency cooperation in a holistic approach to the environment. In this sense overlapping and duplication may not be so much the problem as lack of authorization to take unified action. Informal cooperation may occur among specialists, particularly at field levels, but it may also lack the sanction of the agencies. Such informal activity may therefore lack real power for concentrated long-term action.

Although "memoranda of understanding" made between agencies formalize cooperative arrangements, these agreements usually distract the agencies concerned into defining their responsibilities and limiting any policy innovations or other commitments. A relevant case is the Interagency Wildlife Committee, which was established to handle overpopulation and migration of elk at Rocky Mountain National Park in Colorado. Wildlife specialists from five agencies were involved, but the committee soon found that it could not really make any policy or even policy recommendations for a flow resource on a regional and interagency basis. Its duties consisted of a superficial exchange of information and research.[91] The Clinton administration directive that land management agencies adopt ecosystem-based management approaches that emphasize cooperation across agency and jurisdictional lines and are cross-discipline based may improve this situation.[92]

With failure at the interagency level to address and express concern over values, participating agencies will act out of their own more limiting value orientations while making only token efforts at interagency cooperation. Consequently, unintended "hidden" values and negative consequences may emerge from interagency deliberations. A forest management agency may, for example, authorize logging operations detrimental to the protection of a watershed under the jurisdiction of another management agency, in spite of apparent efforts at interagency cooperation. Wolfinger, Shapiro, and Greenstein summarize the dangers: "Differences between agencies are compromised; unresolved conflicts are papered over by vague generalizations. A false front of harmony is established behind which each agency continues in its own way."[93]

This phenomenon also applies to lateral coordination in interagency clearance systems wherein reports, evaluations, EISs, and policy recommendations are cleared by other agencies involved in the same spheres of operations before being sent to higher authorities. Because agencies do not want their reports to pass on with dissenting opinions, they will usually modify the content of a proposal or its language to meet the objections of other agencies. The end results of interagency clearance may be compromises that "paper over" differences and avoid proper value considerations.

Many values and concepts of the World Conservation Strategy of the International Union for the Conservation of Nature and Natural Resources (IUCN), for example, can still be considered new for many agencies.[94] However, they will not be incorporated into various programs and interagency relations without political and administrative support. Ecosystems and the effects of development seldom observe institutional or political boundaries. Therefore, they must be approached on a cross-sectorial, coordinative basis. In this context of interagency inertia the World Conservation Strategy recommends: "The different agencies with responsibilities for living resources should have clear mandates and such mandates should specifically include conservation; there should be a permanent mechanism for joint consultation on and coordination of both the formulation and implementation of policies."[95] These recommendations grow out of an explicit understanding that ecosystems and the effects of development fail to recognize institutional boundaries.

Interagency relations generally require a form of central control or major mechanism (institutional form) to ensure needed coordination and cooperation for a comprehensive and effective approach toward the environment. This overall approach is necessary to avoid serious conflicts and negative actions by individual agencies relative to the environmental public interest. Growing efforts are under way by many agencies and personnel to achieve interagency relations based on general environmental values and integrated components. These efforts must be supported, however, by sufficient political power to transcend the immediate interests of individual agencies. Finally, changes within the agencies, innovative communications, and new forms of interagency arrangements are needed to facilitate common environmental approaches and unified, coherent actions throughout the administrative process.

Decisionmaking is possibly the most important aspect in the administrative process. The foregoing material illustrates the extreme complexities involved in assuring that appropriate value considerations enter into this process. At all decisionmaking levels the environmental public interest must be considered. In a closed decisionmaking system this is not likely to occur. In order for past achievements in the environmental arena to be maintained and satisfactory future objectives to be achieved, changes will be required. This can only be accomplished through the "substitution of the paternalistic dictates of policymakers, scientists, and special interest groups with a participatory evaluation and decisionmaking process; [and a] reenforcement of quantitative research with qualitative evaluation of scientific information and public values."[96]

The situation with regard to energy policy has improved significantly over the past ten years. A bipartisan effort in Congress produced the Energy Policy Act (EPACT) of 1992. The act required all states to develop new energy efficiency provisions for residential building codes. Manufacturers were required to improve energy efficiency in a wide array of machines and appliances. All federal agencies were required to reduce energy consumption by 20 percent below 1985 levels by the year 2000. One of the more important provisions of EPACT was the requirement that the Department of Energy (DOE) develop a comprehensive national energy strategy that would set all of these and many other energy policy initiatives into motion. The Energy Policy Act also required the Department of Energy to establish financial mechanisms to increase the development of renewable energy technologies in the United States.

The Clinton administration has enthusiastically developed the Comprehensive National Energy Strategy that emphasizes energy security, energy efficiency, and the development of new sources of energy with an emphasis on renewable sources in the United States. It has, through DOE's Energy and Renewable Energy Network (EREN), made significant contributions toward the development of a wide array of renewable energy technologies. Solar power towers are now tied into utility power grids generating more than 20 megawatts of energy.[1] A National Center for Photovoltaics has been established. Alternative fuels are being developed as well as hydrogen-based fuel cells.

The worldwide outlook for energy supplies has improved over the last decade of the century. In addition to research on and implementation of renewable fuels, the traditional fossil fuel situation has improved as well. Exploration for coal, natural gas, and oil has identified new sources that have expanded the amount of "known" fossil fuel reserves around the world. The known reserves of oil have increased so much that oil is now projected to last another one hundred years at current consumption rates. Ten years ago, the projected life of an economically viable supply of oil was only twenty-five to thirty years into the twenty-first century.

Fuelwood still remains the primary source of energy for cooking and heating for a large portion of the developing world. There continues to be a shortage of fuelwood; in fact, a fuelwood deficit exists in many less developed countries. However, recent developments in renewable energy technologies may provide the necessary means to transfer much needed energy technology to the developing countries. Solar and biofuel technologies seem to hold the most promise to provide these countries with economically viable sources of energy.

Energy and Environmental Policy

It is impossible to separate a discussion of environmental and natural resource policies from energy policies. Both the production and consumption of energy resources affect the quality of the environment in multiple ways. Until recently, however, the United States lacked a coherent policy for the long-term supply and maintenance of energy. This revelation is somewhat amazing in light of the fact that the American economy is almost totally dependent upon a continuing, economical source of energy and the political salience of the energy policy issue over the past twenty-five years or more.

The formulation of energy policy is primarily the responsibility of the federal government. But the federal government has historically failed repeatedly to develop a coherent and comprehensive energy policy. The bipartisan consensus that produced the Energy Policy Act of 1992 prompted the Clinton administration to develop the 1995 National Energy Policy Plan that contains a broad-based sustainable energy strategy. The plan emphasizes economic productivity, environmental quality, and national security.[2] Until then, the government's record was not promising.

In 1977, for example, it appeared that President Carter was on the verge of developing a national energy policy, but for one reason or another his efforts fell short. Nevertheless, Carter did succeed in consolidating the various disparate federal offices with energy responsibilities into a cabinet level Department of Energy. At the end of the Carter administration there appeared to be some hope that the United States would develop a policy to ensure a stable, long-term, economical source of energy. But any promise for the establishment of an energy policy died with the advent of the Reagan administration, which opposed both environmental and energy initiatives. One of the first actions taken by President Reagan with regard to energy policy was the termination of federal domestic regulations on domestic oil prices and supplies on January 28, 1981.[3]

Almost immediately after Reagan came to office, he proceeded to dismantle many of the energy programs initiated under the Carter administration. Reagan's first National Energy Plan set the tenor for future policies. The plan optimistically states that "overall the outlook for this country's energy supplies is not nearly as grim as some have painted it." Federal intervention in the allocation and pricing of energy would be reduced, with an "increased reliance on market decisions" because they offer "a continuing referendum which is a far better means of charting the nation's energy path than stubborn reliance on government dictates or on a combination of subsidies and regulation." "Regulatory relief" would characterize future energy policy regulations. As Rosenbaum concludes, "Within three years the Reagan administration had drastically altered national energy policies. Federal spending on energy research had decreased. Energy production received more attention and environmental protection less."[4]

The very nature of the American policymaking system may have kept it from anticipating the 1973 energy crisis and developing a means to guarantee an adequate supply of energy in the future. In a detailed analysis of American energy policy efforts Don Kash and Robert Rycroft indicate:

> The system was organized to make and manage policy for five individual fuels. As scarcity was perceived to be replacing abundance, however, the conviction grew that energy should be managed in a more comprehensive fashion, rather than by fuels. In 1973 the focus of policymaking moved from problem solving in five stable fuel policy systems to a search for issue resolution by the president and Congress. While it was generally agreed that the traditional arrangements for controlling energy no longer worked, there was no agreement on what new arrangements should replace them. Only after an issue resolution process had occurred would it be possible to reestablish a stable problem solving policy process for energy.[5]

Kash and Rycroft believe that the president and Congress work through the issue resolution process to form a consensus on goal definition and identification of the "appropriate political, managerial, or organizational mechanisms" for goal accomplishment. There is a great deal of openness and conflict in this process. Once consensus is reached, a different, more limited set of semiautonomous policy systems operate. In a relatively closed environment a problem-solving policy process is established where vested interests shape the policies that affect them. Unfortunately, when there is no consensus on energy policy in the United States, little real problem solving occurs. Although some consensus occurred with the passage of EPACT, few hard decisions have actually been made to reduce energy consumption. In fact, the short-term oil surplus of the latter 1990s will only serve to erode any consensus that has developed.

Despite the difficulty that government has in developing a coherent energy policy, there is a growing consensus in the United States that energy policy should be directed toward accomplishment of four basic goals: (1) maintenance of adequate energy supplies; (2) conservation of energy through reduction in demand and achievement of greater end use efficiency; (3) reduction of dependency on foreign oil; and (4) protection of foreign energy supplies and energy distribution systems. Governmental attempts to attain these goals, however, remain controversial, with no particular agreement on the means to accomplish them.[6]

Overall Considerations

Energy policy cannot be based only on availability of fuel resources. Other social, economic, and environmental considerations must be taken into account. Walter Rosenbaum poses three critical questions that relate energy policy to these other issues. "First, what should be the balance between public and private benefits from energy policies?" The resolution of basic energy issues will undoubtedly require balancing of public and private costs and benefits. Environmental protection is one of the benefits of "public goods" that cannot be assigned to a given group. This contrasts with "private goods," which selectively benefit specific groups without benefiting a broad spectrum of society. For example, in the raising or lowering of air pollution control standards, public benefits often conflict with private benefits.

"Second, how should the economic costs and benefits of a policy be distributed among those affected by it?" Conflicts over energy policies also involve the distribution of economic costs and benefits among those affected by them. These costs and benefits include externalities or outside effects from energy production and use. Often political conflict results from the failure of energy producers to internalize various social, health, and environmental costs into the costs of production. Those segments of society that are forced to absorb these costs argue for policies to correct such market failure. Difficulties in policy resolution or compromise between public and private energy agendas result in polarization of opposing forces.

"Third, what should be the respective roles of government and private business in future energy development?"[7] The Reagan administration placed an emphasis on public-private interaction in the formulation and implementation of public policies. However, too much influence of private interests in the development of government policies blurs the boundaries between private and public decisionmaking and reduces accountability. Ultimately this is not in the environmental public interest. As Robert Stobaugh and Daniel Yergin suggest, "The fundamental purpose of U.S. energy policy . . . should be to manage a transition from a world of cheap imported oil to a more balanced system of energy sources. . . . Because of continuing imperfections and failures, it is unrealistic to expect an uncorrected 'free' market to solve U.S. energy problems."[8]

Most people do not agree that the market is capable of meeting U.S. energy needs. As Michael Reagan says,

> It is in some degree *uncivilized* (perhaps out of frightful naiveté) to suppose that so basic an ingredient of modern society as national and world energy supply and usage patterns can be simply entrusted to a particular set of market institutions whose compatibility with long-run societal needs is itself in continuing need of "course corrections" through public sector interventions. Energy policy should remain public policy.

In essence, greater consideration should be given to long-term energy supply diversification regulated by the public sector.[9]

The nature of environmental conflicts, the severity of environmental effects, and the substance of environmental policies are in part determined by the directions taken by energy policy and programs. It is important, therefore, to seek consensus in the development of energy policy through the political process, although, as Hans Landsberg suggests, the forging of energy consensus should not be permitted "to founder on the rocks of particulars." Consensus is possible: "First, it is more important to plot persuasively the direction in which we must travel than the exact itinerary and second, we probably know enough to be able to agree on direction."[10] Still, consensus, however general in content, may imply just the lack of visible opposition rather than actual agreement or support. Insofar as the American culture is concerned, the dominant forces affecting consensus are pragmatism, with its mistrust of ideology and stress upon short-range goals, and pluralism, with its tendency to fragment one policy into many policies. James Katz believes that pluralistic decisionmaking "obstructs policy adjustments to meet energy challenges. Thus while an innovative and comprehensive energy policy has become increasingly necessary, the means for its realization have become increasingly complicated and elusive."[11]

The conflict between energy development and environmental protection may be both inevitable and constant. Rosenbaum observes that "government can rarely establish environmental pollution standards without deciding, in the process, what levels of energy production and which energy technologies will be tolerated. The connection between energy management and environmental quality not only undergirds most energy issues; it also explains why their resolution often is so difficult and the struggle to create a coherent national energy policy so bitter." [12]

Two diametrically opposed views affect choices for energy and related environmental policies: one is "expansionist," and the other "limited." The expansionist view is consumption- or growth-oriented, optimistic, and identified with short-term problem solving and technological solutions. This view has dominated public policy throughout U.S. history. In contrast, the limited view emerged as a significant alternative after the mid-1960s in a climate of growing opposition to and disillusionment with expansion and growth. This view argues that the natural world is limited. The possibilities for human domination and exploitation of the planet are finite; a restructuring of human values and behavior is necessary for humans to achieve harmony with nature.

Milton Russell argues that energy policy decisions have emerged from the basic assumptions of the

> expansionist who believes in (1) the desirability of increasing energy production and consumption, and (2) the inevitability of technological solutions to energy problems. Expansionist energy policy is directed toward: (1) making energy cheap, plentiful, and available, and (2) encouraging discovery of new reserves and sources for energy. Such policy facilitates growing energy production and consumption with relatively little constraint imposed upon energy producers. [13]

The limited view assumes energy growth cannot continue because of (1) the finite resource base, (2) environmental constraints and limits, (3) probable degradation in the quality of human life, and (4) risks to human survival. Energy policy articulated from the limited view would endorse conservation of available fuel reserves, reduction of energy outputs, and alteration and reduction of end uses for energy. Such a policy would also have to take into account the need to alter human lifestyles and values. These changes in human patterns inevitably require the alteration of existing human institutions and the creation of new ones. Ultimately, advocates of the limited view would call for policies encouraging evolution toward a steady-state society that is in harmony with the environment. [14]

Given its fragmented and controversial nature, much of what passes for energy policy has been and continues to be based on the belief in unlimited and uncontrolled growth held by the expansionists. In his *Soft Energy Paths: Toward a Durable Peace,* Amory Lovins declares the expansionist policy unrealistic and unworkable:

> It is looking *politically* unworkable; most people, for example, who are on the receiving end of offshore and Arctic oil operations, coal stripping, and the plutonium economy have greeted these enterprises with a comprehensive lack of enthusiasm, because they directly perceive the prohibitive social and environmental costs. Extrapolative policy seems technically unworkable; there is mounting evidence that even the richest and most sophisticated countries lack the skills, industrial capacity, and managerial ability to sustain such rapid expansion of untried and unforgiving technologies. And it seems *economically* unworkable; for excellent reasons, such as free market mechanisms to allocate to the extremely capital-intensive, high risk supply technologies, and the money needed to build them. The inexorable disintegration of current policy thus makes us reexamine its [growth] premises.[15]

Many of the expansionist aspects of energy policy are associated with exponential growth. When the rate of consumption is growing at 7 percent a year, for example, the consumption in a decade will exceed the total of all the previous consumption. The astonishing per capita growth in consumption of resources throughout U.S. history may be at the heart of its energy problems. Exponential growth automatically leads to monumental energy demands for and consumption of limited energy sources. Thus, a lowering of the exponential growth rate and conservation are fundamental approaches for responsible energy policy.[16]

In a report the National Research Council Committee on Nuclear and Alternative Energy Systems said, "The first and dominant 'facet of the solution' relates to the issue of how fast, and indeed whether, our use of energy may need to grow and, ultimately, how much energy our society will require to sustain the way of life that it chooses. *Energy is but a means to social ends; it is not an end in itself.*"[17] In sum, basic social ends underlie energy policy. Many experts agree that the United States could reduce its energy consumption by one-half. Much lower rates of energy consumption are found in the industrialized, developed nations in Europe. Much of the U.S. economy, however, is oriented toward an exponential growth rate in energy consumption.

In considering the belief in growth as the basis for energy policy, Jeremy Rifkin argues that the realities of entropy law apply. "The Entropy Law states that matter and energy can only be changed in one direction, that is, from stable to unstable, or from available to unavailable, or from ordered to disordered." Consequently, entropy is a measure of available energy that can be transformed. When the transformation of energy occurs it is done at the expense of order in the surrounding environment and produces waste.[18]

Rifkin questions the expansion and growth assumptions and the values of energy policies that increase the rate of entropy. Growth in energy use greatly decreases the limited resources available in usable form for sustaining society on a finite planet while also creating high environmental costs and risks. The entropy law challenges the notions of history as progress and of science and technology as creators of an ordered world. Hence, a major element of energy policy should be the dramatic reduction of the rate of growth in energy use and the slowing down of entropy, so energy will not be converted from available to unavailable forms at such a rapid and destructive rate.

High levels of energy consumption and waste make the United States a high-entropy nation. This is as true today as it was twenty or thirty years ago.

Energy Conservation

Energy consumption growth rates have not been as rapid as forecast in the 1970s. From 1979 to 1983, energy consumption growth rates actually decreased in the United States. In large part this was a direct reflection of a declining economy, but a large measure of the reduction is attributed to energy conservation measures. In *The Politics of Energy Conservation,* Pietro Nivola highlights one of the ironies of the energy crisis: "Beneath all the controversy and confusion, the remedy to the energy problem was clear: the nation's chief demand for energy could best be brought into balance with the supply chiefly through conservation. No other course of action promised to meet the country's energy requirements more dependably, quickly, cleanly, and cheaply."[19]

Some Americans have unjustifiably correlated energy conservation with a sacrifice in the quality of life. John Adams et al. note that "in 1983 the American economy used nearly 25 percent less energy to produce each dollar of product than it did in 1973. Yet for many Americans, the myth that energy conservation means sacrificing creature comforts has not been totally dispelled."[20] It could equally well be argued that this commendably greater efficiency in energy use goes only a short way toward addressing long-term energy problems. Although it would appear that some sacrifice of creature comforts will be required in order to reduce the incentive to consume energy, the fact remains that cheap energy prices, regardless of any social equity consideration, encourage consumption. European energy prices are far greater than in the United States, and energy consumption is considerably lower per capita as a consequence.

Despite the effect on creature comforts, the gains made through energy conservation will improve the quality of the environment, because more efficient energy processing results in less pollution. The fact that pollution results from inefficient processing of energy sources like fossil fuels has been known for a long time. More efficient processing would reduce carbon dioxide, sulfur oxide, and particulate matter emissions that contribute to the global warming problems. However, if total energy processing is greater, even though more efficient, pollution of any given kind may be greater.

"Hard" energy paths with high entropy and increasing demand for energy would require ever-increasing consumption of fossil fuels and nuclear power. Conversely, "soft" energy paths that exhibit low entropy and restraint of energy demand would lead to conservation of fossil fuels and the development of long-term renewable energy resources. Also, hard paths would encourage growth of centralized electrical systems vulnerable to terrorist attacks, whereas soft paths would support creation of small, decentralized electrical systems.[21]

The choice between a hard or a soft path for energy policy will help determine future styles of governance. Commitment to a given energy policy could automatically preclude future recourse to other policy options.[22] In an era of limited capital resources, for example, the commitment to construct and bring on line nuclear power–generating facilities could preclude the development of other energy sources.

Diversity, resilience, and practicality are standards for developing energy policy. These standards are to be preferred to endorsement of crash programs based on hard-path technologies. Yet federally funded research and development generally backs hard-path projects. Joel Darmstadter et al. point out a danger in that practice: "The trend is likely to further entrench the primacy of electrical power research, and of nuclear within it, while solar, geothermal, fossil fuels, and conservation R and D are likely to lose ground, at least for a while."[23]

In *Energy Future*, Stobaugh and Yergin state, "Broadly speaking . . . the nation has only two major alternatives for the rest of the century—to import more oil or to accelerate development of conservation and solar energy. This is the nature of the choice to be made. Conservation and solar energy, in our view, are much to be preferred."[24] Any choice among these paths will inevitably be controversial because of complex potential environmental and social costs, disagreements over technical facts, and disagreements over basic values. However, the choice or program should be a politically acceptable one rather than a theoretical or economically optimum one. The Reagan administration chose the path of importing more oil and increasing an unfavorable balance of trade, leading to a serious weakening of the value of the dollar. In spite of considerable conservation efforts under Bush and Clinton, the United States still imports approximately 50 percent of its oil, just as it did in 1973.

More than one-half of the energy used in the United States is wasted, particularly through inappropriate and inefficient applications. In recommending conservation as the primary basis for energy policy, the *Energy Future* report of the Harvard Business School agreed with numerous other energy studies that United States energy policy was characterized by inefficient and wasteful use of energy. "Conservation may well be the cheapest, safest, most productive energy alternative readily available in large amounts. By comparison [with other energy sources] conservation is a quality energy source. . . . [I]f the United States were to make a serious commitment to conservation, it might well consume 30 to 40 percent less energy than it now does, and still enjoy the same or an even higher standard of living."[25]

All new national energy needs for the next few decades can be met through reduction of waste and more efficient use of energy. Because it is capital- and energy-intensive, the hard-energy path is inappropriate. By contrast, the soft path, which is far less capital- and energy-intensive, would be far more likely to permit the United States to implement solutions appropriate in size and scope to its energy problems. The soft path would make possible optimum energy conservation, and long-term human survival.[26]

Politically, conservation agendas lack appeal by virtue of their fragmented and undramatic nature. The possibility of designing sophisticated technological fixes like synfuels and magneto-hydrodynamics is perceived as more appealing. At the policymaking level, the unthinking compulsion to meet energy demands in a growing market economy, regardless of cost, appears to be dominant.[27]

With less than 6 percent of the world's population, the United States uses one-fourth of the energy consumed in the world each year. Per capita consumption of energy in the United States is among the highest in the world.[28] As one of the world's largest oil importers and producers, the United States constantly moves between wanting to fulfill the myth of energy self-sufficiency and wanting to have access to unlimited supplies of energy. While offering an immediate and reliable alternative to U.S. utilization of imported and domestic oil, gas, and coal, energy conservation also eliminates the need to develop nuclear power while providing the time and opportunity to move toward development of alternative, renewable energy resources. With this in mind the Clinton administration has made transportation efficiency and fuel flexibility cornerstones of the national energy policy to improve U.S. energy security as well as environmental quality.[29]

Because of the complex and provisional nature of international oil supply and delivery systems, the world could experience another oil-related energy crisis at any time. Unaddressed resource scarcities and resource-distribution problems in Third World countries inevitably will cause problems. The Persian Gulf War is a dramatic example of potential problems accompanying U.S. efforts to maximize energy availability by protecting energy lifelines to Third World countries. U.S. efforts could necessitate greater reliance on authoritarian measures enforced through employment of the military across the globe to guarantee access of U.S. multinational corporations to world markets and resources. A steady-state society in harmony with the environment is desirable: one that would use nonrenewable resources as little as possible while moving toward use of renewable sources of energy and toward adoption of less energy-intensive processes. Production modes should be adopted that are labor-intensive rather than capital-intensive.[30]

Energy and Environmental Relationships

Energy policies and environmental policies are inextricably interrelated. Energy production, uses, and impacts are all contained within the environment. They occur within ecosystems and interact with other elements within ecosystems. Hence, realistic, long-term energy policies need to incorporate environmental perspectives, just as environmental policies should incorporate energy perspectives.

The formulation of a unified overall energy policy is difficult enough without the added difficulty of relating such a policy to and integrating it with environmental policies. In the energy arena there are the complexities of competing energy development scenarios and energy technologies, as well as the numerous policies and positions emanating from both the public and private sectors. The macro- and micropolicies of each competing energy constituency reflect the immediate values and interests of that group. Numerous, often conflicting, and fragmented energy agendas have led to policy confusion rather than to the articulation of effectively coordinated energy and environmental policies. Almost all public or private sector policymakers claim to believe in protecting the environment.

Unfortunately, this claim is too often made as a public relations gesture rather than a commitment.

Fragmented and pluralistic energy policy approaches have resulted in similarly fragmented policies for environmental concerns related to energy production and consumption. The lack of a powerful and unifying energy policy, however, may assist environmental policymakers. They can point to the deficiencies in value assumptions predicated on unlimited energy growth. Commitment to energy growth generally is recognized as leading to deterioration of the environment and a decline in public health. Rapid expansion in use of fossil fuel and nuclear technologies under a unified developmental energy policy would have severe environmental repercussion, whereas a limited growth or steady-state energy policy would mitigate negative environmental impacts associated with energy use.

Environmentalists typically advocate energy conservation—reducing energy demand, reducing energy consumption, utilizing energy wastes by such means as cogeneration, recycling, and retrofitting existing structures for energy efficiency. Environmentalists also urge commitment to utilization of soft technologies—solar power, wind power, and other alternative renewable energy sources. Environmental interests strongly resist efforts to eliminate or relax existing environmental standards and requirements merely to facilitate new energy production. Politically, environmental advocates often serve as watchdogs, pressuring public officials to keep environmental considerations in mind when contemplating the viability of proposals coming from either the public or the private sector.

Environmental advocates are concerned lest an energy crisis become a powerful psychological weapon for weakening existing environmental standards. In such an emergency people might feel forced to choose between energy production and environmental protection.[31] Even without an energy crisis, constraints such as high unemployment, a declining economy, and mistrust of government all could serve to justify a policy directed toward energy growth and a reduced emphasis on environmental protection. These among other rationales evidently influenced the Reagan administration's movement toward decontrol of fossil fuel prices, deregulation of the energy industry, and a reduced governmental role in the areas of energy and the environment.

Various issues have emerged in political and governmental processes affecting energy and environmental policy. Concern often has centered on the form and degree of energy production rather than on the potential benefits of energy conservation in a free-market economy. Government, moreover, has played a major role in underwriting costs of energy-intensive technologies such as nuclear power. Substantially less governmental commitment has been made to underwrite costs for protecting the environment and public health. In light of such tendencies, proposed energy projects need to be scrutinized carefully.

Comprehensive policies affecting the evaluation of an energy production facility proposal would have to deal not only with production and production-related impacts but also with the uses to which the fuel produced would be put, and to the long-term primary and secondary impacts. The form and degree of production for these uses often are at issue. Factors such as the impacts of coal strip-mining, production facility size, type and effectiveness of pollution control, and impacts associated with transient or new populations could be assessed.

Critical questions need to be answered about energy projects. Is there a need or simply a demand for the petroleum or other energy that might be produced? Are there viable alternatives for producing whatever energy is needed? What are some of the secondary impacts associated with construction of proposed production and processing plants? What effects on the economy may be traced to diversion of capital resources to particular projects? How would an energy facility proposal relate to others if potential impacts of all these questions were studied cumulatively? Answers to questions like these require examination of potential long-term problems as well as basic values. Vested interests of private corporations and governmental agencies as well as the interests of the general public would have to be taken into consideration. Conflicts in values would arise but the debate might be useful, and certainly would be preferable to ignoring long-term values altogether.

Assessment of short-term impacts alone in evaluations favors powerful political and economic pressures that support currently existing hard-technology approaches. These include energy growth, centralization of power production and distribution in regional energy grids, and corporate determination of the public good. Soft alternatives—utilization of alternative renewable energy resources, decentralization of power production and distribution, public determination of the public good—are often treated summarily or ignored altogether.

Governmental bias favoring short-term considerations manifests itself through concentration on symptoms rather than causes. The cumulative effect of this failure to identify and deal with the causes of environmental problems can and often does permit irreversible environmental damage to take place. The public interest would be better served if government took both short- and long-term considerations into account. A similar shortsighted bias is represented in the private sector, for example, by public utility industry spokesmen who argue for greater generating capacity to support growth in the economy. These same individuals tend to ignore the potential negative economic and environmental consequences of bringing excess generating capacity on line. More serious than this neglect, however, is their inability to predict accurately how much power will be needed and in what form. The confident proclamation of the need to develop increased generating capacity obscures the fact that industry experts have been unable to read the future accurately. Former chairman of the Federal Energy Regulatory Board, Raymond J. O'Conner, remarks, "The proliferation of so many conflicting futures underscores massive uncertainty about demand growth, construction costs, alternative technological options, and regulatory policies."[32]

Perhaps government has become so identified with short-term private corporate concerns that the public interest in the areas of energy and environment may not be served. In fulfilling their energy management missions, government agencies such as the Department of Energy and the Department of the Interior may tend to overlook important public concerns. All major federal agencies are affected by energy and environmental considerations in some degree. Further, the views embodied by each agency will differ from the views of other agencies, and this underscores the fact that political conflict will of necessity arise among competing agencies.

Energy policy and energy-related decisions are also determined by what can be termed the "primacy of politics." These decisions, in turn, determine the direction of environmental policy. Politically motivated decisionmaking tends to inhibit the possibility of developing scientifically sound and economically reasonable solutions to energy and environmental problems unless political bodies are careful. David Davis believes that policy is "the net effect of the equilibrium of the forces acting in an arena at any one time. This temporary equilibrium may lack any rational basis. In recent years Congress has shown signs of breaking out of this form of decisionmaking by being more self-consciously analytical. It has deliberately tried to evaluate environmental costs against energy benefits."[33]

A more skeptical view, held, for example, by James Katz, is that the "congressional approach to energy is strongly influenced by local and special interest considerations, while attempts to understand issues or concerns about the national interest play a less important role. . . . In general, Congress's involvement in energy policy tends to be intermittent, short lived, and influenced by the newsworthiness of an energy policy issue."[34] This view is similar to the view that the president and Congress are more involved in issue resolution than problem solving.

Underlying debates concerning energy and environmental issues are value conflicts. At all levels of government, technical-scientific, economic, legal, and social problems are at issue. In each of these areas trade-offs between energy and environmental values and goals occur. Different levels of government may be involved in debate in varying degrees. One level of government may try to raise a problem while another level may attempt to solve it. Although public participation in environmental issues has increased as a result of new laws, decisionmaking on energy and environmental issues remains highly political and rife with value conflicts.

Many of the conflicts associated with growth of energy production are unrelated to the realities of the marketplace. The pressure of growth continues side by side with actual reduction in consumption of liquid fuels and electricity. While growth values remain associated with fossil fuels and nuclear energy, the economy of the marketplace has forced dramatic shifts in demand and consumption patterns. Irrational pressures for energy growth led to a generation of surplus power in the Pacific Northwest under the jurisdiction of the Bonneville Power Administration. As a consequence, consumers were faced with increasing electrical bills that incorporate costs for construction of new and unnecessary power plants. In the report *Indictment: The Case against the Reagan Environmental Record,* the Natural Resource Defense Council and other major environmental organizations pointed out: "The [Reagan] Administration's energy policy has been to eliminate virtually every program that provides direct benefits to individuals and small businesses seeking to conserve energy or use solar energy, while protecting billions of dollars in subsidies for nuclear power, synthetic fuels, and the oil industry."[35] Reagan filled the top administrative and staff positions in environmental agencies with individuals who were antiregulation and proindustry. Consequently, energy technologies associated with fossil fuels and nuclear power were given high priority, while environmental constraints were reduced or eliminated.[36]

Reagan's administrative policies also resulted in the authorization of unprecedented leasing of energy resources. The leasing was made apparently on the assumption that making unlimited supplies of such resources as coal and oil available to energy corporations under speculative market conditions, regardless of actual needs and environmental consequences, would be good for the U.S. economy. At one point in the leasing scenario, then assistant secretary of the interior Gary Carruthers overruled the relatively modest leasing recommendations of a Bureau of Land Management study team. Based on long-term environmental considerations and conclusions drawn from a series of public hearings, the team recommended leasing from 400 to 800 million tons of coal from the Fort Union area. Carruthers ignored these recommendations and suggested leasing from 1.2 to 1.8 billion tons in the northern great plains.[37] Those Montana and

Wyoming ranchers and farmers most directly affected by the proposed coal development complained that excessive coal leasing left too much of the coal resource in the hands of a few large energy development corporations that speculated in energy futures. Under slow coal market conditions with large leases sought by relatively few corporations, establishment of fair market values for the public through competitive bidding was unlikely. A similar political intervention into energy policy occurred when, under pressure from a Republican-dominated Congress, Clinton decided to open up the Strategic Petroleum Reserve for oil production.[38]

Almost all forms of energy development result in negative environmental consequences. Some forms of passive solar technology are a possible exception. The federal government tries to balance energy production and environmental protection. But Department of Energy and Department of the Interior policies favor energy development and consumption. Many of the problems that have arisen can be traced to tensions between energy-promoting and environment-protecting agencies and policies.[39]

Unwise energy development policies can affect the quality of the environment in many ways. For example, leasing programs for geothermal, oil, and gas development in the national forest areas surrounding Yellowstone National Park threaten the environmental quality of the greater Yellowstone ecosystem as well as the park itself. Various secretaries of the interior have scheduled leasing of most of the geothermal resources in the area. These leases risk damage to geysers and other geothermal resources in the park. In fact, the Old Faithful geyser is no longer faithful because of reduced geothermal pressure. The leases also exemplify the problem of inadequate consideration of potential negative environmental impacts upon the quality of the existing environment. Insufficient environmental protection requirements and unrealistically low minimum bid requirements led to leasing of several hundred thousand acres of public land in this area.[40]

More threatening are environmental and public health risks associated with operation and eventual shutdown of existing and proposed nuclear power plants. Operating plants emit primarily low-level radiation decay products that possibly increase the incidence of radiation-related cancers in those living near power plant sites. Operating plants also create problems of storage of radioactive wastes, chiefly in the form of spent fuel rods. Some radiation decay products can remain dangerous to human and other life-forms from 250,000 to 500,000 years. The Department of Energy has yet to develop credible solutions to the permanent storage of radioactive wastes, especially high-level wastes. Also there are radiation risks associated with uranium mining (such as radon gas emissions from mine waste storage piles) and with uranium enrichment. Transportation of radioactive materials is also fraught with dangers. Probably the most dramatic and possibly the gravest potential hazard is from some form of nuclear catastrophe such as a primary core meltdown—the so-called China Syndrome,[41] which gets its name from the notion that a core meltdown would make a hole all the way to China.

The possibility of such a catastrophe captured the consciousness of the American public during the Three Mile Island and Chernobyl nuclear accidents. Regardless of whether they result from primary core meltdowns or from other causes, accidents leading to the release of radioactive materials from operating nuclear plants are a likely eventuality. The steam explosion at the Russian Chernobyl plant, for example, produced consequences over a surprisingly broad area. High levels of food contamination occurred more than one thousand miles from Chernobyl.[42] As long- and short-term impacts are considered, the wide dispersion of radioactive materials are likely to lead to many serious health problems. Both the Chernobyl and Three Mile Island events illustrate the sheer unpredictability of nuclear accidents,[43] as do other unforeseen accidents. "Canada once claimed that its CANDU [Canadian Deuterium Uranium] heavy water reactors were virtually immune to sudden accident, but in the summer and fall of 1983 an unprecedented series of large leaks occurred at several reactors, including one leak where radioactive water was spilled into Lake Ontario."[44]

After the partial nuclear meltdown at Three Mile Island, which occurred on March 28, 1979, ten miles from Harrisburg, Pennsylvania, a Nuclear Regulatory Commission study warned that this accident was only one of 141 nuclear power plant mishaps and incidents that could have led to severe reactor-core damage or meltdowns.[45] Nevertheless, the $60 billion public and private investment in the nuclear alternative has kept policymakers from abandoning currently operating and some proposed nuclear power plants. But growing public opposition to nuclear technology has reduced expansion of the industry. High costs, along with public opposition, have resulted in termination of several nuclear projects during the planning or construction phase. Funding nuclear power projects has grown increasingly complex and difficult. Economic, political, and environmental problems associated with nuclear power have multiplied greatly.[46]

Institutions and policies aimed at control must recognize the ungovernable and unpredictable nature of nuclear power, particularly when in the hands of imperfect human beings. The human and technological perfection demanded in day-to-day operation of nuclear plants appears to be impossible to achieve. And, as nuclear technology evolves, additional difficulties surface. A case in point is the commitment to the breeder reactor. The use of liquid sodium as a heat transfer material has inherent dangers. But the liquid metal fastbreeder reactor also represents a particularly dangerous technology in that civilians might find access to materials eminently suitable for the creation of nuclear bombs. Recent history has shown that it is difficult to control international traffic not just in hard drugs but also in radioactive materials. As Joel Darmstadter declares, "The main argument against the breeder has been that it is fueled by plutonium, the weapons material par excellence, and that possible lack of competitiveness and safety make it even less sensible to incur the risk of spreading this material around."[47] Moreover, beyond the severe health, safety, and environmental risks associated with the use of nuclear energy, there is a growing realization that use of the technology does not appear to affect worldwide dependence on oil in any substantial way.

The Clinton administration has substantially reduced the budget for research and development of nuclear energy. The administration has placed greater emphasis on energy efficiency and alternative energy technologies. Meanwhile, the Department of Energy's inability to solve the problem of disposing of highly radioactive waste continues to plague the nuclear power industry.[48]

Coal is relatively abundant compared to other energy sources in the United States. About one-fourth of the world's known coal reserves are in the United States: 90 percent of all North American reserves.[49] But utilization of coal as an energy source poses severe health and environmental consequences. Strip-mining in such locations as the Northern Great Plains obliterates rangeland. Various attempts to reclaim land often fall short of expectations. In some instances underground water resources are destroyed where the coal-bearing strata are also the aquifers. Burning of coal in energy conversion processes produces substantial sulfur dioxide emissions.[50] Coal-fired power generating facilities have exacerbated the acid rain problem in western regions. At one time acid rain was believed to be a problem of great significance only east of the Mississippi, but now it is recognized to have major impacts in the American West as well.[51] Coal burning, along with the combustion of other fossil fuels and wholesale destruction of rain forests in Latin America and Southeast Asia, are presumed to be hastening the "greenhouse effect," a consequence of the substantial worldwide increase in atmospheric concentrations of carbon dioxide.[52]

The Surface Mining Control and Reclamation Act of 1977 (SMCRA) addresses some of the environmental problems of strip-mining for coal — destruction of productive rangeland, disruption of aquifers and surface water systems, acid drainage into previously useful water systems, and escalation of soil erosion and sedimentation. But the positive effects of this law upon strip-mining as actually practiced have been minimal because federal enforcement has been lax. Cuts in funding for enforcement in the OSM have further reduced the potential effectiveness of the act.

The law stipulates that, after the coal is taken, the mine sites are supposed to be restored to productivity through reclamation procedures bringing back ground conditions similar to those in existence prior to strip-mining. Even if the law was properly enforced, there is still some question as to whether or not, over the long term, implementation of reclamation procedures would result in bringing the land back to productivity. Because of semiarid conditions prevailing in much of the American West, serious doubts still exist about the extent to which reclamation has been and can be successful. Low annual rainfall, periodic droughts, short growing seasons, and poor, thin soils all conspire to make reclamation next to impossible. Native plant life, which has adapted over many centuries to these harsh prairie conditions, cannot simply be replaced by exotic plant species without risking the long-term survival of the soils themselves. Reclamation efforts thus far attempted have included irrigation, introduction of exotic and natural grass species, treatment with mulches and fertilizers, and various other techniques. There is no doubt that some of these often rather costly techniques have succeeded over the short term. However, some of the ranchers in the Northern Great Plains argue that they could grow grass in their pickup trucks with similar treatment. The reclamation standard of long-term survival of previously stripped acreage on a self-sustaining and self-perpetuating basis has not been met effectively. However, some states do better than others. Pennsylvania, for example, improved reclamation performance through a bond forfeiture program. Further improvement occurred when the federal government established the Abandoned Mine Reclamation Fund. OSM still has not addressed problems such as "approximate original contour" when dealing with mountain top removal coal mining.

International Energy Aspects

The majority of industrialized countries import well over one-half of their energy for domestic use, mainly in the form of petroleum from the Middle East. Virtually all industrialized countries lack back-up energy systems should they no longer be able to depend on petroleum. Thus, they are susceptible to any changes in fossil fuel flows from Middle Eastern and other oil-producing countries, many of which are less developed than the oil-consuming nations.[53] At the same time, other less developed countries are forced to compete with industrialized countries for high-priced petroleum products under uncertain world market conditions. Furthermore, lack of application of conservation measures in the less-developed countries has led to low efficiency in oil use.

Some of the key international problems appear to be (1) excessive reliance upon fossil fuels, especially oil; (2) inequities in the distribution of energy resources between industrialized and less developed countries; (3) uncertainties associated with energy delivery systems; and (4) limited supplies of energy resources.[54]

This section provides some conclusions about the future of various energy sources. The likely trend is for pressures to consume nonrenewable energy resources to increase, in light of the probability that per capita consumption of energy worldwide will increase, especially in developing countries. The EIA *International Energy Outlook 1998* forecasts that by 2020 the worldwide consumption of energy will be three times what it was in 1970. Most of the growth is expected to occur in developing countries. In fact, EIA estimates that by 2020 developing countries will consume about 6 percent more energy than the industrialized countries. The developing countries are expected to increase energy consumption by about 3.8 percent annually from 1995 to 2020, industrialized countries by 1.2 percent annually.[55]

Presently, oil is the most often used fuel worldwide. From 1973 to 1984 oil consumption in developed countries declined with economic declines. Oil consumption in developing countries rose almost 50 percent in the same time period and will continue to increase. At current consumption levels, world oil supplies could continue for another one hundred years and then fall off rapidly.[56] Additional oil exploration and development might extend this period.

Information on world natural gas supplies is not as fully developed as for other energy sources. In 1997, 25 percent of U.S. energy consumption was in natural gas. United States natural gas production is second in the world to Russia. Worldwide natural gas consumption is likely to be maintained at current levels for several decades. Unconventional methods of gas production (those associated with oil shale and coal methane) have been developed that will prolong the supply. It is expected to occur in North America, for example, and may assume a significant role in the United States within twenty to thirty years.

World coal deposits could last for thousands of years at current levels of consumption, if they all were economically recoverable. The "proved amount-in-place" sources should last about one hundred years at current consumption levels. The United States has 25 percent of the world's proved total of coal, the former Soviet Union 23 percent, and China 11 percent; the United States is the second largest coal exporter in the world behind Australia.[57] As the availability of oil and gas decline, however, much greater pressure will be placed on coal.[58]

The oil glut of the latter 1990s is only a temporary condition and one that is hastening the depletion of oil reserves. Oil-using nations have little control over oil prices that are set in the international marketplace. However, these countries can raise prices within their borders—which most developed countries have done through taxes. Although excess oil production encouraged OPEC (Oil Producing and Exporting Countries) to lower oil prices, the shift in price encouraged still greater consumption. Some major oil sources—Great Britain's North Sea and Alaska's Prudhoe Bay reserves—will soon decline. There is little doubt that future control of oil production, distribution, and pricing will be more firmly in the hands of OPEC. Reemergence of OPEC control appears inescapable in view of the fact that "Middle Eastern countries have two-thirds of the free world's proven oil reserves; [and] the United States holds 6 percent."[59]

However, Robert Fri reports that three events have significance for the crude oil market: the creation of an active futures market, which has reduced speculative pressures in the market; a more competitive energy market overall, due largely to use of new fuel-switching capabilities; and "netback pricing," a method of estimating price of oil and gas between wellhead and sale, which links crude prices closely to product markets.[60]

To offset fossil fuel shortages, new mixes of energy sources are needed. Conservation measures will also have to be taken in order to eliminate energy waste and improve efficiency of energy use. The United Nations Environment Programme recommends that consideration of energy and environmental relationships be incorporated into the policy planning of nations.[61] The varied and complex global environmental consequences of energy production and use must be taken into account in order to protect the earth's ecosystem and resource base. Among possible negative environmental consequences are

1. reduction of productive agricultural base through diversion of croplands to other uses, erosion, degradation of soils, and pollution
2. deterioration of water quality through pollution
3. reduction of water reserves through diversion of water resources to noncritical uses and waste of existing water resources
4. degradation of the world's atmosphere and alteration of the world's climatic patterns through air pollution
5. elevation of ambient levels of radiation in the atmosphere through failure to control release into the environment of radioactive materials associated with the nuclear fuel cycle

6. reduction of forests, degradation or elimination of watersheds, and elimination of plant and animal species through vast deforestation projects.[62]

In the less developed countries the demand for wood as a fuel has led to severe shortages of wood as well as to elimination of forest areas that could not be taken over for productive agricultural uses. Along with charcoal and agricultural residues, fuelwood as a percentage of total energy supplies in developing countries ranges from 10 percent in wealthier Latin American countries to over 90 percent in some poor African nations. Overall, fuelwood is the primary source of energy for roughly one-half of the world's population.[63] Traditional fuels such as firewood, charcoal, dried dung, and crop residues provide essential cooking and heating for half the world's population who spend much of their time seeking these materials. In many of the less developed countries consumption of wood for fuel is increasing, rapidly outpacing the growth of replacement supplies. To give just one example, approximately 65 percent of the energy for all activities in Honduras is derived from fuelwood. This need is a major reason for the rapid rate of deforestation in Honduras.[64]

Projections indicate that there will be much less fuelwood available in the future than there is today. A study by the UN Food and Agriculture Organization (FAO) reports that by the year 2000 more than 3 billion people will experience a fuelwood deficit (that is, consume more than they can produce).[65] More and more time will be required to gather fuelwood. Currently substantial areas surrounding many cities have already been deforested. Urban families spend 20 to 30 percent of their incomes on wood. Present scarcities and projected shortfalls will restrict use of wood to essentials. Increases in fuelwood prices as well as accelerating deforestation may be expected. Dung and crop residues formerly used as natural fertilizers and fodder will instead be used as household fuels.

Deforestation through extensive logging, fuelwood collection, and burning is already a recognized pattern in the less developed countries. Deforestation in tropical areas is becoming especially critical. In 1984 the FAO estimates that 43,600 square miles of tropical forests are being deforested each year.[66] Of this amount about 29,000 acres are previously undisturbed forests. In a review of the literature on tropical forest problems Robert Stowe cites several reasons for concern over these losses. "First, the global gene pool is reduced."[67] The loss of undisturbed forests has tremendous implications with regard to the potential loss of plant and animal species. Untold numbers of plant and animal species are disappearing in these forests before they have even been identified. In fact, the vast majority of the species extinctions in the next fifty to one hundred years are likely to be in the tropical forests. As these species are eliminated so too are the possibilities of deriving some potentially lifesaving drug or other important substance. The World Resources Institute reports that "more than 50 percent of modern medicines come from the natural world, many from tropical forests, including strychnine, ipecacuanha, reserpine, curare, and quinine."[68]

"The second set of consequences of tropical deforestation has to do with the nature of tropical soils."[69] Tropical soils tend to be extremely thin and highly susceptible to soil erosion. Removal of forests leads to siltation of streams and erratic stream flows that will affect agricultural production. Also these soils are poor in nutrients and are not suitable for most monoculture crops.

"The third major effect of tropical deforestation is upon the climate."[70] Locally, deforestation increases the reflectivity of the land and temperatures become more extreme; hotter in the day and colder at night. This is important because many species of plants can only exist in very narrow temperature ranges. A slight increase or decrease in temperature could eliminate such species. Further, with the loss of ground cover there will be a loss of water vapor and the lands will become more arid. Global climate patterns will also be affected by the loss of tropical forests. Carbon dioxide concentrations will increase and exacerbate the greenhouse effect. That is, global temperatures will continue to rise as a result of increased concentrations of gases like carbon dioxide, leading to melting of polar ice caps, raising of sea levels, and drastic changes in global climate patterns.

"Finally, tropical deforestation has economic impacts."[71] Worldwide demand for the wood products from tropical forests adds further pressures. Typical products are rubber, palm oil, kapok, balsa, and forms of lumber. Industrialized countries like Japan and the United States are chief importers of wood from the poorer countries in Southeast Asia and Latin America. A U.S. interagency task force on tropical forests cites that as of 1980 82 percent of the $682 million in U.S. hardwood imports came from tropical countries.[72] The commercial value of wood products encourages government officials in less developed countries to exploit forests beyond their self-sustaining capabilities. A 1980 FAO study projected that if such trends continued, especially in Latin America and Southeast Asia, 12 percent (150 million hectares) of the remaining closed tropical forests would be deforested by the year 2000. This estimate did not include open tropical forests where fuelwood was already being collected; another 76 million hectares of these lands were expected to be deforested as well.[73] The U.N. Food and Agricultural Organization's report *State of the World's Forests—1997* presents mixed news. There is still a high level of deforestation worldwide, but there are some signs that the rate of deforestation is slowing down slightly in developing countries. From 1990 to 1995, developing countries lost 13.7 million hectares of forest per year, compared with the 15.5 million hectares they lost per year from 1980 to 1990.[74]

Reforestation offers some hope for addressing the fuelwood problem. In order to avoid the consequences of some fuelwood shortages the current rate of tree planting should be increased fivefold just to provide the wood needed in the short term. Analysis by the World Bank and others indicates that wood production, mainly the planting of fast-maturing trees, is the most cost-effective way to meet a large share of rural household energy needs. Although the need for massive reforestation is generally agreed upon, the ecological effects of introducing exotic or nonnative trees into various environments needs to be carefully researched. Early studies have been optimistic. Analysts at the World Bank do not believe that such a program would impair food production or harm the environment.[75] And, using the example of Haitian farmers growing trees to produce charcoal for urban markets, Gus Speth concluded that conservation incentives that emphasize the commercial potential of tree farming would help persuade developing nations to reverse the loss of the world's tropical forests.[76]

Sound global energy policy requires that industrialized and less developed countries face the necessity of changing rapidly to more sustainable patterns of energy production and use. Energy conservation and utilization of renewable energy resources are certainly attractive alternatives to present patterns. Effecting transitions of this nature would help reduce pressure upon less developed countries to adopt inappropriate and exotic energy technologies such as nuclear power to help solve their problems. Some industrialized countries are now in positions that will enable them to demonstrate some of the benefits of their practices and assist in the development of renewable energy resources in the less developed countries.[77]

Too little emphasis worldwide has been given to energy conservation and the development of renewable energy resources. Focusing on short-term market factors, the United States had led the way to excessive dependence upon fossil fuels, chiefly oil. Yet long-term considerations and problems, especially projected shortages of fossil fuels, certainly encourage shifts in global energy policy toward solutions more appropriate to satisfying global needs on a finite planet. There appears to be little international commitment to enacting transitions in global energy policy. Overall industrialized nations are better able to achieve energy policy transitions than less developed countries.

According to the World Bank, less developed countries spend the equivalent of about one-half of the development assistance they receive from all sources to purchase oil. To reduce their increasingly heavy dependence upon oil, the less developed countries need wide-ranging energy programs. Funding should encourage development of a broad spectrum of energy alternatives. In order to meet energy needs in less developed countries efforts should be made to utilize coal and natural gas as well as renewable energy resources. From an environmental standpoint, however, energy conservation and renewable energy resources should also be given higher priority.

Lynton Caldwell predicts, "As the end of the twentieth century approaches, the world faces profound uncertainties regarding its energy future. Changing relationships in the international order are inevitable as the sources of energy change in the decades ahead." Supplies of petroleum are finite, insufficient to last longer than several decades. Coal presents too many economic and environmental problems, including toxic atmospheric contamination after extensive and continuing use. Supplies of uranium for nuclear power are localized and finite. Caldwell believes that the present dependence of some nations upon energy fuels may be of relatively short duration. Utilization of renewable energy resources such as solar power may soon become economically feasible. Caldwell concludes:

> Changing patterns of energy supply and generation during the next half century seem certain to produce changes with implications for the environment and for relations among nations. But it is difficult to forecast with any accuracy just what these changes will be. Their nature will depend much upon the extent to which technology effectively develops substitutes for present energy supplies. In particular, develop commercially feasible systems for utilizing solar energy.[78]

Caldwell's view suggests only a provisional hope for the future. Efforts right now to enact positive transitions in global energy, and environmental strategies could help to make survival into the twenty-first century more attractive. Energy scarcity certainly makes a difference in how air pollution is dealt with, how water pollution is treated, how solid waste is addressed, with the management of parks, with urbanization, and with land use in general. Thus, energy is important for environmental administration generally and for most specialized environmental and natural resource administrators.

Forests, rangelands, wildlife, and water policy are the traditional natural resource areas given consideration in governmental and agency deliberations. It should be recognized that all federal policy in natural resource areas is complex, dynamic, and interrelated. Hence, explicit and comprehensive analysis of natural resource policy is difficult. Yet the multitude of decisions made at various levels with respect to a given resource, in combination with decisions made in other policy areas and programs, add up to a given federal policy for that resource. For selected natural resource areas in the United States some things have become better and others have become worse.

The situation for renewable resource management in the United States has changed considerably during the 1990s. The Clinton administration instructed all land management agencies to utilize ecosystem management principles in the implementation of their legislative authorities. Leadership of the USDA Forest Service shifted, for the first time, from a professional forester to a wildlife biologist. His successor was trained as a fisheries biologist. Forest policy implementation took on a whole new dimension. Instead of focusing on the production of timber as a commodity, forest managers began to scrutinize management decisions from a broader perspective, incorporating ecological and social dimensions as well as economic dimensions.

With regard to specific forest policy issues, timber harvesting on federal lands has declined from high levels in the 1980s to modest levels in the 1990s; from about 13 billion board feet in 1987 to about 4 billion in 1996. A consequence of the reduction in timber harvest on public lands has been a greater reliance on nonindustrial private timberland owners. Despite the reduction in timber harvest, however, the federal government continues to lose millions of dollars annually on timber sales. Furthermore, certain softwood species may be in declining health. Air pollution and global climate change may be affecting the growth rate and health of some tree resources. Severe deforestation is occurring in most less developed countries, which is altering hydrological cycles and facilitating desertification.

The federal government loses money in virtually all sales of natural resources because Congress either refuses to raise fees or blocks increases in fees by the federal agencies. Bruce Babbitt, secretary of the interior in the Clinton administration, in particular attempted to increase grazing fees on public lands. He wanted to raise grazing fees to about one-half of those paid in the private sector, roughly $5.00 per AUM (animal unit per month). The Republican-dominated Congress, however, rejected his proposal and even lowered grazing fees on public lands from $1.82 to $1.35 per AUM. Meanwhile the USDA Forest Service is studying how it can restructure timber sales to more effectively recoup more effectively federal expenditures required to conduct the sales. Congress also has rejected all efforts to change the fee structure of the General Mining Law of 1872. Meanwhile, private individuals and firms continue to stake claims on hardrock minerals on public lands and then sell the minerals without having to pay royalties.

The overall status of rangelands in the United States is relatively poor. But the quality of BLM rangelands has improved modestly over the past twenty years. Over one-half of BLM rangelands were classified as being in fair or poor condition in 1994. As poor as this condition may appear, it represents a considerable improvement over the situation in 1975 when 83 percent of BLM rangelands were in fair or poor condition. In general, too many animals are permitted to overgraze BLM rangelands and not enough attention is given to restoration of damaged lands.

The status of wildlife in the United States has improved since the 1980s for some species and declined for others. Several highly visible endangered species have been delisted or reduced from endangered to threatened status such as the bald eagle, peregrine falcon, and brown pelican. Two species that were declared extinct in the wild, the California condor and the red wolf, were propagated in captivity and subsequently released into the wild. The total number of listed endangered species, however, has at the same time grown to over 1,100 organisms. However, the Endangered Species Act is under extreme scrutiny by a Congress seemingly intent on weakening the act's provisions. One of the most hotly debated issues centers around the designation of critical habitat, which affects land uses. The Clinton administration has developed an innovative policy to give landowners greater flexibility in the use of their land if they agree to set specific areas aside for endangered species. Meanwhile, the populations of many other species continue to decline for a variety of reasons, such as habitat destruction, pesticide contamination, and effects that may be associated with global warming.

A bright spot in the wildlife policy area is the North American Waterfowl Management Plan. Implementation actions pursuant to the plan have increased the production of waterfowl in Canada, Mexico, and the United States. The plan involves numerous public-private cooperative ventures among conservation organizations and private corporations in conjunction with federal, state, and provincial governments. The ventures specifically promote the protection of wetlands that provide breeding and wintering habitat for waterfowl in the three countries.

Water supplies are increasingly threatened by nonpoint source pollution from agricultural and urban runoff. Hog waste has been implicated in the creation of a new threat called *pfiesteria*. This microorganism, which is found in polluted waters, kills fish and appears to affect the memory and vision of humans. Groundwater supplies in the Midwest and arid West are seriously depleted, while the presence of toxic pollutants in surface and groundwater supplies is increasing. Pesticides and agricultural wastes, in particular, are contaminating drinking water supplies.

Renewable Resources Policy

As the opening chapters of this book imply, there are both natural resource managers and environmental managers. Natural resource managers generally have land management responsibilities. Environmental managers are more likely to be involved with the control of pollution. Such a differentiation may be confusing in that both groups of managers deal with the environment and both deal with natural resources. However, natural resource managers tend to be more involved with the formulation and implementation of policies for the management of renewable resources such as forests, rangelands, water, and wildlife, and nonrenewable resources like soil and minerals. Chapters 5, 6, and 7 deal primarily with the roles and responsibilities of natural resource managers in the administration of policies concerning these resources.

Natural resources are those resources that occur in nature without human intervention. They are features or components of the natural environment that are of value to humans. Some natural resources, such as timber, have readily determined economic value. Other resources, such as scenic beauty, have what is referred to as nonmarket value. Renewable resources consist of living organisms and other resources such as soil and water that are closely associated with and affected by living organisms. When they are conserved properly, renewable natural resources are capable of sustaining or perpetuating themselves. Nonrenewable resources are not capable of perpetuating themselves. Some nonrenewable resources, such as fossil fuels, are, so far as we know, still being created by natural forces but at a rate so slow as to be irrelevant to policy or management; likewise soils and organic minerals such as shale. The latter consist of nonliving organic and inorganic materials such as coal, oil, and other minerals that exist in finite quantities and cannot be replaced.

Forest Policy

Forest policy often is considered in combination with range policy. The correlation between forest and range is reflected in U.S. federal laws and policies. The complex interaction between forest policy and other natural resource policy areas such as water, wildlife, and recreation is reflected in the multiple use provisions of various laws and ultimately in the Multiple Use and Sustained Yield Act of 1960 and the Federal Land Policy and Management Act of 1976. Thus, one resource policy area is intertwined and involved with other resource policy areas in a number of ways that extend from politics and economics to ecology and environment.

Forestland is land at least 10 percent occupied by forest trees of any size, including land formerly having such tree cover and capable of natural or artificial reforestation. The term "timberland" refers to forestland capable of producing more than 20 cubic feet per acre per year of industrial wood products.[1] Rangeland is land on which the potential natural vegetation is predominately grasses, grass-like plants, forbs, or shrubs and is less than 10 percent stocked with trees. Rangeland includes savannas, shrublands, most deserts, tundra, and coastal marshes.[2] Extensive forest and range acreage is under federal control, especially in the West, where the federal government owns more than 50 percent of several states. Approximately 66 percent of all forestland and 47 percent of all rangeland is nonfederal. Only 8 percent of the eastern forest is federal, but 69 percent of the western forestland and 61 percent of the western rangelands are in federal ownership.[3] Therefore, a great deal of forest and range policy is subject to federal determination. The principal federal land-managing agencies are the USDA Forest Service, the Bureau of Land Management (BLM), the National Park Service, the U.S. Fish and Wildlife Service, and the Army Corps of Engineers.

Early in the history of the United States the policy was to produce revenue by transferring forest- and rangelands to private ownership.[4] Forestland was of more value if it could be cleared and converted to agricultural production than used as a standing forest. Initially, an era of free timber existed when railroads and other private enterprises in the 1800s were allowed to cut timber on public lands for personal use. By the 1920s, agricultural clearing, logging, and other industrial and domestic impacts had reduced America's original forest cover by more than one-half. A sizable portion (250 million acres) was left incapable of supporting second growth.[5]

This policy of timber disposal changed to a degree around the beginning of the twentieth century, with the reservation of vast areas of public lands as forest and watershed. Later, additional forestlands, primarily in the eastern states, were acquired and added to the public domain. This retention and acquisition policy was due to the belated realization that America's forest and range resources were not limitless. The USDA Forest Service, established in 1905, was charged with management of the forest reserves by the principle of "sustained yield" as envisioned by its first chief, Gifford Pinchot.[6]

As a consequence of the forest reserve policy, the federal government now owns about 20 percent (97 million acres) of all timberland in the United States.[7] The federal government controls a greater share of the commercial forest resource than any other resource on public lands. National forests are the largest federal ownership; they account for 17 percent of United States timberland and 12 percent of timber harvested in 1991.[8]

A pervasive component of public forest and range policy in the United States is extensive public ownership of lands. Logging and grazing are public-interest concerns of government that require public action. This general policy is now well accepted, and questions appear to center on the types and scales of actions that should be permitted. Before the growth of the environmental movement, many private forestry and grazing concerns had a virtual free rein on public lands. These individuals reaped significant benefits at public cost, often harvesting timber and grazing animals at rates substantially below market value. The mission of the USDA Forest Service appeared to have been coopted by the interests of wood- and cattle-producing industries. As environmental awareness grew, more citizens became informed about the negative environmental consequences of poor timber harvest and grazing practices on public lands.

Today, burgeoning societal demands for the diversity of uses and resources provided by public lands have resulted in what Greg Brown and Charles Harris term the "public policy paradox of natural resource management."[9] This paradox refers to the perception held by many individuals and environmental organizations that the federal agencies created to preserve the resource base have devolved into resource commodity brokers.

Under various laws and general statute guidance, many federal lands are to be managed under "multiple use" and "sustained yield" concepts—primarily BLM and USDA Forest Service lands. Both of these concepts, developed before the contemporary concepts of noncommodity and existence values of forests, embrace the principle of *use*. Multiple-use management requires that the government attempt to incorporate or maintain various diverse uses and values, such as protection of wildlife habitat, recreational opportunities, and production of forest products on public lands. No dominant use should have priority over another. Competing and conflicting uses and values often create severe problems in application of the principle of multiple use. On the one hand, timber companies may be permitted to log extensive forest areas to the exclusion or detriment of other alternative uses or values. On the other hand, alternative recreational uses or wilderness preservation may place constraints and impose severe economic costs upon timber harvesting.[10]

Ideally, multiple-use management practices should involve coordinated and comprehensive planning for the most judicious and harmonious use of the land and its natural resources. This is done under the concept of combining two or more uses or purposes with attention to sustainability and nonimpairment of the natural resources and land area.

In practice, multiple-use conflicts often exist over judgments and decisions pertaining to allocations among competing uses and values. In the allocation of resources among diverse uses, such as logging, grazing, mineral development, watersheds, recreation, fisheries, and wildlife, an overemphasis of one often will have a negative impact on the others. In some cases, multiple use decisions of this nature may be made on the basis of political and economic forces, rather than on what might be best for the land and for the environmental public interest. For example, overlogging or overgrazing of an area may occur under a "multiple use" concept that is employed merely as a convenient slogan covering a "dominant use" decision at the expense of other uses and values. Still, value judgments and decisions for the wise allocation of resources viewed in relation to the concept of multiple use are crucial.

Applying more to forest than to range policy, the principle of "sustained yield" prohibits the government from permitting more timber to be harvested than is being replaced by the natural growth of the remaining trees. Under this principle, the USDA Forest Service is prohibited by law from depleting national forests, in contrast to industrial landowners who often have significantly depleted their timberland.

Ideally, the sustained utilization of a forest means to maintain and perpetuate its living resources through wise management. Sustained use should ensure that the various uses and values of the forestland and its living resources will be available to both present and future generations. Under this interpretation, the "sustainability" principle is intended to safeguard ecological processes and biological diversity essential for the maintenance or sustainability of the living resources involved.

In practice, the sustainability principle, as applied to the national forest system, has been interpreted to mean sustained yield of wood products. According to the Multiple Use and Sustained Yield Act, "Sustained yield of the several products and services means the achievement and maintenance in perpetuity of a high-level annual or regular periodic output of the various renewable resources of the national forests without impairment of the productivity of the land." The act requires that national forests be administered for a variety of purposes or values, "not necessarily the combination of uses that will give the greatest dollar return or the greatest unit output."[11] Throughout the Reagan and Bush administrations, however, national forest supervisors were still instructed to "get the cut out."[12]

The BLM is also charged with the long-term multiple-use management of its forest- and rangelands. With the Federal Land Policy and Management Act of 1976 (FLPMA), the Forest and Rangeland Renewable Resources Planning Act of 1974 (RPA) and other related legislation, the BLM and USDA Forest Service have similar land management and planning responsibilities for forest and range areas that they administer. This legislation established the principle and policy that the BLM and the USDA Forest Service should manage their forests and rangelands on the basis of detailed inventories of resources, careful planning, and public participation, with the overall goal of achieving a balance among competing values and uses. The FLPMA particularly requires the two agencies to adopt a systematic planning process, providing for public participation, interdisciplinary approaches, and coordination with other federal, state, and local governments.[13]

Sustained yield and multiple use principles are also contained in the National Forest Management Act of 1976 (NFMA) that establishes guidelines for multiple-use/sustained yield management established under earlier legislation.[14] This act, as well as the RPA, requires the USDA Forest Service to develop land and resource management plans that integrate recreation, range, timber, watershed, fish and wildlife, and wilderness uses into the more than 190-million-acre national forest system.[15] The RPA assessment is required every ten years. Assessment includes detailed information and projections of future supplies and demands for each resource. The NFMA requires the development of an integrated land and resource management plan for each national forest administrative unit. The NFMA also prescribes that the secretary of agriculture should generally limit the "allowable cut" of timber harvested on each national forest to a "quantity equal to or less than a quantity that can be removed from such forest annually in perpetuity on a sustained yield basis." These and other provisions should ensure that national forests are managed to produce a broad range of publicly beneficial goals through multiple use, and a continuous supply of high quality timber for future generations through sustained yield. However, John Gordon, Pinchot Professor of Forestry and Environmental Studies at Yale University, argues that the shift in public views and values from "sustained yield" to something like "sustained forest" has exposed the inadequacy of the classical model, thus emphasizing the need for a new forest management paradigm.[16]

The RPA and the NFMA also involve recognition of the need for long-term planning that is comprehensive and integrated for the various uses and purposes of an entire national forest. Single plans for one purpose, that is, a plan for timber or a plan for wildlife, are not authorized. The requirement of comprehensive and integrated planning recognizes the ecological realities of forestlands that can be managed properly over the long term, but only if the complex interrelationships among various natural resources and uses in a forest community are taken into account.

Short-term and single-purpose planning are too segmented to fulfill legal requirements. Similarly, planning that focuses on local issues rather than national needs fails to serve long-term needs. Current provisions call for new land and resource management plans that are responsive to local concerns while providing for national requirements and concerns.[17] Public participation and coordination with state and local governments are also required during the preparation of all plans. Katherine Barton and Whit Fosburgh's lament still holds true: "While the Resources Planning Act and the National Forest Management Act set up a theoretically neat system of planning from the national to the local level, the process—at least so far—has not worked out as envisioned. One problem is the apparent irrelevance of the Resources Planning Act programs issued so far to the decisionmaking process."[18] A major constraining factor is that appropriations typically are not provided in accordance with RPA program guidelines.

The RPA and NFMA also established land and resource management regulations for limitations on clear-cutting, protection of riparian (and watershed) areas, determinations of lands suitable for timber production, and the rate of timber harvesting more than twenty years ago. However, the debate over timber harvest and forest management practices on public lands has continued.

The effectiveness of legislation and regulations greatly depends on the degree and kind of interpretation and discretion (with underlying value judgments) given throughout the governmental process. For example, in

1998, Congress requested the chief of the USDA Forest Service to provide information to a committee that was considering changing the agency's mission to that of forest custodian.[19] Critics of the USDA Forest Service continue to argue that the agency is too oriented toward the timber industry to examine multiple use critically and should be restructured.[20] The counterargument is made that wood, the most renewable and energy efficient of all modern building materials, is best obtained from public lands managed according to best management practices.[21]

Such debate was somewhat diminished after the appointment of Jack Ward Thomas as chief of the Forest Service by the Clinton administration. Thomas, a wildlife biologist by training, implemented an ecosystem approach to forest management that gave greater consideration to newly recognized noncommodity values, such as biological diversity, on public lands. His successor, Mike Dombeck, reaffirmed the agency's commitment to sustain the health, diversity, and productivity of public lands through "collaborative stewardship."[22]

Present-day public distrust in the ability of the USDA Forest Service to administer its landholdings according to multiple use stems from the years of the Reagan administration, when multiple use was interpreted as dominant use. Approximately 84.6 million of the 96.6 million federal acres of potentially commercial forestlands are in national forests.[23] But low pricing of timber in the 1980s resulted in a failure to meet even the cost of managing timber in the national forests. Consequently, the federal government lost money while causing serious economic damage to private tree owners who cannot compete economically with the subsidized timber industry.[24] Although the USDA Forest Service is supposed to sell timber at fair market value, the sales in the 1980s were generally well below costs. In 1982, for example, the USDA Forest Service announced plans to spend $102,000 to build a road to harvest timber worth $54,000 in an area in the Gallatin National Forest—a dubious decision, despite the fact that the road would be available for future sales.[25] Often the public is so uninvolved in the sales process that the vast majority are not even aware of whether they are receiving fair market value for public resources. The Wilderness Society estimated that taxpayers lost $21 billion between 1975 and 1985 from below-cost timber sales that required the USDA Forest Service to spend more to produce and sell timber than it received in return from logging companies. Some environmentalists charged that the USDA Forest Service, under the guise of facilitating logging operations, has managed to push construction of thousands of miles of new road into previously roadless forest areas in order to forestall any future consideration of these areas as wilderness.[26]

Under pressures from budget-minded congressmen and environmentalists, the U.S. Department of Agriculture ordered the USDA Forest Service in August 1985 to begin revising its money-losing sales of federally owned timber to private lumber companies. The agency was ordered to rewrite its management plans for three large western forest regions where the USDA Forest Service had underwritten deficit timber sales with losses exceeding $100 million in some years. The USDA ruled that there was inadequate economic justification, especially considering government costs for such things as road building, paper work and processing, and personnel time for selling timber in some areas, particularly in the Rocky Mountains and Alaska.[27]

Despite pressure to reduce deficit timber sales and a large backlog of sales to the timber industry, the USDA Forest Service proceeded with plans initiated in the 1980s to increase timber sales. By 1982 the volume of uncut timber under contract to private industry reached 36.1 billion board feet. Some companies even sought congressional relief in an attempt to be released from their contracts to buy USDA Forest Service timber because of the low lumber prices.

The proposed escalation of timber harvesting was still insufficient to the Reagan administration. Relatively quietly, pressures were applied to forest supervisors to work toward increasing levels of timber harvest. Thomas Arrandale reported that "top-level administration officials, with backing from timber industry leaders, were pushing forest supervisors and planners out in the field to study the possibilities for accelerating harvests even more rapidly. Crowell, agriculture's assistant secretary who oversaw the USDA Forest Service, told Congress in 1982 testimony that national forests could produce between 20 billion and 30 billion board feet a year if their timber inventories were harvested at the same rate as industry."[28]

The USDA Forest Service of the 1990s, however, differs drastically from the agency it had been in 1981, when Jimmy Carter left office.[29] Pressure from various administrations, Congress, and the timber industry led to an internal revolution in agency culture, in which a new generation of forest managers aligned with agency wildlife biologists to resist timber targets and reduce timber sales. Thoreau Institute forest economist, Randal O'Toole argues:

> Contrary to popular belief, the reduction in timber sales and other changes happened not in spite of the Forest Service, but because of it. In 1993, agency leaders were far less enthusiastic about timber and far more concerned about practicing true ecosystem management than they were in the late 1970s. Environmental lawsuits, lobbying, and other pressures certainly contributed to this change. But so many other factors were involved that environmentalists really were little

more than a Greek chorus: Urging events onward but not playing the starring role.[30]

Donald Floyd reports that, although timber harvest within the national forest system peaked in 1987 at 12.7 billion board feet, by 1996 it had dropped to about 3.7 billion board feet.[31] Timber sales on the national forests, however, remain extremely cost-ineffective. The USDA Forest Service reports that national forest timber sale losses climbed from $15 million in 1996 to $88 million in 1997, claiming that this increase was due to a change in how it accounts for roads.[32] In 1998, USDA Forest Service Chief Mike Dombeck announced organizational restructuring efforts to address the agency's serious financial and accounting deficiencies.[33] A large number of other statistical indicators, such as expenditures on noncommodity recreation and fish and wildlife management programs, do seem to indicate that the USDA Forest Service has changed.[34] Only time will tell the quality and durability of the change.

Reduced timber harvest on public forestlands has focused renewed interest on private timberland ownerships. As of 1992 forest industry timberland holdings amounted to 14 percent (70 million acres) of United States timberland and one-third of the U.S. timber harvest. Nonindustrial private ownerships, including millions of small parcels, accounted for 58.7 percent of U.S. timberland and 49 percent of the U.S. growing stock timber harvest.[35] About 94 percent of the nonindustrial private ownerships are individuals who, with fewer than 100 acres each, collectively hold 59 percent of private forestland acreage. However, less than 10 percent of nonindustrial private forest (NIPF) landowners have prepared a written forest management plan.[36] Federal and state forest stewardship programs will become increasingly important to reduce depletion of NIPF lands and to engage the landowners in cross-boundary cooperative efforts for sustainable forest ecosystems.[37]

Additional debate persists over specific forest management practices, clear-cutting in particular. According to Thomas Barlow, clear-cutting—the logging and removal of all trees in a given block of forest—may sometimes increase the volume of wood but also "allows public forests to be converted into even-age stands, which benefit timber companies, but not ecosystems."[38] The Society of American Foresters, however, views clear-cutting as a scientifically based management tool, which, when judiciously applied, can create desired conditions of forest disturbance and regeneration.[39]

Clear-cutting is a part of an even-aged silvicultural management scheme that produces trees of uniform size and quality, typically associated with monocultures or the controlled growth of a single tree species. From the perspective of a professional forester, clear-cutting is merely one of several forest management practices necessary for regenerating tree species that are shade intolerant (e.g, Douglas fir, aspen, and most pines). From the perspective of the public, clear-cutting is distressingly unsightly and has become virtually synonymous with environmental degradation.[40] Malcolm Hunter Jr. explains the basis for controversy as follows:

> Despite its limitations, clearcutting is an important and legitimate part of forestry's repertoire of techniques, and there are even circumstances in which it can significantly diversify a forest from a wildlife perspective because of its marked effect on spatial heterogeneity. However, it is likely to always remain somewhat controversial because of its visual impact and its alleged negative impacts on wildlife. Of course, its effects on many species of wildlife can be distinctly negative, at least in the short term, and if these species are public favorites, then controversy is inevitable.[41]

Contemporary conflicts about forest sustainability issues reflect the gaps between the diversity of forest values and the capacity of forestry practices to resolve differences for mutual benefit.[42] The Sierra Club defines "excellent forestry" as

> (a) limiting the cutting of timber to that which can be removed annually in perpetuity, (b) growing timber on long rotations and allowing trees to reach full maturity before being cut, (c) practicing a selection system of cutting wherever this is consistent with the biological requirements of the species involved or keeping the openings no larger than necessary to meet those requirements, and (d) taking extreme precaution to protect the soil, an all-important resource. The advantages of such forestry are overwhelming: (1) low fire hazard, (2) low windthrow hazard, (3) minimum risk from insects and diseases, (4) yield is of maximum value, (5) wood fiber is highest quality, (6) it costs no more, (7) seed source is reliable, (8) soil is protected, (9) landslides minimized, (10) soil fertility maintained, (11) preserves watershed and fish habitat, and (12) preserves natural beauty.[43]

The Society of American Foresters, a national organization for forestry professionals founded in 1900 by Gifford Pinchot, puts forward an alternative definition: "Forestry is the science and art of attaining desired forest conditions and benefits. As professionals, foresters develop, use, and communicate their knowledge for one purpose: to sustain and enhance forest resources for diverse benefits in perpetuity. To fulfill this purpose, foresters need to understand the many demands that forests must satisfy and the potential for forest ecosystems to satisfy these demands now and in the future."[44]

These two venerable environmental stewardship organizations, and the many others like them, are at a juncture in history when it is time for consensus building. Forestry, the source of the concept of sustainability, should be perceived as a process for conflict resolution rather than as a partisan in the conflict of values.[45] Sustainable forestry must supplant the concept of sustainable use, if natural resource managers and stakeholders are to reconcile competing values and respond to new challenges and opportunities.

Forest Health

Pressure from development has intensified at the very time global forest reserves have been put in jeopardy due to environmental health factors. Evidence has grown that air pollution and global climate change may pose threats to forest resources. The National Wildlife Federation (NWF), in its *1985 Environmental Quality Index,* reported "that a systematic and sustained growth decline of some species has occurred in the last 20 to 30 years,"[46] and the rate has not slowed since.

Experts are not yet certain if the basic underlying stress factor is acid rain, ozone, multiple pollutants, or declining moisture levels due to drought and global warming. Regardless of the precise cause, there is a growing consensus that some tree species of eastern higher altitude forests, as well as those elsewhere, are declining substantially. A Forest Health Task Force of highly respected scientists reports, for example, that "ozone . . . has been proven to cause injury and mortality of eastern white pine and injury to white fir, ponderosa pine, and other trees in the mountains of southern California. Acid deposition, pollution-induced nitrogen fertilization, and aluminum toxicity, in addition to the effects of natural stresses are possible causal mechanisms for declines of red spruce in the Appalachian Mountains."[47] The task force further points out that the acidification problem in northeast aquatic systems may be reaching a steady-state situation, while the problem is worsening in the lakes and streams of the southern and central Appalachian mountains.

Mount Mitchell, North Carolina, is one of the more intensively studied locations with regard to the effects of air pollution on forests. During the winter months, the trees at the upper levels are constantly bathed in an acid mist. Photographs of the area over time show tremendous loss of trees (primarily spruce). Although a direct causal relationship between air pollution and the death of trees has not been established, a more foreboding suspicion is developing. Air pollution may be weakening the defense systems of the trees, making them more vulnerable to bacteria and insects. One writer who challenges the air pollution theory points out that the Fraser firs that are dying at the top of Mount Mitchell appear to be dying from aphid infestation and not air pollution.[48] Although the primary cause of death may be the aphids, abiotic stress factors can trigger the initial decline. Recent droughts may be another contributing factor. The problem may be a cyclical phenomenon of twenty or more years, meaning that the problem started several decades back. If immediate resolution were possible (and it is not), currently affected trees would continue to die for another twenty years. Dr. Robert Bruck of North Carolina State University reports that about 7 percent of the red spruce at or near the top of Mount Mitchell are dead and at lower altitudes roughly one-half display symptoms of decline. He does not believe that acid rain per se is the problem, but he does believe that it is a part of a larger problem.[49]

Significant forest declines are also occurring in Canada, Mexico, North America, and the European continent. The term "forest decline" has been defined as referring "to a wide range of symptoms including loss of foliage, discolored or abnormal foliage, reduced growth, branch dieback, and eventual tree death."[50] A number of tree species have been afflicted with this forest decline syndrome, including Ponderosa and Jeffrey pines in California; red spruce and sugar maples in eastern United States and Canada; silver fir, Norway spruce, Scotch pine, and beech in Europe; and the sacred fir in Mexico.[51] The Black Forest in Germany, perhaps the best known forest in the world, has been experiencing *Waldsterben* (forest death) for several years.

Arthur H. Johnson and T. G. Siccama report: "It is probably significant that forest dieback and decline are occurring simultaneously on both continents" and that parallels exist between the North American and European diseases.[52] At first, German scientists blamed air pollution in the form of acid precipitation for the decline of the German forests, but recent studies have thrown considerable doubt on this hypothesis. The Commission for Air Pollution Control of the Association of German Engineers concluded "that acidic deposition, either directly on vegetation or indirectly via effects on soils, is not the primary cause of forest decline in Germany. The group believes, however, that direct effects of gaseous pollutants may be important."[53] William Pierson and T. Y. Chang's detailed analysis of the literature on acid rain supports the conclusion that the "forest decline in West Germany and the eastern United States is stress-related," but the cause is unidentified. They say that acid rain is not the only stress and the degree to which it is a "contributing stress is highly uncertain."[54]

No irrefutable scientific relationship can be established between acid rain and forest decline. But a great number of well-founded suspicions exist. As the American Forestry Associations's "White Paper on The Forest Effects of Air Pollution" points out:

> The quest among some policy makers for complete understanding and a magically irrefutable scientific explanation of the problem — undertaken as a substitute for additional pollution controls — is short-sighted and foolhardy. By definition, the scientific quest for knowledge can never be completed. And yet many decisionmakers and interest groups involved in the air pollution debate are demanding absolute, cut-and-dried explanations of the complex interactions between the forest ecosystem and the bewildering array of natural and pollution-related stresses which affect it.[55]

Long-term forest health and productivity is receiving increased attention in the forestry and natural resource management arenas. The Forest Ecosystems and Atmospheric Research Act of 1988, for example, mandated long-term monitoring of the health of forest ecosystems.[56] However, some regional ambiguity exists in the attempts to define an unhealthy forest scenario. Alaskan yellow cedar has been experiencing a high rate of mortality attributed to effects of global warming, while many southwestern ponderosa pine forests are in decline due to wildfire suppression and pine bark beetle infestations. The focus on forest health protection and enhancement is expanding beyond individual trees and stands to encompass concerns about long-term site productivity and ecosystem functions. Alan Lucier suggests that, with further development of regional definitions, forest health could in fact become a useful criterion of sustainable forestry.[57]

Forest decline combined with the rapid deforestation of tropical forests is likely to produce serious environmental and social consequences in the next few decades. Consequently, the formulation and implementation of policies directed toward conservation of forest resources should be considered a critical element in any future strategy. Perhaps the global climatic impacts associated with the greenhouse effect are the most serious long-term consequence. In the short term there will be energy shortages, increased flooding and soil erosion, food shortages, and a much less appealing place in which to live.

Dennis and Donella Meadows have observed that "at one time one-third of the earth's land surface was forested. Now only one-fifth is. By the year 2000 the world's forest area will have been reduced by half, with half that reduction taking place after 1950." They note that the land area under forest is fairly constant in temperate zones. Yet they also note that the annual loss of tropical forestlands is equal to an area larger than Portugal or Austria. This would imply a halving of all tropical forestlands in 120 years. Rates of deforestation vary greatly from country to country with some countries having very fast rates and others slower rates.[58] Sandra Postel states, "Greater emphasis on social forestry (including agroforestry) reforestation, and forest preservation engender optimism. Yet the deforestation dilemma still cries for more commitment, clarity, and vision."[59]

Range Policy

The USDA indicates that a little more than one-third (803 million acres) of all the land in the United States is classified as rangeland. Fifty-five percent of this land is privately owned. The bulk of the private grassland acreage is in the Rocky Mountains and Great Plains states. Thirty-six percent of remaining grasslands are managed by the federal government, and much of these lands are in the semiarid lands of the Southwest as well as the interior of Alaska.[60] Most of these acres are devoted to the grazing of domestic cattle or sheep at least part of each year. Administration of the federal acreage is shared by the BLM, which oversees 268 million acres, of which 164 million acres are managed for grazing, and the USDA Forest Service and other federal agencies that oversee the remainder.[61] The BLM administers 164 million acres as grazing land under the Taylor Grazing Act and the Public Rangeland Improvement Act.[62]

Public rangelands are in relatively poor condition due to chronic mistreatment. Estimates from the BLM and the Forest Service indicate that less than one-fifth of the public range produces as much as 60 percent of its capacity. The BLM estimates that about 18 percent of the public range is in poor condition, 42 percent is in fair condition, 31 percent is in good condition, and 5 percent is in excellent condition.[63] The 1992 National Resources Inventory indicates that 37 percent of the federal grasslands are in fair condition and 13 percent in poor. A fair condition means that existing grasslands possess only 26 to 50 percent of their potential plant community, and poor 0 to 25 percent. This constitutes a slight improvement from 1986, when 41 percent were in fair condition and 18 percent in poor. Private rangelands were in worse condition in 1992: 44 percent in fair condition and 15 percent in poor.[64]

Historically, little effort was made to manage and protect rangeland until the 1930s, when the Taylor Grazing Act was passed. This act terminated open range policies that had previously allowed land abuse without imposing grazing fees upon transient and other livestock owners. Up to that time, the quality of the land had suffered greatly from chronic overgrazing, lack of soil conservation, and poor range management. The Public Rangelands Improvement Act of 1978 (PRIA) provided additional relief. Under the PRIA the USDA Forest Service and BLM conduct an experimental stewardship program (ESP) to improve the public rangelands through increased cooperation among all rangeland users. Incentives and rewards are given to livestock operators whose grazing management efforts improve resource conditions.[65] Partially as a result of the Taylor Grazing Act a slight upward improvement in range quality occurred. Improvements were brought about through application of state-of-the-art intensive range management, such as proper allocation, distribution, and monitoring of livestock.[66]

Phillip Foss's detailed analysis of the implementation of the Taylor Grazing Act, *Politics and Grass: The Administration of Grazing on the Public Domain,* disagrees with this assessment and holds that little improvement in rangeland quality has occurred as a result of the act, due largely to the domination of the Department of the Interior's policymaking process by the stockmen's associations. Although overgrazing may have been reduced in some areas, the purpose of the grazing act—to *stop* overgrazing and soil deterioration—has not been accomplished.[67]

As mentioned earlier, the FLPMA and NFMA require the BLM and USDA Forest Service to manage federal public lands for multiple use. Rangelands do satisfy that criterion, providing forage for livestock, produce wood products, water, recreation, minerals, habitat for wildlife (including endangered species), wilderness values, and other multiple uses. Although grazing has traditionally dominated all other uses of rangelands, energy development is also facilitated through leasing of public rangelands. Energy development is now becoming an exclusive as well as dominant use for many rangeland areas. Energy developments such as coal strip-mining, however, exclude all other uses of an area through imposition of negative and often irreversible impacts. A large portion of the budget of the BLM is prescribed for energy-related activities, with operations divided into two areas: energy and mineral resources and land and renewable resources.

Grazing—overgrazing, in fact—is by far the dominant use for most rangelands, and the intensity of grazing activity is a major cause of poor range quality. Other factors that modify the vegetation and soils of rangelands may include off-road-vehicle (ORV) use, timber harvesting, use of herbicides or insecticides, insect and disease damage, wildfires, concentrated uses, and weather changes. These, in combination with overgrazing and the land's inherent vulnerability, can cause desertification (the transformation of productive range into desert) and the gradual reduction of the carrying capacity of rangelands for plants and animals.

Grazing allocations are made on the basis of animal unit months (AUMs), based on a charge for grazing one cow and calf, one horse, or five sheep or goats for a month on BLM or USDA Forest Service rangelands. Fifty percent of the money collected from grazing fees is used to finance range investments. The federal grazing fees can be considered a major subsidization of cattlemen and woolgrowers. In 1978 the PRIA set out a fee formula based on the livestock operator's ability to pay, as opposed to a fair market value. In 1980 the fees rose to a "high" of $2.36 per AUM. After 1980 prices dropped steadily, and by 1985 they were $1.35 per AUM. Fees temporarily increased, to $1.98 in 1994, and, then, Congress lowered AUM fees back to $1.35 in 1996.[68] The appraised market value of grazing on public lands in 1986 range from $4.05 to $8.85 per AUM.[69] Meanwhile, private grazing fees range from $8.00 to $18.00 per AUM. Low grazing fees have permitted monopolies on public lands. Grazing permits, which are obtained largely through inheritance or purchase of private ranch property, allow 5 percent of the permit holders to control 52 percent of the grazing, and 11 percent of the permit holders to control 74 percent of the grazing. Although final decisions on grazing allocations and amounts rest with the government, the livestock industry and individuals continue to exert much political influence in this area, particularly through members of Congress.

Efforts to reduce AUMs to permit recovery from overgrazing on public rangelands usually meet subtle but solid resistance. The so-called Sagebrush Rebellion reflected efforts by the livestock and energy industries to transfer all or some of the public lands to state control for sale or lease to private interests to avoid federal grazing reductions and other forms of regulation. Some public lands have ended up being managed solely for livestock interests, with little attention given to other legitimate and compatible uses such as recreation and wildlife. BLM's past applications of herbicides and current efforts to eradicate sagebrush through burning exemplify management bias in favor of livestock production. Neither of these techniques has proven effective in increasing available livestock forage or improving range conditions. Furthermore, sagebrush is recognized as providing needed and valuable food and cover for wildlife.

The Reagan administration initiated policy changes to give ranchers who lease BLM rangeland more control over public land and water resources. For example, ranchers were encouraged to file under state law for private title to the water on public lands. As a consequence of increased private control of water rights and the water of the surrounding rangelands, grazing interests were able to constrain the federal government's ability to manage the lands for other uses.[70] This program was terminated when the courts ruled that cooperative agreements with ranchers unlawfully abrogated the responsibility of the Department of the Interior to manage federal rangelands.[71]

Recommendations for grazing and other resource management policies under the BLM are made through advisory boards composed of private citizens. These boards have varying levels of responsibility. Membership in many advisory boards is drawn from a pool of grazing permit holders and individuals with local economic interests. Many permittees regard themselves as owners having private property rights to the land they lease. Although they are supposed to have strictly advisory functions, many advisory boards exercise substantial control over decisions pertaining to grazing and other uses. Furthermore, advisory board decisions tend to reflect the private interests of board members and permit holders.

In recognition of the importance of diversity, the Public Land Law Review Commission suggests that "membership on advisory boards should be chosen to represent a range of interests and that representation should change as interests in, and uses of, the land change. We believe the appropriate range of representation includes not just the obvious direct interests . . . but the professor, the laborer, the townsman, the environmentalist, and poet as well."[72] Selection of membership from a much broader spectrum of interests would help the public receive fair market value for the uses that take place on BLM lands. Broadening the membership base for advisory groups might also result in greater emphasis on intangible and environmental quality values in the decisionmaking process.

Presently, federal land management agencies such as the BLM appear to be ill equipped to withstand political pressures from strong private interest groups. Mechanisms for the identification and incorporation of the environmental public interest are few and insufficient. Various groups and individuals have tried to highlight problems that result from this inadequacy. In one instance the Natural Resources Defense Council won a suit in which they addressed the problems of excessive influence upon management decisions exercised by grazing interests.[73] As a result of the suit, the BLM now prepares site-specific environmental impact statements before making grazing-related decisions. Other policies and legislation, including the PRIA, provide additional measures designed to control and restrict grazing allocations and impacts. For example, the BLM is to complete range inventories to determine carrying capacities of rangelands.[74] But the Reagan administration, with its western base of political support, initiated actions to change such environmental policies shortly after coming to office. In July 1981 the Department of the Interior circulated a new draft rangeland management proposal to abandon use of the vegetation inventory under the pretext that it was not cost effective.[75]

The broad approach of the Reagan administration to shift as much control as possible over rangelands from government to private economic interests was a sharp contrast with what is needed. Under the Carter administration, the BLM indicated in a publication entitled "Managing the Public Rangelands," that "improving publicly owned grazing lands presents considerable difficulties. Both the narrow economic interests and the tradition of free access to the public lands for private gain stand in the way."[76]

On a worldwide basis, rangelands and deserts are encroaching on formerly more productive land. Overgrazing by excessively large populations of livestock representing both economic values and noneconomic values (social values in Africa and religious values in India, to cite two examples) accounts for this trend. Abused and damaged rangelands have been rendered incapable of supporting both livestock and wildlife. Consequently, land worldwide is subject to increasing erosion and disruption of watersheds.[77]

The end result of such abuse is desertification, which has become a worldwide problem. Approximately 62 percent of the rangelands worldwide have become desertified (that is, the biological productivity has been reduced or destroyed, leading to desert-like conditions).[78] "According to the United Nations Desertification Map of the World, more than two of every five acres (40 percent) in nondesert Africa are at risk of desertification; Asia's at-risk proportion is 32 percent and South America's is 19 percent."[79] The UN Environment Programme reports that "desertification is now perceived as an intensifying worldwide threat. . . . Each year an additional 21 million hectares of agricultural land deteriorates through desertification to a point at which it is no longer economically productive."[80]

Transformation of productive lands into desert has consequences far beyond the impacts on the land itself. Victims of this process often become those who can least afford to suffer the consequences, like those subsisting in the drylands of developing countries like the Sahel in West Africa. The process of desertification may be irreversible. That is, many African countries appear to be in a downward spiral, leading to an inevitable process of desertification. Experiencing food shortages, the Africans overcultivate and overgraze their lands. They borrow money from international banks to make up the shortfall. And when the banks demand their money along with accumulated interest payments, the Africans are forced to plant cash crops to generate enough money to pay the interest. The soils are not particularly well suited to the cash crops, and further drying and consequent erosion occur leading to further erosion. Hence, the national balance-of-payments problems lead to further exploitation of lands negatively affecting agricultural productivity and soil fertility. Ill-suited Western technologies also continue the downward spiral, potentially leading to further poverty and hunger problems. Considered from this worldwide perspective, then, rangeland policymaking indeed has serious consequences.

In the United States, multiple-use management is generally compatible with the grazing of domestic livestock. If appropriately controlled, grazing does not necessarily lead to permanent damage of basic soil and water resources. However, management plans need to be developed with a realization that livestock grazing can restrict other competing uses by altering the natural vegetation and reducing the available forage. In this context, the "carrying capacity" of rangeland should refer to the number of livestock a given area can support without engendering range deterioration. Specific management options can include determination of the actual number of livestock allowed under a given grazing permit ("level of use") as well as of how livestock are to be distributed and when livestock are permitted to graze ("management intensity").[81] Problems having to do with range policy often focus on value judgments relating to level of use and management intensity.

Water Policy

In addition to being a necessity for the accomplishment of various human activities, an adequate supply of water is an essential requirement for the survival of all life-forms. In part, because of the essential role of water in the preservation of life, water policy is complex, diverse, and often controversial. To some extent, the controversy surrounding water issues is made even more intense because of the existence of many competing jurisdictions. In the United States, besides the numerous state and local agencies and governmental entities dealing with water policy and management, eight federal departments and thirty-five federal agencies are involved with water, along with numerous other political interests.

Harvey Doerksen observes that water management involves conflict management, coordination between agencies, and planning under uncertainty. "Three basic tenets emerge from analysis of current water management systems. First, rational comprehensive decisionmaking by almost any definition is impossible. Second, policies are more the result of factors external to public agencies than of decisions made within agencies. Third, conflict in the decision arena is axiomatic."[82]

Dean Mann's classic study, *The Politics of Water in Arizona,* and Helen Ingram's *Patterns of Politics in Water Resource Development: A Case Study of New Mexico's Role in the Colorado River Basin Bill* provide similar findings.

Their case studies illustrate that controversial political maneuvering associated with water problems virtually prohibits long-range planning and decisionmaking in the formulation of water policies in western states, especially in Arizona and New Mexico. This is truly important, since declining water resources in arid western states will require formulation of explicit allocation policies based on long-range forecasts of water supplies in the near future. In her book, Ingram describes how influential representatives and senators managed to push authorization of the Central Arizona Water Project through Congress, despite totally unrealistic assumptions of available water supplies.[83]

Much of the complexity of water policy is associated with the natural characteristics and uses of water:

1. *Pervasive and evasive character:* As a flowing substance, water is everywhere—in, under, and over land—which makes water the subject of various governmental and private claims.
2. *Degree of immobility:* Water is where it is, and water problems and uses are typically very localized with governmental efforts at management occurring at the local or regional level.
3. *Variety and uncertainty:* Water varies tremendously in quality and quantity over time, location, and use. Too much or too little water often exists.
4. *Multifaceted nature:* The same unit of water, depending on its quality and treatment, can be devoted to a variety of uses, for example, household, recreational, wildlife, or industrial purposes.
5. *Systemic uniqueness:* A drainage basin is the only reasonable unit for dealing with water in public management even though no governmental jurisdiction conforms to drainage basins and use in one drainage area may affect use in another.
6. *Susceptibility to treatment:* Although naturally susceptible to pollution, water can be treated and, depending on the pollutants, can be restored or rendered less harmful to varying degrees. For example, flowing waters often reduce or abate the effects of pollution.
7. *Potentially negative effects:* Through erosion and flooding, water can have negative environmental impacts.
8. *A setting for conflict:* Various user demands create conflicts and competition over limited water resources. Intense conflicts, for example, develop between off-stream users, interests in economic development, and in-stream users interested in fish, wildlife, and recreation.

Water is a public resource as far as the government is concerned. The costs associated with water management are met through subsidy and taxation. Traditionally, water has been regarded as a *free* good in commercial terms. The price of water still plays only a small part in the market system. This pattern is on the verge of changing. The days of water as a "free good" are rapidly ending. The governor of South Dakota tried to sell water from a public reservoir to a private slurry company at a price per gallon at the source. Unappropriated groundwater in New Mexico has been valued at $38 billion, which is twice the value of estimated oil reserves. Nebraska found that irrigation water was more valuable than a bushel of corn. As the shortage of water mounts around the world, not only in the United States, the price of water will escalate considerably.[84]

A system of water demand management may be possible in which alterations in demand patterns may be accomplished through responsive pricing. Through the use of responsive pricing and effective management of supply, a water manager would have "the means to assure a more efficient utilization of the resources employed to produce and distribute water and a more equitable distribution of the costs incurred in providing water services."[85]

Little relationship currently exists between the supply and demand for water in the United States. Competitive pricing has been limited. Other factors determine water resource allocation. Much of the allocation of water supply depends on developments that use water—traditionally, agriculture and industry. With private economic interests concerned only with isolated segments of the water process, however, government is required to carry the cost burden of water supply as a public function. Moreover, government invests substantial amounts of public capital in water projects that serve the public welfare only indirectly through local economic development. Allocation of expenditures for water projects becomes highly political, particularly among members of Congress who may wish to bring short-term employment and other short-lived economic benefits to their constituents. The pattern of lavish water projects for western states has slowed considerably in recent years as greater competition for public resources has occurred. Water projects of the future will be smaller than in the past and oriented more to local conditions and objectives. There will be greater emphasis on cost-sharing and public-private cooperation as well as between levels of government.[86]

Typically, water needs are correlated with population, industrial, and urban projections. In some cases this produces a self-fulfilling prophecy. Water expansion acts as a causal factor for overconcentration of populations and urban and industrial developments in areas where excessive demands for water resources already exist. This, in turn, creates greater demands for water in an apparently endless cycle. But water could serve as a crucial control factor in distributing population and industrial growth to avoid overuse and excessive concentration.

Worldwide the greatest use of water is for irrigation purposes. Approximately two-thirds of U.S. fresh water is used for agriculture. Industrial use of water has declined by 36 percent in the U.S. since 1950 because of increases in price per unit.[87] However, agriculture receives federal and state government subsidies such as public works projects and low water rates.

A major proportion of the allocation of water resources and governmental expenditures in the western United States is for agricultural irrigation. About 83 percent of consumptive use of water in the West is accounted for by agriculture.[88] This pattern continues in spite of the marginal agricultural gains and low economic efficiency of such use. Much of this misdirection is due to ambiguous and unsound water laws. A major problem in water allocation relates to prior appropriation rights in the West that act as a stimulus for large governmental projects designed to assist distant users to tap into remote water sources.[89] Jack Hirschleifer, James DeHaven, and Jerome Milliman point out that "water rights are not clearly defined, do not have legal certainty, and cannot be transferred with ease as are rights to other types of property, land, mineral rights, etc. As a consequence, the market processes . . . are either severely limited or prevented entirely from operating in the case of water."[90]

The U.S. Congress has asked the World Bank to reduce negative environmental impacts of development projects in developing countries. Consequently, the World Bank has attempted to develop means to derive economic values for social and environmental variables. One UN dam researcher asked to advise the World Bank on dam projects in Third World countries recommended "an energy-based analysis in which environmental costs of a project are translated into real costs and cranked into the project analysis. If the social and environmental costs of a project are greater than its economic benefit, the final payoff may be nonexistent." Unfortunately, there is little local involvement in planning of dam projects, but there would be a large payoff in educating local people about the costs and benefits and soliciting their participation in the planning process.[91]

Among the social and environmental effects observed in several dam projects initiated in the Third World have been the displacement of 250,000 people for a reservoir site in India's Narmada River, and a diversion of the Yangtze River in China that would extend the range of schistosomiasis, a parasitic disease carried by aquatic snails.[92] Water projects in less developed countries throughout the world are also causing environmental and social problems. The pace of dambuilding has been so rapid that most of the major rivers of the world have now been dammed. Unfortunately, the needs of the majority have been sacrificed for the short-term economic gains of a relatively small number of people. In the process, essential aquatic and terrestrial ecosystems have been destroyed, the viability of sustainable agricultural practices has been damaged considerably, and the lives of untold numbers of people have been irrevocably altered.

In considering water resources development and river basins of this nature, a UNEP annual report declares:

> As a consequence of socioeconomic development, and especially water resources development in the river basins of the world, freshwater ecosystems are becoming more and more complex in their structural, spatial and temporal dimensions. Perceptions of the functions of freshwater bodies (such as rivers, lakes, aquifers) also change during development. Fresh water is not only a renewable natural resource for which no substitute exists, but forms a very important part of ecosystems and landscapes. Contrary to the prevailing practice, which focuses almost exclusively on the function of fresh water as a natural resource, all the related functions should be considered simultaneously. This means that environmental impacts and their management cannot be viewed only in the context of specific water and water-related projects, but should also be considered on a basin-wide scale. This challenge is especially important with regard to international water systems.[93]

A major component of water policy, allocations, and priorities for government projects concerns the cost-benefit ratio. Under interagency agreements a formula is required for justifying a project on a dollar basis with a higher return of benefits than costs. It is assumed that social benefits correlate with economic benefits for an area of the project, and on this socioeconomic basis primary and secondary costs and benefits are assigned. But cost-benefit ratios are sometimes subject to manipulation and used to justify projects for other reasons. Furthermore, costs and benefits associated with environmental quality, such as aesthetics and wildlife appreciation, cannot be quantified in monetary terms, and the true value is of a nonmarket, subjective nature. Cost-benefit analysis also provides a means to avoid having to deal with issues concerning the basic morality of decisions and decisionmaking altogether.

In striving for congressional appropriations, various water agencies compete with one another through the political process. A means of deciding between projects is necessary. One means used is cost-benefit analysis. With the stakes being so high, it is only natural for agencies to manipulate their cost-benefit ratios in order to present one higher than those submitted by other agencies. One agency may challenge another's ratios, especially when both have made studies of the same area. Furthermore, with the tremendous economic rewards associated with a large construction project, rational members of Congress attempt to get projects for their districts. Hence, water project politics are characterized by "logrolling" and "pork barrel" activities, as members of Congress jockey for necessary votes and power to get water projects for their constituents.

One of the more infamous pork barrel projects is the Garrison Diversion Project in North Dakota. The project was authorized by Congress in 1944, 1965, and 1986. (The 1986 authorization substantially changed and deauthorized parts of the 1965 project.) The project is designed to alleviate the semiarid conditions of North Dakota through diversion of water from the Missouri River basin to irrigate over 1 million acres. Despite dubious benefits and substantial negative impacts on the natural environment, North Dakota representatives succeeded through extensive coalitions in Congress in getting the project funded. About 600 million dollars were appropriated for construction of the project that would pollute water going into Canada (from agriculture runoff), in violation of international treaties, and destroy thousands of acres of critical breeding grounds for a large proportion of North American waterfowl.[94] Although the diversion may never be completed because of environmental concerns in Canada, more funds have been requested to complete a scaled-down water delivery system for North Dakota.[95]

The lack of reality in cost estimation is also reflected in the 7.5 percent discount rates (at the time far below the actual costs of borrowing money) suggested for the Garrison Project. By any reasonable accounting standards, unfavorable cost-benefit ratios of less than one-to-one should prove projects uneconomical and provide sufficient justification for cancellation of proposals. However, certain pork barrel favorites, including the Garrison Project, have failed to meet even this minimum standard. Approval of projects under such circumstances amounts to endorsement of substantial subsidies to receiving districts.[96] The Conservation Foundation notes that some formerly disadvantaged farmers have become privileged, and that water projects in general have not provided the help to the poor they were originally planned to serve. "Finally, federal agencies are being forced to recognize that single-mission water policies are often detrimental to environmental quality."[97] Fortunately, in these more restrictive budgetary times, fewer irrigation projects are being approved and funded. It would also appear that water projects are being subjected to closer scrutiny. The NWF has released analyses of federal audits on several major water projects.

The analyses for 1985 illustrated that the projects were costing taxpayers millions of dollars each year. As a result, Congress was prodded into directing the Department of the Interior to audit all such water projects to determine whether or not they were paying their own way.[98] In one USDI audit, conducted in 1994, the inspector general determined that the Animas–La Plata Water Project in southwest Colorado would be financially infeasible. The annual operating expenses would produce a negative cost-benefit ratio even if the project could be built for nothing. A Bureau of Reclamation analysis projected thirty-eight cents return for each dollar spent.[99]

The cost-benefit method for assessing water project feasibility has major, perhaps insurmountable limitations. It is difficult to determine the appropriate yardsticks for measuring costs and benefits. Such elements as scenery, open space, wildlife, recreational opportunities cannot be adequately assigned quantitative or dollar values. And merely requiring that tangible and qualitative value be considered on an interagency basis is no guarantee that these factors will be *adequately* or even properly evaluated. Additionally, the long-term positive and negative impacts of water projects—flood control, and degree and rate of siltation, in the case of dams, for example—cannot be determined with any certainty. Also, it is often hard to tell which are costs and which are benefits. Is the creation of a reservoir to provide irrigation water for agriculture a benefit if this method of water management also results in accelerated water losses through evaporation? Ultimately, whether a particular factor is perceived as a cost or benefit will depend on who is doing the measuring. One immediate consequence of all these limitations is that qualitative factors, having received little or no recognition in the assessment process, exercise practically no influence upon decisionmaking. Especially in the case of water projects, cost-benefit analysis often appears to be employed merely to justify decisions already made. Quantitative measurements and analyses can provide a quasi-scientific disguise for political manipulation.

Under cooperative interagency arrangements, agencies such as the U.S. Fish and Wildlife Service are often contracted to do surveys and other assorted forms of research for major water resource and development agencies, such as the Bureau of Reclamation and the Army Corps of Engineers. Depending on the source of funding, the agencies that undertake the research may develop orientations similar to the water developers. They may begin to develop vested interests in seeing that water projects are approved. John Bruner and Martin Farris suggest that "conventional wisdom promulgates a sense of urgency that can cause such decisions to be made. . . . The value of water used to preserve wildlife sanctuaries and fish spawning grounds, the aesthetic appeal of verdant agriculture, the maintenance of open stretches of rivers are difficult to measure."[100] Hence, to the extent that societal needs can be satisfied by increasing water supplies, they will be included in policy proposals or discounted if they do not.

With the National Environmental Policy Act of 1969 (NEPA) new environmental criteria and requirements have been combined with the cost-benefit ratios for determination of the feasibility of proposed projects. The act requires that methods and procedures be developed that will ensure that unquantified environmental amenities and values will be given appropriate consideration in decisionmaking, along with economic and technical considerations. The act also stipulates that the responsible official and lead agency preparing the detailed environmental impact statement (EIS) consult with and obtain comments from any other federal agency that has jurisdiction over aspects of or special expertise on the environmental impacts of a proposed project. Additionally, consultations are to take place with relevant state and local agencies in accordance with directives issued by the Council on Environmental Quality (CEQ) for public participation and hearings.[101]

Still, EISs prepared by water agencies reveal a strong tendency to emphasize quantitative measures based on assumptions and weights concerning numbers of people or plant species and dollar amounts. A decision based on quantitative data of this nature is often no more accurate than the underlying values and assumptions behind it. Thus, analysis is often inadequate in terms of protecting environmental values and the enforcement of environmental legislation. Often, when mitigation funds are required, their allocation has been ineffective. Further, through the public participation process, water agencies can build support from selected economic interests to back a project.

Agency multiple-use directives are also applicable to water projects. Major problems occur, however, when the agencies have traditional missions for a dominant use that exclude, reduce, or impact other uses. The study of water project proposals depends heavily on agency clientele and political support related to economic uses or interests. Consequently, a water agency works closely with pressure and business groups, state and local officials, and others who have economic interests associated with its dominant and traditional use, purpose, or mission.

The U.S. Bureau of Reclamation, for example, maintains close association with the National Reclamation Association and other irrigation groups in conjunction with its primary purpose of facilitating irrigation. Some of its top agency personnel often serve as officials of national and state chapters of the National Reclamation Association and other water or irrigation organizations on a "two-hat" basis. These and other activities can provide political support and justification for a water project to the detriment of other uses, regardless of actual costs and environmental concerns.

The use of water for irrigation is often a particularly inefficient use of water resources. Sixty percent or more of the water supplied is lost to evaporation and is not subject to reuse.[102] Further complications occur with the pollution of irrigated water caused by the large amounts of herbicides and insecticides that are used in agriculture on irrigated lands.

In building political support and influence, the Army Corps of Engineers has the reputation for being the most powerful and most independent agency in the national government because it answers only to Congress through the appropriations process. Corps projects are intended predominantly for flood control and navigation.[103] However, like the Bureau of Reclamation, the Corps gradually came to regard generation of hydroelectric power as a major use and justification. In regard to its flood control purpose, it commonly assumed that structural protection through dams and reservoirs should be extended against natural flooding that normally and periodically occurs on river floodplains. Corps projects and proposals generally relate to protecting human life and developments that have encroached on river floodplains.

Its flood-control mission, coupled with large federal expenditures for projects, permits the Corps to gain political support from Congress and local areas, particularly from interests deriving direct, short-term economic benefits. This is how Corps projects have been promoted in the past. However, both the Bureau of Reclamation and Army Corps of Engineers have experienced increasing difficulties in gaining approval and funding for proposals and projects. More stringent criteria are being applied. Further, regulations including zoning restrictions against developments on floodplains have grown stricter in recent years, thus challenging the Corps in its interpretation of its responsibility for administering floodplains.

In reference to problems associated with water development projects, the CEQ reports:

> The environmental problems of water projects are much larger in scope than the occasional "endangered species" [snail darter] controversies. For example, flood control, navigation, and drainage projects have eliminated several million acres of productive agriculture and forest land and valuable wetlands and marshes. . . . A most important environmental consideration is protection of flow in existing streams [instream flows] when water projects are built. Failure to do so jeopardizes recreation, water quality, aesthetics, and fish and wildlife habitat.[104]

In recognition of some of these problems, several western states have passed legislation legally recognizing that nonconsumptive, instream uses of water are beneficial. With increased pressure for water use, increasing conflicts will emerge between traditional, consumptive uses and non-consumptive uses.[105] Frequent conflicts occur over how much water must be stored for hydropower operations and how much water will provide the minimum flow needed below the impoundments for protection of recreation, fish, wildlife, and habitat.[106]

Channelization, a process where a river or stream is straightened to allow a swifter, more efficient flow of water, has been practiced by the Soil Conservation Service (which was designated the Natural Resource Conservation Service in 1993) and other federal water agencies as a policy for more than twenty-five years. This practice, however, results in removal of natural vegetation along streams and rivers, thus negatively affecting fish, wildlife, and aesthetics. It also results in soil erosion and an increased tendency of flooding downstream. Much channelization is done through federal funding and projects to avoid natural flooding of private wetlands and floodplains. But channelization serves to turn rivers and streams into drainage ditches and to increase the runoff from the land.

Construction of water development projects is correlated with agency missions and orientations that tend to restrict comprehensive and coordinated planning. Consequently, realistic consideration of environmental, economic, and social aspects on an integrated basis seldom occurred in federal land management agencies before the ecosystem management initiatives of the Clinton administration. Keys to the inclusion of such values in comprehensive planning would be rigorous enforcement of NEPA regulations, "revision of cost sharing formulae" for financing water projects, flexibility in agency leadership, recruitment of talented people to key positions, and an "adjustment of agency orientation."[107] Daniel Mazmanian and Jeanne Nienaber note that the Army Corps of Engineers has moved in this direction, having achieved a new level of accommodation to groups concerned with environmental considerations. Although most agencies periodically come up with some new accommodations to secure their survival, the Corps has changed a great deal in this regard. The Corps is noteworthy for its ability to reconcile "seemingly irreconcilable demands for water resource development, environmental protection, and open planning. After making a decision to change, the agency moved expeditiously and rather successfully to accommodate itself to a changing social and political environment. . . . [O]ne cannot help but note also that the Army Corps of Engineers has once again proved to be a most politically astute organization."[108]

To a degree, accommodation to changing times was formalized in Executive Order (EO) 12113 (January 4, 1979), which required all water management agencies to comply with specific guidelines when submitting water development plans to the Water Resources Council (WRC). The Council's funding and personnel were eliminated by the Reagan administration, but policy responsibilities were reassigned to the Department of the Interior. Under WRC guidelines, water development plans are supposed to include

1. Sufficient documentation to allow a technical review of the analysis by the agency of the ratio of the benefit to the cost
2. Evidence that an adequate evaluation has been made of reasonable alternatives, including nonstructural ones, for addressing the water-related problems of the affected regions and communities
3. An explanation of the relationship of the plan to any approved regional water resources management plans
4. A summary of the consideration given to water conservation measures and a listing of those measures incorporated into the plan
5. Evidence that there has been compliance with relevant environmental and other laws and requirements
6. Evidence that the public and state and local officials have been involved in the plan formulation process.[109]

Even these formal requirements when fully implemented were subject to political manipulation.

For example, the Bureau of Reclamation's Garrison Diversion Project did not meet these criteria, and cancellation of the project was recommended. This and other projects similarly evaluated were nevertheless later approved. Garrison was then revised, but for other reasons.

The devastating floods in the Missouri and Mississippi River basins, in 1993, encouraged the Clinton administration to create the Interagency Floodplain Management Review Committee (IFMRC). The committee's findings and recommendations are referred to as the Galloway Report.[110] The report, for the first time, acknowledged the key ecological services that wetlands and forests contribute in the form of water and nutrient uptake and storage. The committee pointed out that the loss of wetlands and forests to agricultural conversion and other uses contributed to the flooding via increased runoff. The committee also reported that floods are natural reoccurring phenomena that must be accepted and anticipated. Most important, the committee determined that federal water resource project planning guidelines are outdated and that they do not account adequately for the values of ecosystems and ecological services. The committee also identified a need for better coordination between federal floodplain and water management programs as well as with other federal agencies, states, tribal units, and local governments.[111] The report should promote greater consideration of the ecological values that are negatively affected in water projects development.

The Galloway Report also recommended the creation of a Floodplain Management Act that would clearly establish the states as the primary floodplain managers. The act also would provide financial assistance for state and local floodplain management actions. The committee suggested, too, that the Water Resources Council be reactivated to coordinate water activities across the federal government and with the states and tribes.[112] A revival of the WRC would encourage greater consideration of environmental values in federal water projects.

In the United States, water withdrawals (ground and surface) for all uses and purposes average about 407 billion gallons a day.[113] Groundwater volume is estimated at about fifty times the annual flow of surface water. It supplies approximately 25 percent of fresh water used for all purposes in the United States. The major water uses are (1) public, domestic, commercial, and industrial uses; (2) rural domestic and livestock uses; (3) irrigation; and (4) self-supplied industrial uses, including thermoelectric power. Less than one-fourth of the water is consumed with the remainder being returned to surface waters. Withdrawals have increased tenfold since 1900. The increases in water withdrawals are attributed to advances in irrigation, industrial production, energy development, and water-intensive home appliances.[114] In 1950 an average of 1,200 gallons of water per day per person was withdrawn for all uses. In 1990, 1,340 gallons of water per day per person was withdrawn; down from 2,000 gallons of water per day per person in 1980.[115] Although water is considered to be relatively abundant in the United States, there is potential for water shortages in the southern and central high-plains states, the Colorado River basin, Arizona, and California. Further, overall water abundance does not mean that there will be an adequate quantity and quality of water available where and when it is needed. In fact, there is a growing scarcity in the availability of fresh water worldwide.

This scarcity is threatening global food security and the health of ecosystems, as well as social and political stability. Sandra Postel characterizes the inefficient use of scarce freshwater sources as a case of "robbing the future to pay for the present." [116] She argues that a fresh approach is necessary that is focused on using water more efficiently and equitably.

Water shortages are partially caused by wasteful practices and mismanagement. For example, the average American family squanders one-fourth of the water it consumes, and much of the water in older American cities is lost through leaky pipes. The NWF also notes that over one-half of the water diverted for irrigation, the single greatest use of fresh water, is wasted through inefficient techniques. Much of this water is used to irrigate low-value crops.[117] With regard to future considerations, the CEQ states that "consumption of water in the year 2000 is projected to increase from 95 to 135 billion gallons per day. . . . Gradual declines in groundwater levels will restrict western agriculture as withdrawals exceed replenishments, particularly through 'mining' of underground water."[118] In the arid Midwest, they are already mining water at prehistoric levels from the Ogallala aquifer that spreads from Nebraska to Texas.

The demand for fresh water increases with population and economic growth. The future availability of water resources is correlated with population trends. As population increases, the availability of water per person decreases. Consequently, clean water is already a critically scarce resource in many localities and will become so in others. An increasing number of people in the less developed countries do not have easy access to clean and safe water. The estimated annual growth rate in water consumption worldwide for the period 1960 to 2000 is in the range of 2.8 to 4.0 percent, while the population of the world increased 122 percent in roughly the same period, from 1950 to 1995.[119] World population is projected to grow at 1.8 to 2.0 percent (and the world gross domestic product will increase in real terms at 4.2 to 5.2 percent). But water is not evenly distributed, and increased demand is projected for those areas already experiencing shortages in Latin America, Africa, the Middle East, and South Asia.[120]

Various facts illustrate how the world came to experience the scarcity of clean water:

1. One-fourth of the world's water runoff is not useable because of pollution.
2. Worldwide there is considerable unnecessary pollution of surface and groundwater.
3. Worldwide there is considerable waste of water resources.
4. In many locations throughout the world, groundwater is pumped out before aquifers can be recharged.
5. Removal of plant cover through deforestation and other means is encouraging soil erosion as well as flooding and droughts.[121]

The picture, of course, is not totally bleak. There are technological and managerial measures for solving at least some of these problems. Suggested examples are water conservation, surface water management, weather modification, desalinization, transport of water, and water recycling or reuse.[122] It may not be possible, however, to provide sufficient water resources to meet human needs into the indefinite future without better water management efforts that emphasize conservation and wise use.

Available evidence suggests that water supplies and quality will increasingly be disrupted because of damage to watersheds, particularly through deforestation. Besides structural projects (dams or irrigation systems) to distribute water, there are numerous nonstructural alternatives that include preservation of wetlands for flood control and a whole range of water conservation techniques and methods. If the negative consequences of water projects like those of the Aswan High Dam in Egypt are to be avoided, "it is essential to plan water resource management efforts intensively within the context of all other resources and needs." [123]

Tomorrow's water managers will have to draw upon a different set of skills than those who preceded them. Christine Olsensius describes the new rules of the game for water use and management. First, the training that water managers receive will have to be expanded. They will no longer simply manage dams and reservoirs for agricultural purposes. Greater attention will have to be given to fish and wildlife, recreation, and water quality management. Second, these interdisciplinary demands will require the water manager to have the capability to draw information from a wide array of information sources. Third, water management must cut across traditional bureaucratic, geographic, and political boundaries. Fourth, the need for regional planning programs and agencies will increase. Fifth, distinctions between management approaches for surface and groundwater, quantity and quality, and structural and nonstructural need to be reduced with greater emphasis on overall integration of activities. Sixth, increased emphasis will be placed on water conservation and nonstructural alternatives and less on supply of demand. Seventh, long-term water management will increasingly stress reduction in use and recycling of supplies. Finally, priorities for water use will have to be established, because we can no longer afford to let people use water in any manner they deem necessary. [124]

Some of the above suggestions are included in the Water Quality Act of 1987. The law includes provisions for assigning priorities for water management activities and development of water basin management planning for those water priorities targeted for action. The states are expected to integrate "various programmatic approaches to clean water into a single, unified state effort" under the guidance of the EPA. Both point and nonpoint pollution source programs are to be integrated. [125] In response to these provisions, the EPA and the states have adopted a watershed management approach to improve the availability and quality of water in the United States.

In 1996, the Interstate Council on Water Policy unveiled a national water policy proposal designed to serve as a blueprint for the management of freshwater resources in the United States. The policy proposal is based on traditional water management principles, but it also addresses more recent concerns such as risk assessment, economic incentives, and government restructuring. The emphasis of the policy is on more partnerships, more flexibility, and more public involvement. The policy also acknowledges the need for a broad-based collaborative effort between the public and the private sectors in order to ensure an effective and safe supply of fresh water.[126]

Fish and Wildlife Policy

Along with various state, local, and private agencies, numerous federal agencies are involved in the management of fish and wildlife (all non-domesticated mammals, birds, reptiles, amphibians, fish, and invertebrates, living in the natural environment). The U.S. Fish and Wildlife Service in the Department of the Interior originated from the Bureau of Biological Survey, founded in 1905 to study the abundance, distribution, and habits of America's birds and mammals.[127] Programs exist for game and nongame species, wildfowl, endangered species, aquatic species, and fisheries. Although a variety of agency policies exist, an overall wildlife management principle is that habitat determines both abundance and biological diversity of wildlife populations, including their diversity. Consequently, the composition and condition of wildlife populations serve as a barometer of the quality of the environment, because these populations depend on wise and proper use of the environment. Climate, topography, and geology are basic influences on the composition of a plant community and the nature and abundance of local plants. Plant communities, in turn, govern the kinds of wildlife that will be supported in a given geographic area. In this sense changes in the environment by humans may radically alter the composition of wildlife populations. Noting the impacts of human activities on wildlife and its habitat, the Wildlife Management Institute points out that as environmental changes occur, desirable wildlife often begin to disappear from an area, despite appropriate legal protection, indicating that something is wrong in the environment. The impact of the absence of wildlife species also is likely to be much greater than the loss of aesthetic values and recreational opportunities.[128]

In relating the decline in wildlife to the probability of systemic deterioration, Carl Reidel illustrates the basic political dimension of wildlife-related activities:

Intrinsic values are given relative consideration as they are integrated into the total man/nature system. This, I believe, is a major point of departure from traditional conservation. Where the old conservation mourns the loss of an eagle for its intrinsic value alone, or perhaps as a vague symbol of paradise lost, the environmentalist mourns the eagle both for those reasons and in the recognition that its death is a clear warning of systematic disruption. The eagle's death, like the death of the 19th Century coal miner's canary, is a signal that something is wrong in the biological system and quite likely in the political, economic, and social realms as well.[129]

These broader considerations extend to wildlife management agencies as well. Except when inside national parks and national wildlife refuges, wildlife is subject to the laws and regulations of the state in which it is located, regardless of whether it is on federal, state, local, or private lands. The majority of wildlife habitat is in private ownership in the United States. Recent studies indicate that nonindustrial private forest landowners consistently rank wildlife observation and habitat higher than timber income as reasons they own their property.[130]

The manner in which the American public views and interacts with wildlife has changed drastically over time. Today, a larger proportion of the public is interested in wildlife encounters for recreational purposes, nonconsumptive wildlife-associated activities, than in the traditional consumptive activities of fishing and hunting. The *1996 National Survey of Fishing, Hunting, and Wildlife-Associated Recreation* indicates that 62.9 million Americans age sixteen and over participated in wildlife watching (formerly nonconsumptive) wildlife-related activities. These individuals observed, photographed, or fed wildlife. In comparison, there were 35.2 million anglers and 14 million hunters age 16 and over. In the pursuit of these activities Americans spent over $72 billion; $38 billion for fishing, $21 billion for hunting and $29 billion for wildlife-watching activities.[131] Consequently, state wildlife departments are subject to considerable political and economic pressure to produce abundant populations of both game and nongame species of fish and wildlife.

The majority of state wildlife agencies derive all or most of their financial support for the operation of wildlife programs from hunting and fishing licenses. The federal aid funds that states receive from the U.S. Fish and Wildlife Service are derived from excise taxes on fishing and hunting equipment. Consequently, state programs are strongly oriented toward the management and production of game species for the use of sportsmen who supply the most of the revenue. Although habitat and other measures for game species may benefit nongame species, relatively little attention, either in operations or research, has been directed toward non-game species by federal or state agencies. In the 1980s approximately $97 of every $100 spent by federal and state governments on wildlife management (including research) went to less than 3 percent of the species—those used for hunting, trapping, or fishing.[132] However, with the growing wildlife interests of the general public in nongame as well as game species, federal and state wildlife agencies are moving away from an emphasis only on game species.[133] All fifty states now have special tax legislation that provides funds for programs and management of nongame species. Further, the federal government did attempt to develop additional funding mechanisms for state nongame programs under the Fish and Wildlife Conservation Act of 1980. The primary sources considered were excise taxes on equipment used by nonconsumptive wildlife users and punitive taxes related to the destruction of wildlife habitat. The purpose of the 1980 act was to assist states to develop restoration programs for all wildlife. Currently, federal excise taxes on fishing and hunting equipment under the Dingell-Johnson and Pittman-Robertson Acts are used to restore sport fish and wildlife populations in the states.[134] The political power of recreational equipment industries, who do not want excise taxes imposed on their products, and the reluctance of Congress to impose new taxes, is reflected in the fact that eighteen years after passage of the act, Congress still had not funded the Fish and Wildlife Conservation Act.

Wildlife policy is based to a degree on wildlife management needs and requirements. As Reuben Trippensee indicated half a century ago, the "field of wildlife management has many sides and many angles. Fundamentally it is the process of making land and water produce sustained crops of wild animals. The goal is clear and definite, first place, many different classes of animals are involved—migratory species, fur bearers, game species, nongame species."[135] Trippensee also suggested that a basic compatibility exists between wildlife management and multiple-use goals. He believed that this is true in conjunction with the practice of both forestry and agriculture, and that "wildlife management in one form or another may be adapted to nearly all programs of land use is a fact deserving major emphasis."[136] Adaptive wildlife management occurs at the landscape level, transcending specific jurisdictional boundaries. The recent movement toward ecosystem management in federal land management agencies, and in many states, is conceptually related to this line of thought.

Federal land management agencies adopted ecosystem management initiatives after Vice President Gore's National Performance Review recommended that all federal agencies adopt "a proactive approach to ensuring a sustainable economy and a sustainable environment through ecosystem management."[137] However, the written legal authorities for BLM and USDA

Forest Service lands still contain multiple-use guidelines, while U.S. Fish and Wildlife Service guidelines for the national wildlife refuge system indicate one dominant use—the protection and enhancement of fish and wildlife and their habitat requirements. This does not mean that alternative uses do not occur on refuge lands. For example, timber harvesting, grazing, agriculture, and recreation activities do occur on refuges. These activities are to occur only to the extent that they do not interfere with or detract from the dominant purpose of the refuge. The National Wildlife Refuge Improvement Act of 1997 (NWRIA) recognizes compatible wildlife-dependent recreation activities as the priority general public use of the refuge system.[138]

Much of public and private wildlife management consists of activities and measures directed toward conserving and producing sustained populations of fish and wildlife. This includes protection of species and their habitat, habitat manipulation, harvesting, and law enforcement. The Wildlife Management Institute asserts, "Altering or maintaining the environment to favor the needs of certain wildlife species, in fact, is a basic technique of wildlife management."[139] Still another major element of wildlife management is managing people in their various relationships with wildlife and with wildlife habitat.[140]

Underscoring the notion that wildlife management entails a balancing of many interests, Daniel Poole and James Trefethen point out:

> The principal goal of wildlife management is to maintain wildlife populations at levels that are in the best interests of the animals themselves and at the same time consistent with the social, economic, and cultural needs of the people. This is a large order. What may be acceptable deer populations for hunters may be excessive for orchardists or truck farmers harassed by deer predations. Dense flocks of red-winged blackbirds may gratify the birdwatcher but alarm the rice farmer.
>
> The best the wildlife manager can do is strive to balance these conflicting interests and try to build, modify, or control animal numbers at levels where viable populations of each species are assured, where excessive economic damages do not occur, and where the well expressed desire of the public for a diversity and relative abundance of all wild species can be satisfied.[141]

The protection, maintenance, and restoration of adequate habitat is fundamental to wildlife policy. This is especially the case where public or private development activities threaten habitat. For habitat modification and loss pose the greatest threat to wildlife. In the United States the amount of habitat lost or degraded from all types of changes in land use such as urbanization, construction, strip-mining, agriculture, vegetative manipulation, and so forth probably exceeds 2 million acres per year.[142] Unfortunately, conversion of wildlife habitat to other uses is often irreversible. A primary example exists in the loss of thousands of acres per year of Louisiana coastal marshes to energy and water projects. Such loss is significant in that "the importance of the Louisiana coastal habitat to fur bearers, waterfowl, and other migratory birds (particularly seabirds and wading birds) is unsurpassed by any other similar sized area in the U.S." This area winters up to 5.5 million ducks, 70 percent of the continental coot population, 50 percent of the mid-continental population of lesser snow geese, and produces 50 percent of the continental mottled duck population. The marshes are also important breeding areas for 6.5 million rails and gallinules and support about 155 colonies of 8 million wading and shore birds that breed in this area. Finally, this area supports the largest nesting concentration of bald eagles in the south-central United States, as well as two-thirds of the U.S. fur harvest.[143] In 1994, the Interagency Ecosystem Management Task Force recognized coastal Louisiana as one of the most productive wetland ecosystems on earth and designated the region as one of seven ecosystems selected for case study.[144]

Although various laws and policies have provided some measure of protection for wildlife habitat, the destruction of natural habitat continues at an alarming rate. This pattern is particularly apparent for wetlands and riparian habitat. A little more than one-half of the original wetland acreage remains in the United States; since the mid-1950s an average of more than 600,000 acres of wetlands has been lost each year. Connecticut has lost 50 percent of its wetlands. Since 1950, 200,000 acres have been cleared annually in the Mississippi Delta area. Most of California's coastal wetlands have been destroyed.[145] This pattern continues in spite of NEPA and other legislation that call for recognition of the need for good faith efforts to accomplish habitat protection. Legislative support for habitat destruction came in the 104th and 105th Congresses, in the form of riders on bills that eliminated NEPA requirements for timber salvage and grazing operations.

Government can play a positive role in habitat protection at the federal, state, and local levels. Through the Fish and Wildlife Coordination Act (FWCA) and the Endangered Species Act, the U.S. Fish and Wildlife Service takes an active role in the protection of habitat.[146] Under FWCA the agency assesses the potential impact of federal projects on fish and wildlife habitat and makes recommendations for mitigation actions as well as the acceptability of the projects. Under the Endangered Species Act the Service also provides biological opinions on the acceptability of activities on or near endangered species' critical habitat. Other government programs may not be as large, but they, too, should provide the same level of protection for wildlife, as the Conservation Foundation points out:

> Federal policy on wildlife habitat may not be explicit, and may not even be recognized as a policy. But the sum of the individual decisions made from different levels of public lands, combined with pollution control and other programs affecting land use, nonetheless constitutes policy. . . . State and local governments own less land, and their actions affect smaller areas, but for specific species and specific areas they, too, must be concerned with the impact of their activities on wildlife. . . . Government can also stimulate increased conservation and maintenance of wildlife habitats on private lands.[147]

Wildlife conservation and maintenance could be partially accomplished by (1) providing information and technical assistance to private landowners; (2) furnishing financial and tax incentives for habitat protection; (3) buying or obtaining fish and wildlife easements in which private owners of the property permanently give up all rights, including future owners' rights, to develop or change the traditional uses of their land, designating a government agency or conservation organization to enforce this easement; and, (4) having regulations, permits, and zoning regulations that would discourage the conversion of valuable habitat to other uses.[148]

Several of the above-mentioned habitat protection actions are facilitated through federal grant programs of the Fish and Wildlife Service and Natural Resources Conservation Service. Under the Partners for Wildlife Program, the Fish and Wildlife Service provides free technical assistance to help landowners manage property for wildlife. They also will pay up to $10,000 of costs required to preserve or restore riparian habitat. Under the Conservation Reserve Program the Natural Resources Conservation Service (NRCS) pays the going rental rate plus 20 percent to property owners to set aside riparian land and plant grasses or trees for areas classified as highly erodible. For a permanent easement, under the Wetland Reserve Program, the NRCS pays landowners 75 to 100 percent of appraised agricultural value of land and 75 to 100 percent of restoration costs if needed. The NRCS also pays landowners for thirty-year easements or for wetland restoration at lower rates.[149]

In the interest of achieving a comprehensive national wildlife policy, Thomas Kimball recommends:

> First, the policy must encompass all forms of wildlife—game and nongame, endangered species as well as those in abundance, and mi-

gratory as well as nonmigratory species. Second, it should encourage wholehearted cooperation among all governmental agencies and concerned private groups to foster and nurture public awareness of the values and benefits stemming from proper conservation of fish and wildlife. Third, it should encourage these same agencies and groups to coordinate their efforts to produce optimum varieties and numbers of fish and wildlife. And, fourth, the policy should recognize both the responsibility of state wildlife agencies to manage fish and resident wildlife on federal lands, except where contrary federal laws exist, and the authority of the federal government to manage wildlife habitats and regulate public use of its land.[150]

The Fish and Wildlife Service formally announced a similar division of responsibilities between the federal and state governments in 1983 (43 CFR, part 24).[151]

Thomas Fitch argues that there is a need for comprehensive wildlife policies and programs that are based on ecosystems and all species rather than just on game species. Comprehensive approaches of this nature would also be appropriate for environmental impact statements and for consideration of protection for rare and endangered species. In observing that public attitudes toward wildlife have changed substantially over the past fifty years, he says that the public now values and appreciates both game and nongame species and the ecosystems in which they reside.[152]

This change in public attitude was reflected in the Fish and Wildlife Conservation Act of 1980 mentioned above, which provided the basis for the adoption of broader wildlife policies and perspectives by both national and state governments. This law included provisions for furthering cooperation between national and state governments for conserving nongame species and their habitats and for planning for the protection of species indigenous to specific states. The act was designed to provide federal assistance to state agencies for the development of comprehensive wildlife management plans for both game and nongame species.[153] However, as noted, as of 1998 the federal government has not developed appropriate funding mechanisms to support the act, for various political reasons.

John Loomis and William Mangun describe the many efforts taken to fund the act. For example, Congress almost funded the act at the outset through an excise tax on birdseed, but political influentials defeated the proposal. Subsequently, the Fish and Wildlife Service conducted an analysis of potential funding mechanisms at the request of Congress. The Office of Management and Budget attempted to keep this analysis from being sent to Congress for consideration because several of the recommendations in it went against President Reagan's position on new taxes.[154] As a consequence of the failure to fund the act, the development of comprehensive wildlife policies has been delayed.

In his book on the need for comprehensive wildlife programs, Fitch considers ecosystems to be relatively stable and self-sustaining units of which game and nongame species are integral and significant parts. Ecosystems can provide a variety of resources to humans as well as relatively stable populations of wildlife. These wildlife and other resources can disappear, however, if the structural or functional components of ecosystems are disrupted by humans. Fitch goes on to say that "management induced introductions, overpopulations, or extirpation of wildlife species can destabilize and degrade ecosystems. The long-term survival of a variety of nongame species may be adversely affected by wildlife management strategies that are designed to maximize the short-term harvest of a few game species." He advocates management activities that grow out of recognition of the need to preserve the integrity of ecosystems because such management strategies are more economically feasible and they provide the basis for developing ecosystem level management expertise by surveying and inventorying wildlife in relation to ecosystems.[155]

In addition to an orientation toward biological diversity, ecosystem and nongame approaches to wildlife policy and management provide a more reliable data base for EISs and for monitoring species that are at, or might reach, threatened or endangered population levels.[156] The ecosystem approach intentionally attempts to preserve biological diversity for a variety of species. The species approach identifies and provides specially designed conservation measures to ensure survival of one species or a group of associated species that are in need of protection. A combination of the two approaches is often used. For example, a natural ecosystem reserve may be used in combination with conservation measures to prohibit hunting or trade of a species to protect rare and endangered species.[157]

The lands and waters of the United States support an estimated 2,900 species of vertebrates (mostly nongame). This total is comprised of 430 mammals, 1,000 birds, 300 reptiles, 175 amphibians, and more than 1,000 fish. Most of these species require specific habitats, some of which may be very narrow in range. By 1920 an estimated 22 species of mammals, birds, and fish had disappeared from the United States. Among the lost species are the eastern elk, passenger pigeon, and the plains wolf. Indiscriminate hunting and habitat destruction hastened the extinction of these species.[158] Today, as Steven Yaffee correctly points out, the endangered species and other populations that often seem to interfere with development projects are "more than likely a mere indicator of ecosystem loss or damage."[159]

Since 1880, over 160 plants and animals have become extinct in the United States. This number does not include 140 species extinctions in the tropical forests of Hawaii since 1850. In order to reduce the possibility of more extinctions, the Endangered Species Act was established by the U.S. Congress in 1973. The Endangered Species Act requires the listing of plant and animal species identified as threatened or endangered, the designation of habitat critical to their survival, the establishment and conduct of programs for their recovery, and the establishment of domestic and international agreements and assistance to conserve endangered and threatened species.

The Endangered Species Act appears to have slowed down the rate at which species disappear in the United States. However, 5 species are presumed to have become extinct between 1982 and 1984. The number of species that are listed as threatened or endangered has steadily increased in recent years. From 1980 to 1995, the number of threatened species increased from 49 to 206 and the number of endangered species increased from 234 to 754.[160] Furthermore, over 60 percent of the listed species are either declining or are in an unknown status.[161]

The formal adding of a species to the list is a regulatory act that sometimes lags behind biological realities.[162] For example, in the process of providing American citizens "regulatory relief" early in the Reagan administration, the listing process was slowed considerably. After the Reagan administration, political pressure from concerned citizens and influential congressmen forced the Fish and Wildlife Service to increase the number of listings considerably, primarily the listing of plants, which now constitute more than one-half of the species. However, in 1996, the Republican majority in Congress stopped this trend and placed a moratorium on the listing of additional species.

Plants constitute the majority of the current federal list of endangered and threatened species, displacing vertebrates, particularly birds and mammals, and there is increasing emphasis on freshwater clams and mussels. In 1984 only 69 listed species were from the plant kingdom, while the U.S. Fish and Wildlife Service had identified 2,560 native plants as candidates for listing.[163] By 1995, 526 plants were listed.

A significant difference exists between proposal as a candidate species and assignment of actual protection under the Endangered Species Act. Some of this can be attributed to resistance from some quarters to giving endangered or threatened species status to any species that might cause delays or serious problems for development in habitats critical to their survival. The Endangered Species Act contains provisions for the conservation and recovery of (1) an endangered species, which is in danger of becoming extinct throughout all or a significant part of its natural range;

or (2) a threatened species, which is likely to become endangered in the foreseeable future. In addition to protection, the act requires that efforts be made at recovery of the threatened or endangered species. By 1997 there were 653 approved management plans for recovery of 1,082 listed species. This represents roughly 60 percent of the species that have been listed. One problem, however, with the recovery plan process is that "no systematic effort yet exists at the federal level to monitor the implementation of recovery programs." [164] Moreover, not all species are treated the same. Harvey Doerksen and Craig Leff point out that "in recent years, about 85% of all recovery funds were allocated to only 8% of all listed species, and 40% of all species received no funding at all!" [165]

Some recovery plans have produced successful results. The whooping crane population has increased tenfold since 1941. Because of a dramatic increase in the population of the American alligator, classification of this species has been changed from "endangered" to "threatened." [166] The bald eagle, peregrine falcon, brown pelican and the infamous snail darter also have been removed from the endangered species list, due to successful management of the former and fortuitous location of additional numbers of the darter.

Condors have also been brought back from near extinction. The battle over the condor, however, is typical of the problem of competing jurisdictions hindering endangered species recovery where federal, state, and private wildlife management agencies are concerned. With the disappearance of six birds in 1985, the number of wild California condors in the wild was estimated to have dropped from fifteen to fewer than nine. The Department of the Interior announced that the remaining wild condors would be captured and placed in a captive breeding program run by two zoos. The National Audubon Society unsuccessfully sought to block the capture on the grounds that the federal agency was not developing a recovery plan or securing vital condor habitat.[167] In 1986, the U.S. Fish and Wildlife Service, nevertheless, captured the remaining wild condors, concluding an effort that spanned several decades to preserve a viable condor population in its natural habitat. The captive breeding program has, in fact, been nominally successful, and several California condors were released back to the wild in California as well as the Grand Canyon in Arizona. At the end of 1996 the population of California condors including both wild and captive individuals was 120.[168]

Section 6 of the Endangered Species Act permits the U.S. Fish and Wildlife Service to enter into cooperative agreements with the state fish and wildlife agencies. The purpose of this section was to stimulate state agencies to develop endangered species programs in return for financial assistance. Most states responded and developed programs according to federal standards in order to obtain the additional funds. Many states took this opportunity to develop nongame management programs in conjunction with the endangered species programs. Because programs for plant and animal species can be separated there are only twenty cooperative agreements for plants while there are fifty agreements for wildlife.[169]

A major problem with the Section 6 program began in 1982 when the Reagan administration made no appropriation request for cooperative agreements under the section. Many state agencies had come to rely on the federal assistance, and their endangered species efforts were severely restricted with the cut in funding. The logic for the cutback appears to be related to the controversial nature of endangered species protection activities. Both federal and private projects can be halted if they threaten a listed endangered species either directly or indirectly by jeopardizing critical habitat for the species.

The U.S. Fish and Wildlife Service has developed a new procedure that allows Service personnel to be somewhat more flexible with landowners who feel threatened by the Endangered Species Act's critical habitat provisions. The Service can enter into a "Habitat Conservation Plan" agreement with landowners. The "Safe Harbor" agreements protect the landowners' rights to use their property in the manner in which they desire, even in the event that an endangered species moves onto the property after the agreement is made.

The World Conservation Union (IUCN) also has an international system for classifying species as endangered, vulnerable (threatened), or rare. A species is "endangered" if near-term causal factors jeopardizing its survival continue to operate. A species is "vulnerable" (or "threatened") if causal factors likely to jeopardize its numbers or geographical distribution continue to operate. A species is "rare" if it has a world population that is small and "at risk" but not yet endangered or vulnerable. It should be noted, however, that rare species, as the result of sudden or unanticipated changes, could become extinct. The IUCN has a Conservation Monitoring Center that maintains "red data books" on the status of plant and animal species. The World Resources Institute and International Institute for Environment and Development report that the most valuable feature of these books is that they stimulate local interests to identify and list species for regional red data books.[170]

Species are declining on a worldwide basis. It was estimated that 15 to 20 percent of the world's species would become extinct between 1980 and the year 2000. These extinctions would be due in part to toxic environmental pollutants but mainly to loss of wildlife habitat. Over one-half of the projected extinctions would result from clear-cutting and degradation of tropical rain forests that contain the greatest abundance and diversity of species.[171] An estimated 25,000 plant species and in excess of 1,000 vertebrate species and subspecies are threatened with extinction. These figures do not include invertebrates such as mollusks and corals.[172]

Some biologists estimate that one to three extinctions occur daily and that the rate will increase to one per hour several years from now. Of the 5 to 10 million species in existence (some estimates range as high as 40 million) throughout the world, most have never been named or studied. Over 1 million are expected to be lost during our lifetimes.[173] A conservative estimate for the overall extinction rate among all known and unknown species is 1,000 species per year. However, based on the "manhandling of the environment" during the last part of this century, a more realistic figure may be approximately 40,000 species per year, or 100 species per day.[174]

Paul and Anne Ehrlich state, "Extinctions that are occurring today and that can be expected in the future are likely to have much more serious consequences than those of the distant past." They believe that current extinctions will destroy a greater proportion of the biological diversity of the world than previous extinctions. They also note that human activities are starting to shut down the processes by which biological diversity can be regenerated or replaced. The Ehrlichs consider the great variety of plant and animal species in all their biological diversity to be working parts of ecological systems that provide people with irreplaceable free "public services." Among the more important of these are clean air and water, the maintenance of genetic diversity for agriculture, climate regulation, and control of pests. Thus, when people exterminate species they are basically "popping the rivets" on their own spaceship through the loss of essential, free public services.[175]

Under provisions of the Endangered Species Act, areas (or ecosystems) on which endangered and threatened species depend for their survival can be designated critical habitats. Federal agencies are required to ensure that their activities do not adversely modify or destroy the habitat or interfere with the continued existence of the species. More than one hundred critical habitats have been designated for over thirty-four species that include the whooping crane and gray wolf. Designation of critical habitat is based on biological factors and can cover private as well as public lands. Critical habitats, however, are not closed to most human uses, but supposedly only those activities that threaten the well-being and survival of species.[176]

The designation of critical habitats is a controversial area of wildlife policy. Efforts to establish or maintain critical habitat on public lands often come into conflict with developmental activities. Mineral development and motorized vehicle use are typical activities creating conflict. Human disturbances in and around Yellowstone National Park continue to put pressure on critical wildlife habitat. A number of conservationists have suggested that the park boundaries are unrelated to the greater Yellowstone ecosystem and that management based on the larger ecosystem would better protect wildlife species.

The ecosystem management approach initiated by the Clinton administration takes the Endangered Species Act into consideration. The Fish and Wildlife Service and the National Oceanic and Atmospheric Administration (for marine mammals) issued a Federal Register Notice of Interagency Cooperative Policy for the Ecosystem Approach to the Endangered Species Act on July 1, 1994. The policy is designed to "integrate the mandates of Federal, State, Tribal, and local governments to prevent endangerment by protecting, conserving, restoring, or rehabilitating ecosystems that are important for conservation of biodiversity."[177]

A promising interagency program for grizzly bear reintroduction is an example of making partnerships work in an ecosystem approach to endangered species recovery. From an estimated 50,000 individuals prior to settlement, grizzly bear numbers in the lower forty-eight states had been reduced to fewer than 1,000 in 1975. Encroachments on prime grizzly bear habitat had involved oil and gas exploration and drilling, timber harvesting, geothermal construction, and grazing domestic livestock. Habitat destruction, along with conflicting policies of federal land use agencies and poaching, posed serious threats to grizzly bear survival. In 1991, the Interagency Grizzly Bear Committee endorsed the Bitterroot Ecosystem in central Idaho and western Montana as a recovery area and authorized the U.S. Fish and Wildlife Service to pursue grizzly bear recovery. A citizen management committee, headed by the National Wildlife Federation, forged a unique coalition between the timber industry and environmentalists.[178]

Predator control is also a controversial area in wildlife policy. Predators are animals that kill and eat other animals. Many of the listed endangered and threatened mammals and birds are predators, such as the grizzly bear, gray wolf, and American bald eagle. In the United States poisoning, hunting, and trapping have been traditional ways of eliminating animals considered to be predators. Yet little is known about the effectiveness and impact of predator control programs. In general the history of predator control has been one of indiscriminate removal of "bad" species or "varmints," as defined by selected interests with complete disregard for ecological considerations.

Supposed nontarget species, however, are often eradicated along with offending predators or pest species. For example, efforts to eradicate prairie dogs virtually eradicated the now endangered black-footed ferret. Fortunately, U.S. Fish and Wildlife Service listing and recovery efforts brought the black-footed ferrets back from near extinction and they have been released back into the wild.[179] Serious reductions in natural predators such as coyotes, mountain lions, and bears may upset natural controls over populations of prey species such as rodents and deer as well. If employed at all, predator control measures should be directed at specific individuals rather than at whole populations.[180] Federal control programs have been substantially reduced since the mid-1960s, reflecting changing public attitudes toward wildlife.

International Wildlife Policy

In many less-developed countries, wildlife still provides subsistence and nutrition on a renewable basis, especially in rural areas. Robert and Christine Prescott-Allen observe that "[t]he contribution of wildlife to developing country diets is paradoxically both minor and vital. However, the minor role of wildlife is still vital because directly or indirectly wildlife supplies much of the animal protein consumed as well as being a source of trace elements, nutritional variety, and from time to time famine relief. Wildlife including fish can provide welcome flavor especially to poor diets dominated by starchy staples."[181]

Wild animals also constitute an important source of income for rural communities in many nations engaged in world trade in wildlife and wildlife products. Unfortunately, some international wildlife trade in the form of organized commercial enterprises has become a major threat to many species of wildlife. Scarce wildlife products are being taken from the wild in less-developed countries and sold in industrialized countries to supply expanding markets. Representatives of eighty countries met in 1973 to establish the Convention on International Trade in Endangered Species (CITES) to control the overharvesting of endangered and threatened species. Under this convention, it is illegal to import or export a listed species in whole or in part. This includes products like furs and jewelry made from such species. In order to import or export an animal or animal product on the CITES list into the United States, an individual must have a permit from the U.S. Fish and Wildlife Service. It should be acknowledged that it is not the purpose of CITES to eliminate trade in animals or plants but to reduce the possibility of trade in species becoming a major factor in the endangerment of species. Michael Bean points out that in order to accomplish this, CITES "establishes a system of trade controls that vary in their restrictiveness, depending on the degree of jeopardy each species faces. The trade controls imposed by CITES apply only to species listed on one of three treaty appendices. Species may be added to or removed from appendices 1 and 2 by a two-thirds majority of the member countries at their biennial meetings." [182]

Wildlife serves still other functions internationally. It brings tourists to less developed countries, contributing substantially to foreign exchange earnings and, if managed well, helping to protect natural areas. As Robert Mendelsohn observes:

> Ecotourism is emerging as a popular approach to protecting natural areas throughout the world. However, to be sustainable, ecotourism must provide reasonable economic benefits locally and nationally, and must avoid damaging the natural areas on which it depends. By itself, ecotourism is not the answer to environmental destruction, but if carried out properly, and in concert with other protective measures, it can contribute to sustainable and economically rewarding use of natural areas. [183]

At the other extreme, worldwide wildlife is recognized to have high symbolic, ritual, and cultural importance. The spiritual lives of many people are enriched through contact with wildlife. All of these contributions taken together should encourage governments to manage wildlife resources on a sustainable basis and to undertake substantial measures for habitat protection. [184] Ultimately, a worldwide recognition of the need to maintain biological diversity and to preserve all species, regardless of their immediate value to humans, is required. In addition to protecting known species, maintenance and restoration of biological diversity also requires protection of unknown species through the protection of natural ecosystems and habitats. The international Convention on Biological Diversity formulated at the United Nations Conference on Environment and Development in Rio de Janeiro in 1992 is a direct reflection of such thought.

The Convention on Biological Diversity's overall purpose is to slow down the loss of species and maintain species diversity through such efforts as increased identification and monitoring efforts and protection of ecosystems. President Clinton signed the treaty in 1993, but Congress did not ratify it. [185]

The shift toward protection of all species and biological diversity is an improvement in international wildlife policy. The shift represents a significant growth in understanding of the place and importance of wild species and wildlife habitat in the biosphere. The contribution of individual species to systemic stability is beginning to be recognized. Aldo

Leopold's vision of the importance of all species native to Wisconsin could well be applied to those species native to any particular region: "Of the 22,000 higher plants and animals native to Wisconsin, it is doubtful whether more than 5 percent can be sold, fed, eaten, or otherwise put to economic use. Yet these creatures are members of the biotic community, and if (as I believe) its stability depends on its integrity, they are entitled to continuance." [186]

But defense of living species is not, of course, entirely a matter of systemic integrity. There is a fundamental value orientation that also argues for species protection. According to Charles Elton: "The first [reason for conservation] which is not usually put first, is really religious. There are some millions of people in the world who think that animals have a right to exist and be left alone, or at any rate, that they should not be persecuted or made extinct as species. Some people will believe this even when it is quite dangerous to themselves." [187]

Part of the awareness of the need to protect plant and animal species and encourage biological diversity stems from disenchantment with the anthropocentric view of the universe. A different view of human beings is that they no longer appear to be so important as they once seemed. "This non-humanistic view of communities and species is the simplest of all to state: they should be conserved because they exist and because this existence itself is but the present expression of a continuing historical process of immense antiquity and majesty." [188] Humans are beginning to realize that rather than being situated at the exact center of the universe—a position of supreme importance—they are one species among many. All species function together to make the biosphere self-sustaining and self-perpetuating.

Natural Resource Policy Impacts across Policy Areas

Various other federal policy areas often overlap natural resource policy areas and can have either positive or negative effects on natural resource policy areas. For example, energy policy requiring inundation of a given flood plain to provide water for energy conversion will remove wildlife habitat from use. Energy policy would affect and, in a sense, become wildlife policy for that area. Policies in other areas, such as national economic policy, also influence specific natural resources. A strong federal development policy, for instance, might promote development to the exclusion of other concerns, thus restricting the possibility for conservation measures to be taken.

National policy areas represent large, but very incomplete parts of what can be considered a society's course of action or objectives (values) for a particular natural resource area. Other important parts are manifested as vested public or private interests that interact with articulated policies for a given natural resource area. For example, export-import policies that guide domestic and overseas operations of an internationally based corporation have impacts on both national and international forest policies. A given policy does not exist or operate in a vacuum. Each policy must be considered in the broader context of other policies that are in operation from the local level to the global level.

Conservation initiatives implemented over the past ten years have contributed to the most significant improvement in soil erosion reduction in the United States over the past century. The various initiatives began with the Food Security Act (FSA) of 1985 (also known as the Farm Bill of 1985). Through the FSA, Congress authorized creation of the Conservation Reserve Program (CRP) and the "sodbuster" and "swampbuster" programs. The CRP is credited as being the primary reason for the reduction of soil erosion from farmlands in the United States from 1982 to 1992. The estimated amount of soil erosion declined from 3.1 billion tons in 1982 to 2.1 billion tons in 1992.[1] The Conservation Reserve Program is implemented by the Natural Resources Conservation Service. It provides subsidies to farmers who retire highly erodible lands. In order to receive the subsidies farmers must plant trees or grasses on the land and not use the land for ten years.

The sodbuster and swampbuster programs are tied into the agricultural subsidies program administered by the Consolidated Farm Service Agency (CFSA). Under provisions of these two programs, farmers who attempt to expand their allotments by plowing marginal lands or filling in ponds or other wetland areas may lose their rights to receive subsidies from the CFSA.

The Federal Agricultural Improvement and Reform Act of 1996 (FAIR) weakened the sodbuster and swampbuster provisions by granting farmers more flexibility in complying with the restrictive provisions of these programs. But FAIR also expanded and extended the conservation easement programs established under the 1990 farm bill, such as the Wetland Reserve Program (WRP) and the Environmental Quality Incentive Program (EQIP). FAIR also provided funding for retirement of frequently flooded croplands. Collectively, all of the conservation easement programs that began with the original farm bill in 1985 have contributed to the substantial reduction in soil erosion from farmland that has occurred in the United States.

Soil policy is important to natural resource management to the extent that most plant life is dependent on the nutrients supplied by the minerals in soil and the manner in which soil serves as a substrate for plants and animal organisms. Mineral policies are also important, in that appropriate policies ensure the ores necessary for energy processing and industrial development. But the acquisition of minerals either through deep or strip-mining causes considerable disturbance to ecosystems.

Soil Policy

Federal, state, and local policies in the United States to control the erosion of soil have not been adequate, but appear to be improving slightly. The indicator of inadequacy lies in the annual loss of soil. The 1982 USDA Natural Resources Inventory (NRI) indicates that 6.5 billion tons of soil were being lost annually on nonfederal lands.[2] This loss would amount approximately to a layer of soil two inches thick across the entire state of Missouri. The 1992 NRI indicates that wind and water erosion produced a loss of 4.6 billion tons from nonfederal croplands, pasture, and range. From U.S. cropland, alone, 2.1 billion tons of soil were lost to erosion in 1992; compared to 3.1 billion tons in 1982.[3]

Soil loss is important because soil is a crucial and living support system for plant and animal life. Soil is a natural resource that is synthesized through biogeochemical cycling from a variable mixture of broken and weathered materials and decaying organic matter. It covers the earth in a thin layer and serves as a natural medium for the growth of plants. Soil is technically a renewable resource, but because of the time required for its renewal it should be considered a nonrenewable resource. "Soil fertility" refers to the quality of a soil that enables it to provide nutrients in adequate amounts and balances for the growth of specified plants. As an essential for life, soils differ widely in types and capabilities. These differences affect soil productivity and susceptibility to environmental and human influences.

Because of different soil capabilities and adverse effects from human influences, the government formulates policies designed to maximize the yield from the soil while minimizing the negative effects of human activities. U.S. soil policy began in the 1930s, primarily as a result of drought conditions and the massive losses of prime agricultural lands in the Midwest. In 1933 the Soil Erosion Service was created as a temporary agency in the Department of the Interior to carry out provisions of the National

Industrial Recovery Act related to soil erosion prevention. In 1935 the Soil Erosion Service was transferred to the Department of Agriculture. Subsequently, Congress passed the Soil Conservation Act of 1935, which established the Soil Conservation Service (SCS) in the USDA for the development of a program of soil and water conservation. In 1993 the Soil Conservation Service became the Natural Resources Conservation Service. In the same period, the Agricultural Stabilization and Conservation Service (ASCS) became the Consolidated Farm Service Agency.

The recognition that soil and water problems are interrelated had merits, but later policy implementation was complicated by interagency conflicts. Frederick Steiner suggests that such conflict grows out of the multiple goals of soil conservation policy in the United States. The policy is designed to serve goals related to "control of erosion, the retirement of marginal land, land-use planning, reduction of water pollution, flood control, and watershed management." At the same time the policy is supposed to meet other goals designed to benefit farmers. Steiner, talking about the SCS and ASCS before they changed their names, observes that "[t]his public versus private dichotomy has led to specific agenda goals which have aggravated interagency conflicts." The SCS had primary responsibility for the "public goals" and the Agricultural Stabilization and Conservation Service had the major responsibility for the "private goals" through its cost-sharing and supply-control programs. The EPA was involved in implementation of programs directed toward both sets of goals. The primary difficulty according to Steiner is that the SCS abandoned its role as regulator and began competing with the Agricultural Stabilization and Conservation Service to see which is the greater benefactor of the farmer.[4]

The NRCS basically relies on voluntary compliance with its guidelines through the state and local soil and water conservation districts. Consequently, American soil conservation policy lacks teeth. This helps to explain the fact that although the United States has spent billions on soil erosion control efforts through federal programs, these programs have only had nominal success in reducing cropland losses, according to the General Accounting Office (GAO).[5]

A major concern in soil conservation policy is the classification of soils for land use and planning. In cooperation with other federal and state agencies the NRCS conducts soil surveys and makes classifications relative to public and private lands. Soil in a given area is surveyed and classified according to type and quality. This work includes national appraisal of the quality and quantity of agricultural and water resources. Interactions among soil characteristics and land uses are assessed in predicting the behavior of various widely distributed soil types. Although the survey classification is oriented toward agricultural productivity, it is valuable for other private and governmental activities. Soil surveys are used for urban and county planning, watershed and water project activities, wildlife management, and construction projects. Environmental impact studies conducted under the National Environmental Policy Act also include soil surveys.

Soil conservation, water conservancy, and drainage districts are legally constituted units of local government established to administer soil and water conservation work within their boundaries. There are about 3,000 soil conservation districts that include about 99 percent of the farms and ranches and more than 95 percent of the agricultural land in the United States. In many instances, soil conservation policies and activities require attention from various governmental units at the county and metropolitan levels. Although the states prescribe a variety of legal procedures for their operations, each soil conservation district is self-governed and has the authority to enter into working arrangements with other governmental agencies and private concerns to carry out soil and water conservation programs. A major weakness with the programs of the NRCS and other federal agencies concerned with soil conservation is that they merely provide technical and economic assistance through the local district offices.[6] The GAO found that SCS staff (and so, probably those of its successor, NRCS) were constrained in the enforcement of conservation program regulations because they lacked clear authority to require state and county offices to follow their guidance.[7]

Although the means has been established to develop a potentially effective nationally coordinated approach through the districts, it has not occurred. The soil conservation associations that regulate the districts are composed of officials who represent the private interests of farmers and ranchers. This complicates efforts to incorporate a coherent national perspective on soil and water conservation into state and local zoning and planning policies. As a result, soil survey inventories continue to demonstrate serious soil loss and land abuses at the state and local level. Because of economic and social pressures from their peers, state and local officials tend to make poor land use decisions predicated on short-term gains with negative implications for long-term environmental quality. Unplanned and unzoned development projects lead to the daily removal of hundreds of acres of land from productive use.

The highly publicized Food Security Act (FSA) of 1985 is another combined voluntary compliance and economic benefit approach of the federal government to reduce soil erosion. Robert Gray of the American Farmland Trust heralded the act as "the most fundamental change in soil conservation policy of the past 50 years." The provisions of the act provide incentives and penalties directed toward the removal of highly erodible cropland from production. Gray, however, expressed serious reservations about the implementation of the act:

> USDA is under enormous pressure to increase farm income. Crop prices are low, and land values continue to fall. If 45 million acres of the most highly erodible cropland are enrolled in the conservation reserve at $60 per acre, more than $10 billion will reach farmers in the next five years. But if the reserve is managed to funnel a share into every county throughout the nation, regardless of a county's erosion problems, it will become an expensive, ineffective program and run the risk of being terminated as budget cuts are made. . . . To protect the validity of these programs, a long-term monitoring and evaluation program is needed.[8]

Past economic benefit programs like the "payment-in-kind" (PIK) programs have not succeeded in meeting conservation goals. Individual farm owners have been permitted to reap economic benefits while violating the conservation intentions of the laws. As Gray highlights, the success of the Conservation Reserve Program will depend on a vigilant monitoring program to assure that appropriate lands are selected for reservation and that they remain in that status. In spite of Gray's concerns, the NRCS reports that the CRP appears to be a highly successful program. The 1992 Natural Resources Inventory estimates that the CRP accounted for an estimated 369 million ton reduction in soil loss from croplands.[9]

One feature that differentiates the 1985 act from previous efforts is that it required the planting of trees on 5 million acres. Former secretary of agriculture John Block points out that "At a rate of 500 trees per acre, that amounts to more than 2 billion new anchors for our fragile soil." The Conservation Reserve Program serves multiple conservation and environmental purposes beyond agricultural needs. Initial estimates indicated that the program would reduce soil erosion by 750 million tons each year it is in operation. Stream and river sedimentation would be reduced by 200 million tons annually. The application of pesticides would be cut by about 60 million pounds annually. More and better habitat is also provided for fish and wildlife. And eventually landowners will be able to harvest the trees.[10] In actual practice, the Natural Resources Conservation Service estimates that the average annual soil erosion on Conservation Reserve Program lands declined from 12.5 tons per acre in 1982 to 1.5 tons per acre per year in 1992. Between 1982 and 1992, 36.4 million acres were enrolled in the CRP.[11]

One of the primary tools for the various soil conservation easement programs is the classification of soils. Linda Lee and Jeffery Goebel question whether the national land classification system used by the NRCS will permit proper classification of "highly erodible cropland." The land capability class-subclass system provides designations of the erosion potential of lands based on physical factors from a universal soil loss equation. Lee and Goebel demonstrate that the land capability system "does not provide a uniform erodibility ranking across regions or a precise categorization of highly erodible soils."[12] From a policy perspective, improper classification of lands could substantially lessen the positive environmental contributions of the Conservation Reserve Program. However, Lee and Goebel suggest that combinations of classification systems or alternative systems may provide better classification. This would also minimize any negative impacts that might result from misclassification.

In order to reduce the possibility of misclassification of soils and to improve the overall effectiveness of soil classification, the NRCS established a new data management facility. The National Soils Data Access Facility (NSDAF) project provides access to various national soils databases for purposes of analysis. Soil attribute databases, soil geographic databases, and the National Soil Information System (NASIS) can be accessed and downloaded through the Internet.[13] The system is designed to manage and maintain soil data from collection to dissemination.[14]

The Food Security Act of 1985, as amended in 1990 and 1996, represents a recognition of the profound effects that agricultural practices have on environmental quality and the long-term sustainable use of land and water for the production of food and fiber. In the United States, farming and ranching directly affect over 40 percent of the land area, or more than 960 million acres. Poor tilling and harvesting, excessive use of pest controls (herbicides and insecticides), excessive reliance upon chemical nutrients, overgrazing, devotion of large tracts of land to monocultural (one-crop) production, and other unwise agricultural practices cause pollution and erosion of soil, degradation of water quality, and generally negative effects on the environment.[15]

Moreover, wetlands and other wildlife habitats are irreversibly lost when converted to cropland use. The North American Wetlands Conservation Act of 1989 (P.L. 101-233) required the U.S. Fish and Wildlife Service to make an assessment of wetlands lost in the United States between the 1780s and the 1980s. The Service estimates that of the 221 million acres of wetlands in the lower forty-eight states in the 1780s, only 104 million acres remained by the 1980s. This constitutes a 53 percent loss from the original acreage. Only 5 percent of the land surface in the lower forty-eight states could be classified as wetlands. Ten states have lost over 70 percent of their original acreage, with California losing the most: 91 percent.[16] Of these wetlands, only about 10 percent are under federal and state protection with the remainder in private ownership.

Wetlands provide a rich habitat for a wide variety of plant and animal species, and loss of wetlands seriously affects wildfowl populations. The sponge-like actions of the water flow of wetlands helps cleanse water by filtering out pollutants and impurities. Wetlands reduce erosion by trapping sediments and functioning as storage areas for floodwaters. They also produce cash crops that include blueberries, wild rice, marsh hay, and cranberries.[17]

The Natural Resources Conservation Service provides financial and technical assistance to private landowners to reduce the destruction of wetlands and where possible to restore them. The results are measurable. NRCS reports that the Conservation Reserve Program has helped to increase nesting success for North American waterfowl by one-third since 1985, due to farmer and rancher participation.[18]

Another major concern for soil conservation is the current rapid conversion of prime agricultural lands for urban development, highways, and water projects. From 1950 to 1992, the total amount of farmland declined from 1.16 billion acres to 950 million acres. At the same time, highly specialized, mechanized, labor-efficient, and capital-intensive farms of larger size have been displacing medium-sized farms.[19] But regardless of size or location of farms, soil erosion is one of the more serious environmental problems for agriculture everywhere.

Present rates of soil loss cannot be continued in many industrialized countries without serious implications for agricultural production. Soil losses from wind and water erosion must be reduced substantially in order to sustain crop production indefinitely at present levels.

David Pimentel argues that erosion is reducing agricultural productivity so rapidly that about 30 percent of the world's arable land has had to be abandoned over the past 40 years.[20] More than 140 million acres (one-third of the cropland base of the United States) are eroding at high rates that seriously threaten productivity.[21] Massive soil loss in the United States is also of concern internationally, since about one-third of all cropland is devoted to producing food for foreign markets. Further, the National Wildlife Federation observes that, despite four decades of soil erosion control, the United States experienced 35 percent more erosion in 1981 than in the Dust Bowl days of the 1930s. Approximately 8 billion tons of soil eroded in that year. The cause was not drought. Rather, poor land use decisions involving overgrazing, plowing of grasslands, removal of windbreaks, and plowing in the fall all contributed to the erosion problem.[22] Blaming farming practices for the deteriorating Plains landscape, William McGuinnies says, "Another bust is going to come. The only question is when. Large acreages of native grasslands on sites not suitable for farming are being plowed." He reports that the washing and blowing away of top soil is occurring two and one-half times more rapidly than in the Dust Bowl days.[23]

Farmers and landowners are generally aware of soil conservation practices. Moreover, subsidies or technical assistance for controlling erosion are often provided for soil conservation activities. The Conservation Foundation suggests, however, that the major reason for the continuing and growing problem of erosion is cost. Measures to control erosion almost always impose some costs on the farmer. Contour plowing is more expensive than plowing long, straight rows. Leaving hedgerows to control wind erosion and leaving stream banks and other lands to control water erosion mean reducing the amount of land in production. Other economic factors such as interest rates, farm tenure, and short-term profitability can also lead farmers to decide against erosion control.[24]

The Conservation Foundation indicates that some policy options along economic lines might be (1) researching soil conservation measures and innovations relative to farm productivity, (2) making conservation measures cheaper through economic incentives, and (3) making erosion more expensive through laws and "cross compliance" or withholding of government benefits.[25] Because of obvious and apparently high up-front costs, long-term considerations receive too little emphasis where soil conservation is concerned.

The 1985, 1990, and 1996 farm bills have partially addressed some of these concerns. The 1985 Food and Security Act (FSA) enacted highly erodible land (sodbuster) provisions that required producers who farmed such land and who received farm program benefits to follow an approved conservation plan or face loss of farm program support benefits. Eligible lands were more clearly defined by the 1990 Food, Agriculture, Conservation, and Trade Act (FACT). And producers were given greater flexibility by the 1996 Federal Agricultural Improvement and Reform Act (FAIR), in order to decrease impact on farm operations. Swampbuster provisions also were included in the 1985 FSA, which discouraged farmers from destroying wetlands, again, under risk of losing farm benefits. However, this restriction was significantly weakened by FAIR in 1996 through a change in the definition of what constitutes a "prior converted wetland." The 1985 bill also created the original Conservation Reserve Program to convert highly erodible lands to tree production or grasslands. The 1990 FACT created the Wetlands Reserve Program (WRP), which authorized the establishment of permanent and long-term easements to protect wetlands. The 1996 bill (FAIR) expanded eligibility to include land that maximizes wildlife benefits also, but replaced Fish and Wildlife Service consultation with consultation with the state technical committee. A Forestry Incentives Program (FIP) and a Wildlife Habitat Incentive Program (WHIP) were also created by FACT.[26]

In an experiment that appeals to the economic self-interests of farmers, the state of Wisconsin created an experimental program to motivate landowners to implement soil conservation practices through a property tax credit. If landowners adopted measures to control soil erosion within the experimental site locations, their property taxes were reduced. All of the implementation expenses were assumed by the landowners. After three years the program succeeded in increasing the amount of cropland adequately protected from erosion from 49.8 to 85.6 percent in the treatment areas. Also, the amount of soil loss in the treatment townships declined 72.4 percent. From a policy perspective, this has advantages in that the program could be instituted and managed by local governments with very little assistance from state and federal governments. The landowners have the flexibility to choose the conservation methods that best suit their economic situation. Funds for the program could originate in a property tax assessment that acts as a penalty for nonparticipants and a reward for participants.[27]

On the international level, the CEQ and the Department of State point out that "political difficulties cannot be overemphasized. Often solutions to soil problems will require resettlement, reduction of herd sizes, restrictions on plantings, reforms in land tenure, and public works projects that will fail without widespread cooperation from the agrarian population." The immediately apparent economic and social costs will far outweigh distant benefits.[28]

Soil erosion poses both on-farm and off-farm problems. The former mainly relate to lost agricultural productivity, while the latter pertain to the effects of erosion on the environment downstream or downwind. Scientists estimate that it takes from 300 to 1,000 years to form one inch of topsoil. At a "tolerable" rate of erosion (which may vary from one site to another) of five tons an acre per year, only thirty-three years are required to lose an inch of topsoil, the soil layer that is the highest in organic content, nutrients, and water-holding capabilities.[29] In the United States the national average annual loss from water-induced surface and rill (small, carved) erosion is over 4 billion tons a year. Almost one-half of the cropland currently cultivated consists of soil classified as having high potential for erosion. Steep slopes, easily erodible soils, and other causes combine with water and wind to cause severe soil erosion problems.[30]

On an international scale, soil erosion is a particularly severe problem, especially in those countries where tropical forests are being depleted at a rapid rate. Tropical forest soils are fragile, of poor quality, and unsuitable for agriculture. Almost all nutrients are found in living vegetation rather than in the soil. With deforestation the thin layer of topsoil quickly washes away. The soils rapidly deteriorate and lose their fertility, leaving hard, packed clay or sand that is likely to support crops or grazing for only very short periods of time. Thus, the very productive land of tropical forests becomes virtual desert when its soil is exposed. Tropical forests shelter the soil from rain, stabilize it, and prevent erosion. In tropical forest areas soil erosion renders lands unfit for agricultural and other uses. Many tropical countries that are not willing to give up any of their land are watching thousands of tons of soil wash away each year because of deforestation.

Faced with croplands eroding faster than topsoil is being formed, greater reliance for agricultural productivity is being placed on chemical fertilizers and new crop varieties to offset these natural soil losses or deficits. This reliance presents particularly severe economic difficulties for less-developed countries. Not only is it costly to maintain agricultural productivity on eroded lands under these methods, but there is evidence that productivity maintained by artificial means may drop at some point by rapid amounts. Coupled with the increased use of pesticides on cropland everywhere, the end result may be the growing of crops on infertile soil that is saturated with chemicals. An ounce of natural and fertile soil, for example, contains millions of bacteria, algae, fungi, protozoa, and small invertebrates like worms and arthropods that play essential roles in soil fertility and structure. Yet pesticides and other chemicals destroy these beneficial organisms and their complex habitats, which results in soil depletion, fertility decline, and soil erosion.[31]

Furthermore, concerns are growing about the role of pesticides as endocrine disruptors. Research indicates that synthetic chemicals, most notably the organochlorines of pesticides, have affected the production of hormones in alligators, fish, and gulls, altering their reproductive systems. Theo Colborn, a zoologist with the World Wide Fund for Nature, is quoted as saying, "We're seeing this same pattern of anomalies linked to endocrine disruptors across a whole suite of animals and humans around the world."[32]

Off-farm problems of soil erosion involve sediment that ends up in streams, rivers, lakes, and reservoirs. Soil sediment is the largest single water pollutant (or resource out of place). It exceeds the sewage load by 500 to 700 times. About one-half of all sediment comes from agricultural land.[33] Sediment can increase flood damage, interfere with water transportation, and destroy fish spawning beds and other wildlife habitat in streams and rivers. It can also settle into the bottoms of lakes and reservoirs, thereby reducing their capacity to prevent floods or to store water. Serious problems also arise from the chemical pesticides and fertilizers contained in the soil sediment. Pesticides can be toxic to various forms of life, including human and fish life. Fertilizers can cause acceleration of eutrophication of lakes and reservoirs, particularly through stimulation of algae blooms.[34]

Another policy concern is pollution of soil and water through improper use of fertilizers, as well as pesticides. These problems are in large measure due to growth of monocrops such as wheat, rice, or corn that extend over large areas and that involve little diversity or crop rotation. To sustain these unnatural monocultures, vast amounts of chemical fertilizers and pesticides are used that produce depletion and erosion of the soil. However, methods are available for mitigating these effects. One such method is farm planning based on the Sustained Yield Agro-Ecosystem, which uses the ecosystem as its planning unit while determining and working within the long-range carrying capacity of the land. In this process, "(1) the chemical nutrients removed by crops are replenished in the soil, (2) the physical condition of the soil by land utilization type is maintained, which means that the humus level in the soil is constant or increasing, (3) there is no buildup of weeds, insects, and pathogens, (4) there is no increase in soil acidity, toxic elements, and (5) soil erosion is effectively controlled."[35]

Other approaches to reducing soil depletion and erosion involve Integrated Pest Management (IPM), minimum tillage, and organic farming. Basically, IPM is a system of controlling pests through a combination of techniques that reduces the use of pesticides. The system emphasizes the use of biological controls, modifying environmental conditions, monitoring carefully, and utilizing chemical pesticides only when needed as part of an overall control strategy.[36] Minimum tillage is still another mitigating alternative that, although often requiring more herbicide control, can provide effective controls over soil erosion by reducing the times when the soil is disturbed.

A greater understanding of the economic and environmental effects of using alternative farming methods has grown out of the increased use of organic farming and other combinations of methods involving fertilizers and pesticides. Organic farming, for example, is regarded as the best agricultural-production system for controlling erosion, minimizing water pollution, and conserving energy. Maintenance of diversity in production systems is needed in agriculture.[37] Encouraging diversity of genetic strains also improves resistance to pest attack.

Soil conservation policy is closely related to agricultural policy, as well as to economic motivations for conservation measures from the private sector. In spite of conservation efforts, fertile topsoil continues to be eroded by wind and water at an unacceptable rate worldwide. More conservation-oriented tools, education, and technical assistance need to be offered to farmers to curb the loss of topsoil where poor management practices have led to declining crop production and rising costs. These tools should be backed up by binding multiyear contracts between farmers and national governments. The contracts should stipulate effective conservation practices in order for farmers to receive benefits, including federal financial assistance. The tools should also support conservation incentive programs, integrating these with commodity support programs and efforts to encourage soil conservation.[38]

Results of the National Agricultural Lands Study suggest that the federal government should establish a comprehensive national policy for the protection and conservation of the nation's agricultural land base. For example, federal agencies should review their programs that affect the use of agricultural lands to ensure that federal action does not unnecessarily encourage conversion to nonagricultural uses. Federal agencies need to initiate actions to mitigate the negative impacts of federal activities on prime agricultural lands.[39] The continued lack of strong, comprehensive policies for soil conservation in the United States will significantly reduce its ability to provide foodstuffs in sufficient quantities for both America and other parts of the world.

Mineral Policy

Public lands managed by the USDA Forest Service and the Bureau of Land Management are available for mineral exploration and development in accordance with the General Mining Law of 1872, Mineral Leasing Act of 1920, and Surface Resources Act of 1955, as well as other laws pertaining to mineral leasing. The Department of the Interior through the BLM and Minerals Management Service (MMS) administers most of the leasing and mining of minerals on federal lands. Under the Outer Continental Shelf Lands Act (OCSLA) MMS manages the leasing programs for oil, natural gas, and deep sea minerals. MMS is responsible for leasing in the Exclusive Economic Zone (EEZ), beyond state jurisdictions out to the two-hundred-mile limit on the Outer Continental Shelf.

Generally, the Mineral Leasing Act and the Surface Resources Act give the federal government more control over mineral development on federal lands than the General Mining Law. "Under the leasing laws, rights to minerals are granted at the discretion of the secretary of the interior, who has clear authority to deny the development of leasable minerals and to place conditions on the lease to protect surface resources. The lessee does not get rights to purchase the overlying land and is required to pay an annual rental fee and a percentage of his or her proceeds to the federal government."[40] National policy with regard to mineral exploration and development on national forestlands is contained also in the Mining and Mineral Policy Act of 1970 and the National Materials and Minerals Policy, Research, and Development Act of 1980. Both laws encourage mineral exploration and development while minimizing adverse environmental impacts.[41]

Emphasizing the crucial need to develop an effective solid minerals policy, the NWF consistently has attacked legislative and administrative efforts to establish mining as a dominant use on public lands. A critical necessity if such advocacy is to be successful would be reform, modernization, or perhaps elimination of the General Mining Law of 1872.[42] For so-called hardrock minerals, the General Mining Law stipulates that all valuable mineral deposits on public lands shall be open to exploration and purchase. Lands upon which minerals have been found to be economically viable (a very loose standard) are to be open to occupation and purchase. The NWF argues that a new national policy is urgently needed to control exploration, development, and use of nonrenewable mineral resources. The NWF recommends that all future mineral exploration should be done under leasing arrangements, whereby surface values of the land may be protected for various uses such as grazing, recreation, timber, and wildlife. Unreclaimed mined lands and consequent destruction of wildlife habitat would be prohibited.[43]

With respect to locatable minerals, the USDA Forest Service and the BLM are responsible for managing and mitigating negative impacts associated with mineral exploration and development. These agencies are also expected to address impacts associated with exploration and development of leases for fossil fuels such as oil, gas, and coal. Both agencies have the authority to impose constraints upon hardrock and leasable (fossil fuel) mineral developers. But USDA Forest Service regulations, in particular, emphasize the primacy of national energy and strategic mineral needs over other interests.[44]

In reference to the Nonfuel Minerals Domestic Policy Review (a mineral policy study by a national commission), the CEQ identifies several pertinent issues pertaining to minerals management:

> Federal government planning and decisionmaking involving nonfuel minerals is uncoordinated. There is no formal process of effective participation by nonfuel mineral policy makers in developing broader

national goals that significantly affect minerals supply and consumption. The data gathering and analysis capability required to support nonfuel mineral policy making in the federal government is deficient or nonexistent. There are serious deficiencies in agency forecasting and environmental impact data gathering and analysis capabilities. Efforts to control environmental pollution problems associated with the mining and processing of nonfuel minerals have been successful in limiting some pollutants. . . . Current federal nonfuel minerals research and development programs may not be adequate to address the current and future generic national problems that confront the U.S. mineral industry.[45]

There are other major problems with government mining controls. The National Academy of Sciences cites "reclaiming disturbed land for anticipated postmining uses, balancing mining against other land uses, and protecting land from abuses under provisions of existing mining laws." In response to these problems the Academy suggests "education and technical assistance, economic incentives, regulation for specific postmining effects, and public ownership of surface rights."[46]

Many land abuses are traceable directly to the General Mining Law of 1872, which authorizes mineral development but does not require mitigation of environmental impacts associated with mining activity. All serious efforts to revise this law have been blocked in Congress through the intense lobbying pressures of powerful mining interests, intent upon preserving the General Mining Law. The Western mining state lawmakers concerned about greater economic benefits for their states continually win out over other members of Congress who want to get more money from those who use federal land for their own economic benefit.[47]

We do not "consume" most of the mineral resources that are used. Materials may become dispersed or oxidized but the rate of loss is not high, particularly for metals. It is possible to reclaim or recycle many minerals without incurring exorbitant energy costs. In many instances, the energy costs of converting raw materials into usable form are much higher than the costs of recycling materials. It is quite likely that recycling resources eventually will be seen as a viable alternative to development of both renewable and nonrenewable resources. Escalating costs of raw materials, conversion of raw materials into usable form, pollution and waste disposal may force changes in our ways of doing things.[48]

The recycling and recovery concept recognizes that nonfuel minerals are essentially nonrenewable. That is, the quantity and availability of these minerals is neither increased nor maintained over time. Each mineral unit used, therefore, diminishes the total stock available unless at least a portion of that stock can either be recycled or recovered.[49] But, following the physical law of conservation of mass, materials are not really consumed in any ultimate sense. Rather, they are converted from usable to residual form.

Dennis and Donella Meadows note that "every year (globally) the human economy mobilizes about 1 billion tons of minerals from earth and discards about the same amount as solid waste. Because the deposits from which these minerals are taken are becoming less concentrated, minerals are obtained with an increasingly high cost in energy, capital, labor, and environmental damage." They indicate that the world's known mineral reserves have remained fairly constant, with discovery rates similar to production rates. But the discoveries have been poorer grades of minerals in lower concentrations. This results in the expenditure of much more energy, labor, and capital with more extensive environmental damage. There is a corresponding rise in price. The existence of such conditions provides a strong argument for careful use and thorough recycling of the earth's mineral resources. By 1985 some countries were already recycling as much as 70 percent of their discarded iron and 30 percent of their aluminum, and even these percentages could be increased substantially.[50]

Change in the mining and mineral industries occurs in response to economic, social, and technological changes. For example, the importance of an ore is related to the social and economic demand for the metal that it produces. The direction of technological change also may affect all phases of activity from exploration, through development and production, to use. Technology or the "state of the art" also affects social needs and wants relative to minerals. Consequently, technology may be the most important and most variable determinant of mineral policy.

The U.S. mining industry faces a very competitive international market that forced many companies to close or sell properties in the 1980s.[51] Competition for American companies occurs from overproduction by nations that rely on export of mineral commodities to support their economies and service their international debt, high domestic labor costs compared with those of less developed countries, and the switch from metals to plastics, glass, and fibers in industrial processes. Nevertheless, the value of nonfuel minerals produced in the United States during 1997 was estimated at $39.6 billion, an increase of $16.5 billion from 1985. For nonfuel minerals there has been an overall increase in value for several consecutive years.[52]

Pressure to import minerals for domestic use or stockpiling is further increased by the strategic minerals policy of the United States. This policy mandates stockpiling of rare or scarce minerals for defense-related purposes. The National Defense Stockpile has the strategic minerals necessary to sustain a conventional war for three years while maintaining the economy. Analysis of the evolution and prospects of national materials policy, however, indicates that stockpiling seems to be decreasing in relative importance, with a greater emphasis on using materials productively rather than hoarding them against a possible future shortage.[53] In fact, as a result of the decline in the possibility of global war with the breakup of the Soviet Union, the Defense Department announced in 1991 that it was going to reduce the stockpile by about 40 percent.[54]

The world may be running out of all major nonrenewable minerals that are necessary for highly industrialized societies. Jeremy Rifkin argues that the United States is responsible for consuming much of the remaining stock of the world's precious minerals, as the above figures support. He points out that many experts predict that the various economies of the earth will have exhausted about one-half of the world's useful metals in seventy-five years or less. Rifkin does not consider seabed mining for nonferrous metals like manganese nodules to be feasible for satisfying exponentially growing demands for these materials. Furthermore, seabed mining is likely to have significantly dangerous environmental impacts.[55] Still, the Minerals Management Service has initiated environmental impact statements for the leasing of minerals at a number of sites on the Outer Continental Shelf (OCS) under authority of the Outer Continental Shelf Lands Act of 1953 and subsequent amendments. Potential leases include polymetallic sulfide minerals in the Gorda ridge area and the cobalt-rich manganese crusts near the Hawaiian Islands and Johnston Island.[56]

In an assessment of the U.S. and world mineral positions between 1985 and the year 2000, the U.S. Bureau of Mines estimates that there should be sufficient supplies of minerals to support rising standards of living for most of the world through 2000. This estimate is predicated on an assumption that no major political or social upheavals occur that would interfere in the production of strategic minerals worldwide.[57]

The trends for nonfuel minerals show constant increases in demand and consumption that will double from 1980 to 2000. Less developed countries in Latin America, Africa, and Asia are projected to share only a small percentage of nonfuel mineral uses similar to current patterns. These regions contain three-fourths of the world's population. But the one-fourth of the world's population that inhabits industrialized countries is projected to continue to absorb more than three-fourths of the world's nonfuel mineral production. Although projections reveal no dramatic mineral exhaustion problems, further discoveries will be needed to meet expected demands.[58] Also, the creation of synthetic minerals has increased domes-

tic supplies of certain strategic minerals. Synthetic minerals created so far
include graphite, quartz crystal, silicon carbide, electrolytic manganese
dioxide, and magnesia from seawater.[59]

A momentous event occurred on March 10, 1983. On that date, President Reagan made the "Exclusive Economic Zone (EEZ) Proclamation."
With this proclamation the United States established "sovereign rights"
and control over economic resources in the offshore area that encompasses some 3.9 billion acres surrounding U.S. territories. In comparison,
onshore acreage of the United States amounts to 2.3 billion acres. Much of
the reasoning for the proclamation can be found in the critical and strategic mineral resources that exist in the delineated areas. Three essential
hardmineral resources are found in the EEZ: manganese nodules, cobalt-
ferromanganese crusts, and polymetallic sulfides.[60] All three produce strategic minerals that are produced primarily in three countries of the world
at the present time—South Africa, Zaire, and the former Soviet Union.[61]
Consequently the EEZ could be extremely important to the United States
in reducing dependence upon foreign imports, once seabed mining techniques have been established.

By the year 2050 the United States is projected to exhaust extractable
quantities of tin, columbium, fluorite, sheet mica, strontium, mercury,
chromium, and nickel in known locations. It will also lead to increased
reliance on foreign imports of most key minerals. This will raise prices and
enhance the bargaining leverage of the mineral exporting countries. Such
leverage will be similar to the power exercised by OPEC nations that possess ample oil reserves.[62]

In the United States federal lands are by far the major sources for both
fuel and nonfuel minerals. More than 90 percent of the federal lands are
located in eleven contiguous western states and Alaska, and these states
have produced in excess of 90 percent of most domestic nonfuel minerals,
including copper, mercury, silver, nickel, molybdenum, potash, and lead.
The Public Land Law Review Commission (PLLRC) recommended that
"public land mineral policy should encourage exploration, development,
and production of minerals on public lands . . . [and maintain] a continuing invitation to explore for and develop minerals on public lands."[63] The
PLLRC also recommended that the principle of "maximum economic efficiency" apply to lands classified as being of principal value for their minerals (in direct conflict with the principle of multiple use of public lands).
The PLLRC further recommended a "dominant use zoning" principle that
would essentially turn over mineral lands to the mining industries and
would recognize no other rights or uses.[64] Some of the top officials of the
PLLRC also held positions with the mining industry.

Even the National Park Service and other agencies that have wilderness lands are to accommodate mineral exploration and development where it can be permitted. Thirty-three of the National Park System units have at least one mining activity occurring on them. As of 1996, 817 mineral development operations were ongoing in national parks; 15 for hardrock metals, 28 for sand, gravel, and soil, and 709 for nonfederal oil and gas.[65]

The special case of the general mining law of 1872

In the United States mineral policy on public lands to a large extent grows out of the General Mining Law of 1872. When the PLLRC in 1965 developed recommendations for legislation and policy pertaining to mineral development on public lands, the commission endorsed continuation of the mining law with modest amendments reflecting mining industry preferences. While noting some of the adverse environmental impacts of mining, the commission indicated that public lands generally should be open to exploration, development, and production, regardless of where the minerals are located and regardless of environmental costs.[66] This recommendation was made, despite the fact that most land management personnel are unsatisfied with the mining law because it does not make adequate provision for management of mining operations by agencies responsible for administering public lands.

The CEQ makes the following observation about the mining law:

> Under the 1872 Mining Law, still in force today, prospectors and miners of gold, uranium, iron, copper, and all other hard rock minerals have free access to the vast acreage in the public domain. Unlike all other commercial users, miners can enter federally owned lands without a permit or license. They can claim any marketable body of ore that they find, and they can dig the ore and sell it without paying any royalty to its owners, the nation's citizens. . . . The over century-old mining law does not allow adequate control of environmental damage caused by mineral development. Nor does it permit orderly planning for use of public lands. . . . Over 60 percent of public lands are open to mining. The 1872 Mining Law had one purpose, to promote mining.[67]

The 1872 General Mining Law expresses the frontier sentiment and development of the West. However, as the frontier closed and land demands increased, along with conflicts, all other uses, including logging and grazing, were subject to controls and restraints. Coal and other fuel minerals are now leased under the 1920 Mineral Leasing Act and other similar legislation. Only hardrock mining remains exempt from substantial controls and restraints. Over the years legislation has been introduced to update and revise the 1872 law with little success. Unsuccessful provisions in recent legislative attempts include measures to obtain fair market value for surface lands, royalties, ensure protection standards, and a reclamation fund for abandoned mines.[68]

One of the most contentious issues in the reform efforts is a proposal to change over to a permiting system for hardrock minerals, to replace the existing location-patent system that may have been more appropriate in the 1800s. Under the location-patent system, miners are given full ownership to the land once a "patent" is issued. With a permiting system the federal government would retain ownership of the surface and mineral rights and maintain discretionary control over mineral development activities.[69] These and other provisions are designed to keep discretion over mining on public lands in the hands of the public owners' agent, the federal government.

The 1872 General Mining Law provides that locators may establish rights to federal land mineral deposits merely by discovery without prior administrative approval, that is, by merely filing a claim. The locator may then acquire legal title and ownership to the land within his or her claim(s) if the locator proves the claim to be economically viable and after the locator secures a federal "patent" at a small fee ($2.50 to $5.00 an acre). Even without a patent, the locator may produce minerals without paying royalties to the government. Qualifications as to what constitutes a claim are ambiguous, and, prospectors exploit the locational provisions of the 1872 General Mining Law in order to obtain public lands for other purposes such as vacation cabin sites, private fishing areas, land speculation, and even illegal activities like marijuana production.[70] Many abandoned or invalid claims existed for over a century because the land management agencies lacked the personnel and resources to check claims. However, FLMPA changed the rules of the game in 1976. The act required hardrock mineral mining claimants to file notices with the federal government initially and each year afterward. The nominal filing requirements enabled the federal government to begin to obtain control in this area for the first time. The mining industry reacted quite negatively to these requirements and attempted to have them removed by the courts. However, in United States v. Locke, 471 U.S. 84 (1985) the Supreme Court clearly acknowledged federal government regulatory powers. Even though the Lockes had maintained a mining operation for over thirty years, because they failed to meet the December 30 filing date by one day, the Supreme Court ruled that they should lose their mining rights. However, Congress subsequently passed a bill to restore mining rights to the Lockes. In 1993-94, the imposition of the filing requirements reduced the number of claims by more than one-half, to less than one million.[71]

Although the USDA Forest Service does have some authority to prohibit claims and mining on sensitive sites or where serious conflicts occur with other uses, this power is rarely used. Nevertheless, the USDA Forest Service is technically responsible for surface impacts that result from mining. Generally the USDA Forest Service will automatically approve mining activities while negotiating some specific site agreements with the mining party to mitigate or "reduce" the mining impacts. Approval and consequent negotiations will occur, despite the severity of mining impacts and whether or not mining activities would eliminate other important uses. Although mining is an accepted use of public lands, the present mineral policy under the law of 1872 permits too many unnecessary abuses of public lands and the environment.

Among the negative impacts are (1) creation of waste dumps and slides, (2) deterioration of quality and quantity of surface and groundwater, often through pollution, (3) loss of soil structure and organic matter in surface soils, (4) elimination of vegetation at the mining site without possibility of reclamation, (5) pollution of fisheries, (6) disturbance of habitat for threatened and endangered species, (7) release of ore dust and other particulate matter into the atmosphere, (8) infusion of transient workers into local communities, often requiring increased social and law enforcement programs, (9) changes in life-style of rural communities, (10) increased demand for water and sewers, (11) increased demand for and deterioration of existing roads associated with high traffic volumes, (12) interference with game habitat and migration patterns, (13) increased recreational use and demand for recreational services, (14) visual alteration of landscape, (15) noise pollution from mining activities, and deterioration of cultural resources.[72] This is just a partial list of the many negative features associated with mining activities on federal lands.

In order to control and mitigate the impacts of mining exploration, location, and development on public lands, more authority and discretion are needed by federal agencies in their allocation of leasing, granting of claims or patents, and planning. Enhanced authority is necessary for the protection of diversity in surface uses of federal lands. Such protection is impossible for land management agencies attempting to function under the existing 1872 mining law. The General Mining Law of 1872 contains none of the environmental protection requirements of modern land management legislation. Furthermore, the mining law takes precedence over subsequent environmental legislation. Serious ecological alterations and pollution have resulted from the 1872 act, which lacks land protection and reclamation provisions. The Sierra Club, Wilderness Society, and other environmental organizations support legislation that would do the following:

1. Balance mineral use with other resource values through agency land use plans with provisions for departmental authority to grant or deny the privilege to mine on public lands under a system leasing mineral development rights, rather than giving those rights away
2. Provide for departmental and public review of mining proposals at both the exploration and development phases of each project
3. Provide that mitigation and reclamation requirements be spelled out as lease conditions with guidelines provided by regulation
4. Provide for rate of return to the public for use of its lands and minerals.

Insufficient funding for outdoor recreation and wilderness programs continues to be a major problem. Congress refuses to adequately fund programs for recreation activities on federal lands. Congress also has not been willing to provide full appropriations for the Land and Water Conservation Fund (LWCF), which is the major revenue source for outdoor recreation in the United States. The LWCF was created by the Congress in 1964 to provide funds for the federal government to acquire land for parks and refuges as well as to provide grants to state and local governments for outdoor recreation projects including land acquisition. The money for the LWCF is derived from royalties on the sale of oil and natural gas on the Outer Continental Shelf. Although Congress is authorized to appropriate up to $900 million per year for LWCF-associated projects, it has appropriated only a fraction of that amount each year, typically about one-third. In its first budget request, the Reagan administration requested no money be appropriated for LWCF.

The lack of funds for acquisition of new lands has had significant implications for both the federal government and the states. Both levels of government are experiencing increased demand for outdoor recreation opportunities from a better educated populace with more expendable income and more leisure time. With a greater percentage of the United States population now living in urban areas, there is a natural propensity for that population to want to escape from the confines of the city and to go out into the countryside to experience nature. Consequently, there is serious overcrowding at many national parks and forests. A similar pattern occurs with state parks and natural areas. Carrying capacity often is exceeded, resulting in depreciated recreational experiences for many participants. With increased use there is increased wear and tear on facilities. Significant deficits exist with regard to cyclical maintenance for the repair and replacement of damaged facilities and equipment.

In an attempt to find additional revenue sources for outdoor recreation on federal lands Congress authorized a temporary recreational fee demonstration program with the 1996 Omnibus Budget Rescission and Appropriation Act (P.L. 104-134). The program is designed to permit federal agencies to experiment with new fee programs at various national parks, national forests, national refuges, and BLM recreational units. Admission fees have been increased, new and increased user fees have been charged for camping, hiking into wilderness areas, boat launching, and even bird watching. Some critics argue that through the imposition of such fees the federal government is setting aside pristine federal lands for the benefit of the wealthy to the detriment of the poor. Whatever the fee structure that does result from the fee demonstration program, equity considerations need to be given considerable weight. One positive aspect about the fee demonstration program is that for the first time a portion of the collected fees will be used at the recreation areas where they are collected and not sent to the general fund in the Department of the Treasury. In the first full year of the demonstration period, FY1997, the National Park Service obtained an additional $45 million in fee demonstration receipts. In FY1998, demonstration fees increased to $132.5 million. For all four land management agencies participating in the fee demonstration program (National Park Service, Fish and Wildlife Service, BLM, and USDA Forest Service) the respective overall increases in fee demonstration funds were $138.8 million in FY1997 and $169 million in FY1998.[1] At least 80 percent of the funds are to be used at the sites where the funds were collected. Given the initial successes of the program, the increased fees are likely to remain in place on these sites as well as additional fees being imposed at other federal facilities.

Outdoor Recreation and Wilderness Policy

In addition to managing renewable and nonrenewable resources, natural resource managers often have outdoor recreation and wilderness management responsibilities. Outdoor recreation and wilderness policies serve various psychological, social, and physical needs of humans. However, outdoor recreation and wilderness policy choices affect the acquisition and utilization of natural resources for other purposes. For example, in planning the future use of resources a manager must make a decision on the amount of land, water, and forests necessary to meet future demand for recreational purposes. A decision made to set aside forest areas for wilderness purposes will proscribe the use of those resources for timber harvest.[2] Conversely, a forest area that has recently been clear-cut will not be suitable for wilderness designation and most outdoor recreation for decades. In making decisions among competing uses natural resource managers have to make sure that a proper balance exists between uses, while protecting the ability of renewable resources to renew themselves.

Outdoor Recreation Policy

For many years the Bureau of Outdoor Recreation (BOR) was the primary federal agency with the responsibility for coordinating and advising other federal agencies on matters having to do with outdoor recreation. This included planning and administering a national program for recreation and providing financial and technical assistance to state and local recreation agencies. In 1977 the Carter administration abolished BOR and combined historical and recreation functions into a Heritage Conservation and Recreation Service (HCRS). The Reagan administration eliminated this HCRS and transferred most of its functions to the National Park Service, including management of the Land and Water Conservation Fund—which was the primary source of funding for outdoor recreation in the United States. With the loss of the BOR and its successor, the federal responsibilities for outdoor recreation planning have been greatly reduced.

As a result of increases in population, income, and leisure time, the demand for outdoor recreation in the United States has risen almost continuously throughout the twentieth century, most rapidly since World War II. One measure of interest in outdoor recreation is demonstrated by the number of visits to national parks, national forests, and state parks. Marion Clawson reports that the average annual increase in number of visits to national parks grew from 1.1 million in the period 1924–41 to 13 million annually in the period 1971–81. Similarly, visits to national forests rose from 0.3 million to 6.5 million annually from 1967 to 1980. Although the rate of increase in visits to state parks was not as significant, the number of visitors also continued to climb. In 1995, the USDA Forest Service had 345.1 million visitors, followed by the National Park Service with 269.6 million, and state parks served another 745.6 million. In 1986, the President's Commission on Americans Outdoors (PCAO) indicated that the rate of growth in demand for outdoor recreation was around 3 to 4 percent annually.[3]

Although two-thirds of the land in the United States is owned by the private sector, a great deal of resource-based recreation does occur on public lands. The PCAO reports that in 1986 there were 778.4 million acres of public recreation land in this country. The federal government owned 90 percent (707.7 million acres), state governments owned 8 percent (62 million acres), county-regional governments owned 0.7 percent (5.7 million acres), and municipal governments owned 0.4 percent (three million acres). There were a total of 109,718 recreational areas—67,685 federal, 19,884 state, 20,375 county-region, and 1,774 municipal.[4]

Outdoor recreation involves many activities. Typical outdoor recreation activities are picnicking, swimming, sightseeing, hiking, backpacking, boating, bicycling, and outdoor sports. According to a market opinion research survey, the attributes that people consider, in rank order, when selecting a park, beach, or other outdoor recreation area are natural beauty, amount of crowding, restroom facilities, parking availability, available information, picnic areas, cultural events, fees charged, concessions, organized sports, and guided activities.[5] In order to be responsive to societal needs, then, resource managers should attempt to develop plans of action that incorporate many of these considerations.

According to the *1994–95 National Survey on Recreation and the Environment* conducted under the leadership of the USDA Forest Service, 94.5 percent of the U.S. population sixteen years of age or older participate in outdoor recreation activities. This figure has remained relatively constant from 1960 to 1994.[6] Sixty-six percent of Americans participated in walking for recreation, and 44 percent participated in swimming, the two most popular outdoor activities.[7] The *1996 National Survey of Fishing, Hunting, and Wildlife-Associated Recreation* reports that 62.9 million people age sixteen and over (31 percent of the U.S. population) enjoyed a variety of wildlife-watching activities such as observing, feeding, or photographing wildlife.[8]

The most dramatic increases in participation from 1982 to 1994 were in birdwatching (124 percent) and backpacking (72 percent). From 1960 to 1982 the participation rate for bicycling tripled, though it has since leveled off, with annual increases of only 1.6 percent. The national recreation survey also indicates that there were also major increases in participation rates for camping and jogging.

Most outdoor recreation activities occur close to urban areas, basically because three-fourths of the U.S. population is urban. Although these urban recreational zones are well used, they do not necessarily represent the optimum quality of outdoor recreational experience. It is also interesting to note that outdoor recreation often occurs on private property. The national recreation survey indicates that those engaged in hunting, fishing, and other activities related to wildlife spent 67 percent of their recreation days on private land where the majority of wildlife habitat is located.[9]

Outdoor recreation ranks high in economic importance in most states, especially in the West, where the majority of public lands are located.

The *1996 National Survey of Fishing, Hunting, and Wildlife-Associated Recreation* indicates that more than 77 million members of the American public spent more than $101 billion in wildlife-related recreation. More than $60 billion was spent for equipment, and more than $30 billion for food, travel, and lodging associated with wildlife use. These participation and expenditure figures translate into economic and political power that affects natural resource policymaking arenas.

Outdoor Recreation Resources Review Commission

The Outdoor Recreation Resources Review Commission (ORRRC) was established in 1958 to assess the state of outdoor recreation in the United States. In 1962, the commission issued a series of reports with major recommendations. The publication of its report *Outdoor Recreation for America* is probably the most outstanding event in the history of outdoor recreation resources management. Among the recommendations of the commission were

1. The establishment of a national outdoor recreation policy
2. The development of guidelines for the management of outdoor recreation resources
3. The planning of improvements in outdoor-recreation programs to meet increasing needs
4. Establishment of a BOR in the national government
5. Establishment of a federal grant-in-aid program to provide assistance to state and local governments for recreation planning and development and funds to federal agencies for acquisition of land for recreational purposes.

The report also recommended acceptance of a federal commitment to assist in bringing recreational opportunities to the American people: "It shall be the national policy, through the conservation and wise use of natural resources, to preserve, develop, and make accessible to all American people such quantity and quality of outdoor recreation as will be necessary to assure the physical, cultural and spiritual benefits of outdoor recreation." [10] The BOR and the LWCF created out of the efforts of the ORRRC were "major forces on the national scene as far as outdoor recreation is concerned" from the early 1960s.[11] As already noted, BOR became the HCRS in 1977, and its functions were transferred to the National Park Service in the Reagan administration.

The ORRRC clearly established a foundation for a comprehensive outdoor recreation policy. One recommendation was for a framework that divided responsibility for creating quality outdoor recreation opportunities among local, state, and federal agencies. This framework still allowed room for private individuals to exercise initiatives in the development of outdoor recreation options.[12] In regard to the role of the federal government the commission said:

> The Federal government should be responsible for the preservation of scenic areas, natural wonders, primitive areas, and historical sites of national significance; for cooperation with the states through technical and financial assistance; in the promotion of interstate arrangements, including Federal participation where necessary; for the assumption of vigorous cooperative leadership in a nationwide effort; and for management of Federal lands for the broadest recreational benefit consistent with other essential uses.[13]

Congress fully endorsed the commission's report and made efforts to implement it. Concern about the flagging commitment of the federal government to implement the commission's recommendations and about the state of outdoor recreation prompted Laurence S. Rockefeller, former chairman of the commission, to organize a new study. The resulting 1983 report indicates that while more and more Americans enjoy the great outdoors, the future of outdoor recreation is in danger because of shrinking governmental commitments to parks, conservation programs, and open spaces. The report points to budget cuts at all levels of government that have greatly slowed efforts to expand protection for wilderness areas, parks, trails, and rivers over the previous two decades. It asked Congress to create a new bipartisan national commission to study long-range needs and to create policies for protecting outdoor recreation in an era of fiscal restraint. The President's Commission on Americans Outdoors grew out of this recommendation.

The federal role and responsibilities toward recreation are also stated in the National Environmental Policy Act. Two major provisions of NEPA that apply to recreation are (1) Section 101, which recognizes that each person should enjoy a healthful environment, and (2) Section 102, what requires that federal agencies orient and administer programs toward environmental quality, values, and policy considerations in accordance with the act.[14]

A major need in outdoor recreation policy, especially in light of continually growing demands, is the establishment of greater consideration of environmental quality values in management decisions. In the past, too much attention has been given to meeting recreational needs and demands of the masses with too little attention to environmental impacts of their crowding behavior. A given land or water area has a carrying capacity or environmental saturation point for people as well as for other forms of life. To exceed the carrying capacity in particular areas will produce negative effects on these areas and ecosystems, reducing recreational quality and potential for other uses.

Until recently, outdoor recreation resource management agencies such as the National Park Service focused on accommodation, adjustment, and development to meet demands of increasing numbers of people, in spite of the sacrifice of environmental quality. A growing pattern among some federal land management agencies is to restrict and regulate the human/land ratio in terms of the carrying capacity of a given area. Under this pattern of regulation some federal land agencies have developed schemes to regulate density of recreational use, depending upon the quality and quantity of land available for such use. But accommodation of environmental quality considerations in recreational management is relatively recent. The National Recreation and Park Act of 1978 directed the National Park Service to establish carrying capacities for each park unit. A major stumbling block to full implementation of management policies based on carrying capacities is the National Park Service Act of 1916, which directs the agency to preserve the natural and cultural resources of national parks while providing for visitors' enjoyment of these resources.

Traditionally, the concept of carrying capacity has been thought of in ecological terms. However, recreational use of any natural area also will involve natural resource, social, and managerial factors. An alternative approach that considers all of these factors is the "limits of acceptable change" concept.[15] Under this concept, limits of acceptable change would be developed and expressed as management objectives, which take into consideration the fragility of various resource bases and weigh their use against the needs and wants of people to determine the level of acceptable use. In addition to biological limitations, people's perceptions and opinions as to acceptable limitations need to be taken into consideration as well as legal directives and agency missions.[16]

In order to reconcile these different management objectives the National Park Service recently developed a carrying capacity-related planning scheme called Visitor Experience and Resource Protection. The process is based on the selection of indicators and standards of quality of the visitor experience that subsequently are monitored over time. A social science survey of visitors is the methodology used to identify these variables. The VERP planning scheme is being refined at Arches National Park, Utah, and may serve as a model for the entire national park system.[17]

Although state and local lands receive the bulk of recreational visits because of their locations near population centers, federal public lands are experiencing the greatest rate of increase in numbers of recreational visitors. With the exception of a small percentage of federal public lands established for an explicit purpose, such as national parks or national wildlife refuges, most federal lands are managed under multiple use guidelines. In addition, the Clinton administration has directed all of the federal land management agencies to implement ecosystem management principles as well. Recreation is legally mandated as one of several uses on Bureau of Land Management and USDA Forest Service lands. The Public Land Law Review Commission notes, "Some of the sharpest public policy issues in recent years have arisen as a result of real or alleged conflicts between various recreational values and other uses of public lands, or between one and another type of recreational use."[18] The changing demographics of the American population and their demands for more recreational opportunities present many potential multiple use conflicts for BLM and USDA Forest Service managers. Ironically, recent studies show a far more rapid increase in USDA Forest Service appropriations for noncommodity line items like recreation than for traditional commodity line items like timber production.[19]

Under the multiple-use concept, value conflicts often have involved economic interests and groups favoring a dominant use that reduces or removes recreational and other uses. Historically, few agency personnel had professional training in recreation resources management, so they tended to favor their areas of specialization such as timber or range management. Growing demand for recreational opportunities has increased the potential for value conflicts in multiple-use management, especially when recreational values are threatened, when recreational areas are degraded or lost through allowance of incompatible dominant uses. Greater demands have been placed on federal lands with increased amounts of available leisure time and money, overcrowding of state and local parks, and improving transportation. Various studies predicted large increases in land-, water-, ice-, and snow-related recreation activities. There was an increase of 37 percent in visitor days on national forests between 1971 and 1981.[20] Visitors to national forests, more than 345 million in 1995, now outnumber the number of visitors to National Park Service recreation areas. However, Marion Clawson predicted in 1985 that the exponential increases in recreational demand would level off: "The rate of increase in outdoor recreation activity for the next 25 years is likely to be much less than the rate in the past 25 years—more on the order of 4 percent annually than of 10 percent annually."[21]

In addition to conflicts between recreation and other uses, an increasing number of conflicts have developed among various types of recreation uses on public lands. Because of negative environmental impacts, certain types of recreational use may diminish and others may be eliminated. The growing use of off-road vehicles (ORVs) such as motorbikes, snowmobiles, all-terrain and sport utility vehicles on public lands has brought on severe conflicts with other uses, nonmotorized recreation, and the environment. The impacts include damage to vegetation, erosion, disturbance of wildlife (particularly when animals are in breeding or wintering condition), and noise pollution. Noise pollution and associated disturbances by ORVs create particularly severe disturbances to people who are seeking nature-oriented experiences through hiking, horseback riding, cross-country skiing, and other nonmotorized recreational uses.

Although much of this discussion pertains to environmental quality considerations, these and other ORV impacts may be too severe for public lands that have the majority of their areas open to various forms of ORV use. In the Council on Environmental Quality's 1979 *Off Road Vehicles on Public Lands* report, the USDA Forest Service stated, "In the land use planning process, the question should not be, should we close an area to ORV use? But—can ORV use, in some form, be permitted in the area? One of the primary questions . . . is generally—how much resource impact can we live with in providing for recreation activity such as ORVs?"[22] Users of ORVs have organized into pressure groups that sometimes successfully resist efforts to establish protected wilderness areas simply because they might want to traverse these areas in or on their machines.

Moreover, agencies as well as offices within agencies differ greatly in their positions and degrees of planning and regulation of ORVs. An inadequate level of control has actually been imposed for various reasons. Relatively little federal regulation, planning, and enforcement have been directed at reducing conflicts between the use of ORVs and other forms of recreation, despite the government's broad powers. However, with increasing recreational pressures and conflicts more governmental involvement and controls will occur. The National Park Service has banned snowmobiles in Glacier, Yosemite, Sequoia-Kings Canyon, Lassen Volcanic and Rocky Mountain National Parks. Like other areas of recreation policy, ORV policy, remains generally ambiguous, but more restrictions seem likely, given the noise and fumes. In fact, park rangers at West Yellowstone reported "an unusual blue-brown pall of smog" in 1993 and complained of headaches and nausea caused by the steady stream of snowmobiles.[23]

Unfortunately, the observation that Phillip Foss made in 1962 concerning the lack of direction in recreation policy still has meaning today:

> None of these, or any other Federal agency, started out to provide public outdoor recreation, but they have had recreation and recreationists thrust upon them. Recreation has been an incidental, and

almost an accidental, by-product of the "primary" national policy on recreation, and few of the agencies have any real recreation policy. Agency practices have been established usually without adequate research and minor planning and oftentimes as defensive measures against the recreationists. Lack of anything resembling a national recreational policy is therefore at the root of most of the recreation problems of the Federal government.[24]

As in many other areas of environmental policy, one of the problems in formulating a unified and comprehensive outdoor recreation policy is the complex pluralistic nature of public and private interests that often are in direct competition. One consequence of policy failure is the significant increase of the private sector in the provision of recreation. Meanwhile, most federal agencies continue to deal with recreation in terms of their own vested interests and missions. The end result continues to be fragmented policies and programs for both recreation and environmental quality protection. This includes both planning and actual arrangements for accommodating growing recreational needs.

A perennial need in outdoor recreation is for public acquisition of open space and natural areas before the lands are purchased and developed for other purposes by the private sector. The PCAO reported that although annual spending for outdoor recreation at all levels of government amounted to about $9.8 billion, spending has not kept pace with inflation. In 1962 the ORRRC recognized the need for federal assistance to state and local governments for outdoor recreation through a specific recommendation for the establishment of a financial assistance program for the states for recreation land acquisition and facility development. In response, Congress enacted the Land and Water Conservation Act (P.L. 88-578) in 1965 to provide a means of federal funding from offshore oil receipts for support of state and local recreation planning, acquisition, and development programs on a matching basis, and for the acquisition of lands and waters for federal recreation areas. Through the Land and Water Conservation Fund (LWCF) the federal government has become the principal stimulus for government spending on outdoor recreation in the United States. However, appropriations for LWCF fell rapidly after 1978 when the highest funding occurred ($905 million). The Department of the Interior requested no funds be provided for outlays for the LWCF in fiscal year 1982, and only carry-over funds from previous years were available.[25] More recently, despite growing needs for land acquisition and construction of recreational facilities to serve the rising numbers of visitors, Congress has not spent authorized funds. Although the act authorizes expenditures of $900 million per year, from 1987 to 1997 appropriations for federal land acquisition did not exceed $341 million. Furthermore, grants to the financially strapped states were virtually nonexistent from 1994 to 1997.[26] Prior to 1980, "the LWCF faithfully financed purchases of wilderness lands and urban parks and paid for building hiking trails, bike paths, fishing piers and public gardens." One-half of the funds went to federal agencies and one-half to state and local agencies. By 1981, James Watt had cut funding by 95 percent; the Clinton administration eliminated state and local funding in 1994.[27]

The Conservation Foundation notes, "Some 190 million acres in the United States are protected in state and national parks and wildlife refuges. The expansion of the system has slowed markedly since 1980 with no corresponding increase in private acquisitions for conservation purposes."[28] Government natural resource agencies tend to be relatively ineffective and slow in acquiring essential natural lands before they are consumed by the private sector. Consequently, the Nature Conservancy and similar organizations often have to supplement the efforts of government agencies. But sometimes, in an effort to maintain natural areas for the public, these groups end up in conflict with government agencies.

The slow pace of acquisition of open space and natural areas by government has resulted in serious loss of needed open space and recreational lands to other uses and irreversible developments. Also, waiting too long for acquisitions often has resulted in prohibitive price increases. In 1971 there were approximately eleven acres of land available for each American. This is about one-third of the land available at the beginning of the century. Projections indicate that there will be only six and two-thirds acres per person available by the year 2000.[29] In spite of the emotional, spiritual, and intellectual needs served by open space and natural areas, governments worldwide have been slow to plan for dedication of such lands. Much of this delay is associated with the intangible quality of benefits and values associated with recreation in competition with more visible economic and developmental considerations. Such difficulties spawned the creation of the field of resource economics that attempts to assign economic values to natural resources through contingent valuation methods or travel cost models.[30]

Concern for outdoor recreation policy in all nations apparently varies with the degree of urbanization and technology. For example, industrialized nations emphasize outdoor recreation much more than less developed countries. This is natural, since key factors in recreation demand are income, leisure time, and mobility of the population. People who live in poor rural areas or who have little to eat have less means to seek outdoor recreation. However, there will probably be increasing demands for outdoor recreation space with the improvement of people's circumstances in less developed countries. Many of these countries have opportunities to save open space now while the demand is less for alternative uses. As demands and competing uses increase in the future it will be more difficult to obtain open space for recreation. Growing urbanization in less developed countries will be accompanied by increasing needs for public open space areas near urban centers. In industrialized countries recreational demands have caused general overcrowding in national parks and open spaces in the countryside. This, in turn, reduces the possibilities for many forms of quality recreation. Past president of the National Parks and Conservation Association, Paul Pritchard, recommends a "four percent solution" to deal with increasing demands for outdoor recreation. Since the average increase in the number of people visiting parks each year is currently 4 percent, he suggests that we should increase parkland by 4 percent each year. However, between 1980 and 1988 the United States added only one national park—Great Basin—to the National Park System.[31] The most significant event in the 1990s related to national parks was the passage of the California Desert Protection Act of 1994 (P.L. 103-453), which established Joshua Tree National Park and Death Valley National Park (both formerly national monuments). The 104th Congress added five new national park units. By 1999 there were 376 national park units with only 56 full parks.

Industrialized nations have made efforts to aid less developed countries in opening or expanding national park and open space recreation programs. Dasmann states:

> The interest of people in the wealthier nations in preserving and in visiting the natural treasures and outdoor resources of the developing nations has been an unexpected boom to the economics of some of these countries. . . . Nevertheless, the support for the preservation of outdoor recreational space in developing nations must come, to a large extent, from outside their boundaries until such time as the economic welfare of their own people has improved. This is a contribution which must be made by those who understand the value to all mankind that these recreational resources represent.[32]

This effort on the part of industrialized nations should also extend to the conservation of wildlands, which can be used for ecotourism. Harold Eidsvik, past chairman of the International Union for the Conservation of Nature and Natural Resources Commission on National Parks and Protected Areas, points out that "unless conservation appears on the agenda of both developed and the developing countries, it will remain a non-issue. If conservation does remain a non-issue, it will result in protected areas disappearing—a loss both for us and for future generations."[33]

Ecotourism is a new concept in which tourism and aspects of environmental education and interpretation supposedly are combined. This concept holds promise for the future, especially in tropical forested areas of some less developed countries. For example, in 1988, approximately 15 million tourists came to Central and South America to see flora and fauna of natural areas. Tropical forest ecotour destinations report steadily increasing visitation over the last decade. Belize, in particular, received 99,000 visitors in 1987, increasing to 215,442 in 1991.[34] Visitors undertaking such tours typically include scientists and students who, for example, "invade" Costa Rica every year, using this Central American country as a living laboratory and educational center. Patricia and Robert Cahn report that "if world records were kept for developing a national park system in the shortest time and against the most severe odds, it would probably belong to tiny Costa Rica." From having virtually no protection of its tropical forests in 1969 to 1978, by 1987 "its national park service had a million-dollar budget, several employees in the field, and eight percent of the country was protected in 20 national parks and biological reserves." And in 1987 land acquisition was initiated for the Guanacaste National Park, which is projected to be 173,000 acres.[35]

Under the best of circumstances, ecotourism can provide financial and other forms of support for efforts to protect tropical forests and other natural areas. Under less fortuitous circumstances, ecotourism may itself have negative environmental impacts. In any case, there is at least the possibility that ecotourism in less developed countries can educate visitors while assisting conservation and preservation efforts in host countries.

Wilderness Policy

Many nations have developed wilderness policies and implemented a variety of programs (including the creation of national parks) to protect and preserve wildlands. By 1994 the United States had 630 wilderness areas, totaling almost 104 million acres of federal land as "wilderness" in the National Wilderness Preservation System established by the Wilderness Act of 1964.[36] Of these designated lands, over 43 million acres are in the National Park System, over 20 million in the National Wildlife Refuge System, over 34 million in the National Forest System of the USDA Forest Service, and over 5 million acres are on BLM lands.[37] The vast majority of the wilderness area is in eleven western states and Alaska. Twenty-five states maintain wilderness or nature preserve areas. Some private lands are also kept in this condition.[38] Elsewhere in the world, in the form of national parks and nature reserves, wilderness areas are given differing degrees of protection.

Interest in wilderness activities in the United States dates back to the turn of the century. Roderick Nash identifies a "wilderness cult" of individuals and organizations interested in experiencing untamed nature at the end of the 1800s and early decades of the 1900s.[39] In 1921 the noted forester Aldo Leopold was reported as having made the "first explicit and practicable suggestion for a wilderness policy."[40] Leopold's apparent desire was to establish the means for future generations to experience nature as it was found by those who settled the United States. A portion of the first designated national wilderness area in New Mexico, the Gila Wilderness Area, was named after Leopold.

Craig Allin's *The Politics of Wilderness Preservation* expands upon Nash's historical coverage of wilderness preservation by addressing the political and administrative issues. He describes the relative quickness of Congress in enacting the Wilderness Act of 1964 and the Alaska National Interest Lands Conservation Act of 1980 to protect millions of acres of wilderness lands. Allin concludes that:

> If Americans adhere to the concept of spaceship earth, then wilderness will be cherished for its naturalness and for its ability to preserve ecological systems. It will be preserved as a symbol of the natural order to which we must adapt our civilization. If, on the other hand, Americans see science as savior, the anticipated technological fix will make it unnecessary to give serious consideration to plowing up our protected wilderness. The economic gain potentially available by doing so will appear insignificant compared to our ever-increasing ability to accomplish what we want by technological means.[41]

Formal wilderness policy in the United States emanates from the Wilderness Act of 1964. Its basic aim is to preserve the natural beauty, solitude, and environment of untouched, wild, natural areas, and to ensure that these areas remain subject only to natural evolutionary processes.[42] Through the National Wilderness Preservation System large areas of wilderness are protected from any form of development. But new developments such as settlements, industries, roads, timber cutting, and other human activities continue to encroach on remaining large tracts of undisturbed lands. The 1964 Wilderness Act indicates that nonmechanical recreational activities such as camping, hiking, backpacking, horseback riding, and cross-country skiing are considered important but secondary uses of wilderness areas. Preservation is the primary reason these areas are set aside.[43]

U.S. wilderness policy according to the Wilderness Act of 1964 is designed to secure for the American people of present and future generations the benefits of an enduring resource of wilderness. For this purpose there is hereby established a National Wilderness Preservation System to be composed of federally owned areas designated by Congress as "wilderness areas," and these shall be administered for the use and enjoyment of the American people in such manner as will leave them unimpaired for future use and enjoyment as wilderness, and so as to provide for the protection of these areas, the preservation of their wilderness character, and for the gathering and dissemination of information regarding their use and enjoyment as wilderness.

The act defines wilderness as follows:

A wilderness, in contrast with those areas where man and his own works dominate the landscape, is hereby recognized as an area where the earth and its community of life are untrammeled by man, where man himself is a visitor who does not remain. An area of wilderness is further defined to mean in this Act an area of undeveloped Federal land retaining its primeval character and influence, without permanent improvements or human habitation, which is protected and managed so as to preserve its natural conditions.

Further, the act makes the following provisions for administration of wilderness:

Except as otherwise provided in this Act, each agency administering any area designated as wilderness shall be responsible for preserving the wilderness character of the area and shall so administer such area for such other purposes for which it may have been established as also to preserve its wilderness character. Except as otherwise provided in this Act, wilderness areas shall be devoted to the public purposes of recreational, scenic, scientific, educational, conservation, and historical use.[44]

Over the last few decades increasing pressure has come from urban dwellers for recreational use of backcountry areas. This pressure stems in part from a general lack of adequate recreational facilities near urban centers. Demands have also come from ORV users to "open up" wilderness areas through road development and official sanction for off-road motorized uses. Motorized recreation is not legally permitted in officially designated wilderness areas. However, in some instances heavy use by legal activities, in particular horseback trail riding, has resulted in overuse of wilderness areas that can tolerate only limited impacts.[45] Ironically, many of the urban residents increasingly responsible for creating negative impacts on wilderness areas are among the strongest supporters of wilderness protection.

Corporate entities with interests in mining, geothermal, oil and gas leasing, and other forms of resource development have long opposed the setting aside of any wilderness areas. The same corporate entities also attempt to undertake developmental activities in already protected areas. As a result of developmental pressures it is now a common requirement for mineral resource surveys to be completed before Congress takes any action on wilderness protection measures.

Even the Wilderness Act itself fails fully to protect designated wilderness areas. Under appropriate legal circumstances, mining and grazing are still permissible uses within designated wilderness areas. Grazing use in a designated area is based upon level and kind of grazing use permitted prior to designation. Mining use is carried forward in designated areas under the provisions of the General Mining Law of 1872, subject to certain constraints imposed by NEPA and the Wilderness Act. Environmental impact assessment procedures, for example, encourage mining operators to undertake measures intended to reduce negative environmental impacts. The Wilderness Act terminated exploration for mineral deposits in established wilderness and wilderness study areas as of December 31, 1983. But this prohibition did not affect mining claims and mining developments that were in operation before that date,[46] although mining is recognized to be a use incompatible with wilderness. As Olen Matthews et al. state, "Although (mining) claims that were valid on that date (December 31, 1983) may be developed in the future, the Wilderness Act states that development should be substantially unnoticeable. Can mining be unnoticeable?"[47] Development of oil and gas leases is just as incompatible with wilderness management.

Under the Reagan administration, strong efforts were made to open up protected wilderness areas such as the Bob Marshall Area in Montana and undesignated backcountry areas along the Rocky Mountain front to oil and gas exploration and development. Congress passed legislation banning oil and gas drilling in designated wilderness areas, but the undesignated areas remain vulnerable to this kind of intrusion.

Despite its limitations, the Wilderness Act of 1964 has, nevertheless, managed to partially fulfill the intentions of its framers who wanted

> to assure that man does not change every acre within the United States, that some places shall be kept where nature is dominant and man comes only as a visitor—where man does not build his material things, where man does not change the face of the earth, where man does not interfere with the natural course of the waters of the earth.

In short, wilderness shall be those designated places where processes of nature continue without interfering or interruption by man.[48]

Having served as the basis for guaranteeing at least partial protection for many pristine acres in the United States and its territorial possessions, the act has served the cause of wilderness preservation. This is not to suggest, however, that the Wilderness Act lacks critics—quite the contrary. Even among those individuals and groups most anxious to preserve wilderness areas, severe criticisms of the act have surfaced. Among the more serious and telling accusations leveled at the act is that it fails to address the need to protect entire ecosystems rather than mere fragments of ecosystems. If the high-mountain country with its rocks, lakes, and vestigial glaciers is protected but the valley country with its forested slopes, low meadows, and free-flowing streams is left unguarded, the potential for destructive human intrusion constantly threatens the integrity of the entire living mountain-valley system that, under ideal circumstances, should function to sustain and perpetuate itself.[49]

Although designated wilderness areas should be managed so as to preserve their natural and undisturbed condition, multiple-use concepts, nevertheless, apply to wilderness management, particularly in those areas administered by the USDA Forest Service. Service personnel acknowledge that the Wilderness Act defines a management direction that excludes some uses and restricts others. However, the act does not change the multiple-use mandate for the USDA Forest Service. Thus, uses permitted or prohibited under the act are still managed under the umbrella of the multiple-use concept in designated wilderness areas. Some of the permitted uses are watershed management, research, grazing for domestic stock and wild animals, and wildlife habitat for rare and endangered species.[50] Wilderness habitat is particularly necessary for the protection of species such as the grizzly bear and the wolverine, which are threatened with extinction. These species must have large tracts of undisturbed land upon which to carry on their struggle for survival.

As noted in the Wilderness Act, each land management agency within the framework of existing legislation and policies is responsible for managing its wilderness lands in such a way as to preserve their wilderness character while serving other public recreational and educational needs. Obviously, management for achievement of such disparate goals can engender conflict. In a survey of designated wilderness area managers, the following were identified as major problems: (1) user abuse (such as littering and vandalism) from visitors lacking a wilderness ethic; (2) user conflicts (for example, between mine operators and ranchers); (3) legal conflicts having to do with competing jurisdictions; (4) illegal vehicular use within wilderness boundaries; (5) badly designed boundary designations based upon political expediency rather than topographical features; (6) intrusion of man-made objects (telephone poles and power lines, for example) within wilderness boundaries; (7) difficulties with disaster control and fire management; (8) deterioration of the visual landscape and trail systems; (9) lack of funding; and (10) lack of adequate data on the wilderness resource.[51]

In the United States, controversies over wilderness have arisen out of serious value conflicts. The wilderness classification process—in which agency recommendations are presented for public consideration before final approval by Congress and the president—has been the focus for many such controversies. For example, the USDA Forest Service had to scrap RARE II (Roadless Area Review Evaluation II) after having failed with an earlier effort—RARE I—several years earlier. Both RARE I and RARE II were USDA Forest Service efforts to inventory roadless lands throughout the entire national forest system with the intention of making management recommendations for these lands. The RARE I and RARE II processes were supposed to identify which lands were appropriate for wilderness protection and which were appropriate for further study. In the case of RARE II the USDA Forest Service ran afoul of NEPA. As the result of a legal challenge, a judge in California ruled that RARE II had failed to provide adequate environmental studies, and, in particular, had failed to take adequate account of the possibility of wilderness classification as a viable management alternative for some of the roadless lands under review. The USDA Forest Service was required to start the roadless-area review process all over again, reevaluating all of its potential wildlands.

In addition to the Wilderness Act of 1964, wilderness preservation occurs through the National Wild and Scenic Rivers Act of 1968, Alaska Native Claims Settlement Act of 1971, Eastern Wilderness Act of 1975, and other acts. The National Wild and Scenic Rivers Act was created to protect rivers that possessed "outstandingly remarkable" scenic, recreation, geologic, fish and wildlife, historic, cultural, or other similar values in free-flowing condition. The three types of rivers classified under the act are wild, scenic, or recreational.[52] Only the wild rivers that are to be inaccessible by road and undeveloped truly apply to wilderness management. But the scenic and recreational classified rivers are also important to wilderness areas in an ecosystem sense, even though they may be partially developed.

The Wilderness Society has expressed concern that the Wild and Scenic Rivers System has been ignored by the federal government. Unlike the Wilderness Act, Congress did not specify that rivers on federal lands would be surveyed and then designated for inclusion within the system. Instead, Congress must authorize each designation. Possibly in response to public concern, four rivers were designated in late 1984, but few new rivers have been designated for inclusion in the system in recent years. The Wilderness Society suggests that "the reason behind this fitful and retarded growth lies in the inadequacies and confusions of the law that created the system, but an equally significant obstacle has been the absence of any clear, organized, and coherent constituency for its growth." Consequently, the Wilderness Society recommends that the federal government pursue the following goals with regard to wild and scenic rivers:

1. Passage of an omnibus wild and scenic rivers bill
2. Completion of already authorized acquisitions and easements to provide ridge-to-ridge "buffer zones" on either side of designated rivers
3. Initiation of oversight hearings on the Wild and Scenic Rivers System
4. A mandate for a national rivers study program
5. Establishment of a permanent administrative body for river policy and planning at the federal level
6. Development of meaningful state wild and scenic rivers programs.

Other difficulties associated with the Wild and Scenic Rivers System according to the Wilderness Society include the absence of a comprehensive management plan and constraints in the law that limit acquisition to an area one-quarter mile from the ordinary high-water mark of both sides of a river, thereby reducing necessary flexibility in management actions.[53]

The Alaska Native Claims Settlement Act of 1971 was highly promoted by preservationist interests. As a result of this act, the secretary of the interior was directed to withdraw more than 80 million acres into the four conservation systems—the National Forest System, the National Park System, the National Wildlife Refuge System, and the National Wild and Scenic River System. The so-called Section 1002 d(2) withdrawals under ANCSA for these systems took precedence over state and native land selections. The federal agencies involved in the management of these lands are directed by Congress to develop comprehensive conservation plans for the responsible management of resources within the withdrawal areas. Douglas Wellman observes, however, that "these immense gains for wildland recreation were achieved at the price of releasing to non-wilderness uses all remaining USDA Forest Service lands being studied under RARE II, canceling the Carter administration's proclamations and withdrawals, and limiting future executive withdrawals in Alaska to 5,000 acres."[54]

Since the majority of public lands are located in the western states, much less consideration has been given to eastern wildlands. The greatest centers of population, however, are in the eastern states. In response to concerns that the less than pristine forests of the east would not be preserved, Congress passed the Eastern Wilderness Act of 1975. This act "designated as wilderness 15 national forest areas totaling 206,988 acres in Alabama, Arkansas, Florida, Georgia, Kentucky, North Carolina, New Hampshire, South Carolina, Tennessee, Virginia, Vermont, West Virginia, and Wisconsin."[55]

In recognition that some wilderness areas were particularly threatened from development and not adequately protected or fully studied for wilderness designation, Congress passed the Endangered American Wilderness Act of 1978, designating another seventeen areas totaling 1,305,307 acres in Arizona, California, Colorado, Idaho, Montana, New Mexico, Oregon, Utah, and Wyoming. The wilderness designation process continued when Congress passed the 1978 Boundary Water Canoe Area Wilderness Act in 1978. This act protected an additional 1,075,000 acres of wilderness in Minnesota national forests.[56] After the 100st, 102nd, 103rd and 104th congresses failed to protect additional wilderness areas in Utah, President Clinton used his authority under the Antiquities Act of 1906 and issued Presidential Proclamation No. 6920 to preserve 1.7 million acres as the Grand Staircase–Escalante National Monument. However, members of the Utah congressional delegation have challenged the president's authority to issue such proclamations under the act.

Wilderness controversies grow out of the pitting of development-oriented, short-term objectives against less tangible, long-term objectives and considerations. Conflicts between immediate and future considerations surface at all levels from local communities to the halls of Congress. Philosophically, conflicts arise between those taking an anthropocentric view and those taking a biocentric view. Individuals and groups subscribing to the anthropocentric view tend to endorse land management options that facilitate human use of wilderness areas. Individuals and groups oriented to the biocentric view are likely to favor land management options that will serve to protect the integrity of complex ecosystems.

Humans belong in wilderness areas only so long as they do not disrupt the functioning of these ecosystems. It is inevitable that those taking the biocentric view also identify with long-term values, since they are committed to management options encouraging the survival of wilderness areas capable of sustaining and perpetuating themselves into the indefinite future with little or no human interference.

Without using the "wilderness" label, many countries have set aside wild land or natural areas that are variously named and exist under varying degrees of protection. Typically, these areas are set aside for the preservation of wild species, the protection of natural ecosystems, and the provision of public recreation that is not supposed to impair these ecosystems or alter the character of the areas set aside. The IUCN Commission on National Parks and Protected Areas publishes *The United List of National Parks and Equivalent Areas* that lists data on the size, nature, and management status of numerous parks and nature preserves in various countries from Australia to Zimbabwe.

Raymond Dasmann says that national parks in many countries exist only as long as no real economic interests crystallize. When these parks stand in the way of development, they simply are not recognized. Thus, logging and other developmental uses occur in many national parks throughout the world.[57] For example, pressures from Maasai grazing interests in Kenya had by 1987 caused the protected core of the Maasai-Mara Game Reserve to shrink by 20 percent.[58] Jeremy Harrison, Kenton Miller, and Jeffrey McNeeley noted in 1982 that "some 2611 protected areas covering nearly 4 million square kilometers have been established by 124 countries, but management must still improve considerably before the full benefits of such areas can be delivered to society."[59]

National parks are relatively large land and water areas that contain representative samples and sites of major natural features and plant and animal species of national or international significance. Such features and species are of special scientific, educational, and recreational interest. National parks can contain one or several entire ecosystems that are not materially influenced by human exploitation or occupation. It is more typical that these parks contain portions or fragments of ecosystems. Generally, parks are protected and managed in a natural or near-natural condition by the governments of the countries containing the parks. Visitors enter national parks under special conditions for inspirational, educational, cultural, and recreational purposes. Some developing countries have dedicated a large percentage of their total land area to national parks and other forms of nature reserve. Costa Rica, for example, has approximately 10 percent of its total area in national parks, and some other less developed countries have similar percentages of protected land. In many industrialized nations, including the United States, national parks make up less than 1 percent of the total land area.

Outside the United States, establishment of national parks and equivalent natural areas requires more national and international effort and funding. Otherwise, outstanding wild areas of the world with unique and diversified ecosystems and species will be lost for present and future generations. Among other efforts, the World Heritage Trust channels public and private money to purchase and provide adequate protection for outstanding examples of the earth's wild country and wildlife. Through an arrangement with UNESCO, nations may dedicate their outstanding natural cultural areas to become part of the World Heritage System for the long-range benefits of all humans. Funds are then provided to these nations to assist them in acquisition and proper management of the areas selected.[60] This and other efforts are needed in working toward a world wilderness policy that encourages nations to acquire and protect wilderness areas and their ecosystems before they are irreversibly degraded or destroyed.

Federal Lands Policy

Of the 2.3 billion acres of land in the United States approximately 28 percent or 650 million acres are public lands managed by the federal government. Four-hundred fifty-eight million of these acres are managed under the concept and legal requirement of "multiple use."[61] About one-half of all federal land is in Alaska. Roughly 90 percent of the remainder is in eleven western states. With the exception of the thirteen colonies and the states of Hawaii and Texas, historically all the land in the United States was at one time public. The original thirteen states gave up any claim on public lands outside their borders and new states joined the Union under an enabling act that required them to disclaim all rights to unappropriated public lands within their boundaries.[62]

Thus, the federal government once owned more than three-fourths of the continental United States, or 1.5 billion acres acquired through secession, treaty, or purchase. Overall, with the exception of some largely unsuccessful efforts at public land disposal by the Reagan and Bush administrations, public land policy has changed from advocacy of almost complete disposal in the earlier days to advocacy of almost complete reservation in recent years. In the American past, especially in the mid- to late-nineteenth century, numerous laws promoted the disposal of the public domain through various methods such as grants and sales to individuals and corporations, including railroads. Most of the better and more productive public lands were claimed under these early laws and removed from public ownership.[63] Over 200 million acres were given to corporations under these disposal laws. This amounted to more land than was earned by homesteaders. More than a hundred years ago, however, a reaction developed against abuses, corruption, and waste associated with the disposal law that led to a shift toward conservation and the retention of public lands.

The establishment of Yellowstone National Park in 1872 and the passage of forest reserve legislation in 1891 marked the beginning of the concept of federal protection and retention of environmentally significant land. Public land policy gradually became one of federal retention, with reservations or withdrawals of public lands allowed for various purposes, such as national forests, national parks, and national wildlife refuges.[64]

Although it was assumed that most of the remaining unappropriated public lands or "public domain" would go into private ownership, land abuses and conflicts about overgrazing on public lands called for regulating livestock through the Taylor Grazing Act of 1934. However, this act also provided that public lands had to be classified before being disposed of by the national government. This requirement served to curtail disposal effectively. As a result, this act, along with the growing public interest in better management and recreation for public lands, strengthened the concept or value of retaining all public lands rather than disposing of them.

The September 19, 1964, law that established the Public Land Law Review Commission states the "policy of Congress that the public lands of the United States shall be: (1) retained and managed or, disposed of, all in a manner to provide the maximum benefit for the general public."[65] When it completed its report in 1970, the PLLRC recommended to the president and Congress that "the policy of large scale disposal of public lands . . . should be revised and that future disposal should be of only those lands that will achieve maximum benefit for the general public in non-Federal ownership, while retaining those whose values must be preserved so that they be used and enjoyed by all Americans."[66]

A major result of the PLLRC's efforts was the passage of the Federal Land Policy and Management Act (FLPMA) of 1976, which was debated through three Congresses. The act is recognized as the first definitive and comprehensive statement on public land policy and management. In the "Declaration of Policy" section, FLPMA states "Sec. 102. (a) The Congress declares that it is the policy of the United States that—(1) the public lands be *retained* in Federal ownership, unless as a result of the land use planning procedure provided for in this Act, it is determined that disposal of a *particular parcel* will serve the national interest."[67]

Under the Reagan administration, however, massive public land disposals and sales were inventoried and planned through initiatives from Secretary of the Interior James Watt.[68] These public land disposals, however, are subject to the authority of FLPMA, which permits the disposal only of *particular parcels* after land use planning procedures have been undertaken. Overwhelming public and congressional resistance to large-scale land disposals or "privatization" in fact prevented most of the land sales and exchanges from occurring.

The FLPMA also requires that public lands be managed under the concept of multiple use, which is defined as follows:

The "multiple use" means the management of the public lands and their various resource values so that they are utilized in the combination that will best meet the present and future needs of the American people; . . . the use of some land for less than all of the resources; a combination of balanced and diverse resource uses that take into account the long-term needs of future generations for renewable and nonrenewable resources, including, but not limited to, recreation, ranges, timber, minerals, watershed, wildlife and fish, and natural scenic, scientific, and historical values; and harmonious and coordinated management of the various resources without permanent impairment of the productivity of the land and the quality of the environment with consideration being given to the relative values of the resources and not necessarily to the combination of uses that will give the greatest economic return or the greatest unit output.[69]

The USDA Forest Service is well known for utilizing and for handling conflicts over the concept of multiple use. The passage of the Multiple Use and Sustained Yield Act of 1960 for national forests provided official recognition of the concept. Through actual or implied authorization other resource agencies such as the BLM also operate under this policy. With increasing conflicts and demands for energy, mineral, and timber resources and growing concern for protection of environmental values and nonexploitative uses, a new era for public lands began with formalization of multiple-use management through various legislative acts in the 1960s and 1970s. These laws gave specific legal authority for multiple-use management as a means of achieving a balance among competing uses and values through resource inventory, planning, public participation, and other management measures. These laws also gave official sanction and recognition to environmental values and noneconomic uses, as contrasted to the largely economic and exploitative uses of the past.[70]

Along with the growth and formalization of the multiple-use concept, the idea that environmentally significant public lands should be set aside also has gained wide public support. By 1964 three major public land preservation systems had been instituted: the National Park System (1872), the National Wildlife Refuge System (1903), and the National Wilderness Preservation System (1964). All three systems now protect more than 16 percent of the public lands located in all fifty states. Although these systems are managed under their respective primary purposes of national park preservation, wildlife conservation, or wilderness preservation, they are also considered, under some circumstances, to operate under multiple-use policy.[71] For example, among other uses, all three systems provide watersheds, protection for rare and endangered species, genetic pools, recreation, geological and archeological sites, wildlife, fish, cultural resources, and research opportunities. Some wildlife refuges and wilderness areas permit grazing use.[72] All of these uses, along with the production of wood, forage, minerals, and other uses are also included under multiple-use legislation for public lands outside the three systems.

The Department of the Interior through its agencies, the BLM, the U.S. Fish and Wildlife Service, the National Park Service, and the Bureau of Reclamation, administers more than 430 million acres of public land in the United States. The BLM has the largest share, with 266 million acres.[73] As part of the Department of Agriculture, the USDA Forest Service administers an additional 191 million acres.[74] The Department of Defense and other agencies administer another 287 million acres.[75] These federal departments and their various agencies issue directives concerning classification, inventory, and planning for public lands. These directives may have more importance than they first appear to have. For example, a basic policy question for public lands is how they should be classified as to general type and category of use. Classification thus limits the range of uses to which public lands may be subjected.[76]

The West, where most federal lands are located, retains many elements traditional to frontier culture. For example, westerners possess a culture that encourages independence from governmental control and a belief that individuals and groups have the right to exploit natural resources existing in the public domain. Many westerners believe federal lands are state-owned areas subject to development on an unrestricted basis. Although some of these frontier attitudes may be changing, the majority of support for preserving environmental quality and permitting noneconomic uses on these lands comes from urban areas. Western members of Congress, many of whom were put in office partly as the result of financial support coming from resource extraction and energy development industries, dominate the committees on interior and insular affairs in both houses of Congress. Consequently, these congressmen have a great deal of influence upon legislation, policy, and agency decisionmaking insofar as it affects western federal lands.

In a summary of the impacts of quasi-frontier attitudes, Phillip Foss points out that "the history of public land policy has been a history of conflict since the founding of the Republic. . . . [O]pposition to federal (and sometimes state) management of public lands is nothing new." And, although these conflicts and controversies have been present throughout western American history, they recently have been aggravated because of a proliferation in the 1960s and 1970s of new legislation, policies, and regulations aimed at the protection and enhancement of the natural environment.[77]

The so-called Sagebrush Rebellion, representing the vocally expressed opinions of an extreme minority of westerners, is an example of a western reaction to the prospect of continuing restrictions on access to and use of federal lands. The object of this movement was to gain transfer of land from federal to state or private ownership. Supporters included individuals engaged in ranching, oil, gas, coal, and hardrock mineral development. The illusion of strength was bolstered by the actions of several state legislatures that, led by Nevada in 1979, passed nonbinding and legally irrelevant resolutions backing the Sagebrush Rebellion.[78] Unable to achieve substantial support even in the West, the rebellion died for lack of interest.

As previously noted, major public land controversies often center on energy development, particularly in its extraction and leasing aspects. Energy development activities such as coal strip-mining may be considered a dominant use, excluding and too frequently permanently removing other uses even from the realm of possibility. At the same time that such dominant uses are being facilitated, federal agencies are required by law to protect environmental values and encourage multiple uses in the public administration of all public lands. For example, with respect to the protection of environmental values, Section 8 of the FLPMA directs that "the public lands be managed in a manner that will protect the quality of scientific, scenic, historical, ecological, environmental, air and atmospheric, water resource, and archaeological values; that, where appropriate, will preserve and provide food and habitat for fish and wildlife and domestic animals; and that will provide for outdoor recreation and human occupancy and use."[79]

Still, Paul Culhane's analysis of the USDA Forest Service and the BLM, *Public Lands Politics: Interest Group Influence on the Forest Service and the Bureau of Land Management,* reveals that both agencies are managed in accordance with appropriate professional standards. Furthermore, he suggests that the user or special interest groups served by these agencies do not assert any undue influence on agency policies. In contrast to Phillip Foss's analysis of the implementation of the Taylor Grazing Act by BLM, Culhane states that any notion of "agency capture theory" with regard to both the USDA Forest Service and BLM is not appropriate. Instead, he asserts that both agencies balance the influences of timber harvester, cattlemen, woolgrowers, and environmentalists one against the other. Like Foss, however, many others may object to Culhane's observations; especially in light of the poor condition of the BLM grazing lands.[80]

According to the BLM, the FLPMA—which also applies to the USDA Forest Service—mandates environmental protection and serious consideration of long-term values when public land decisions are made:

Protection of environmental values is explicitly and clearly a major criteria for all management activities. . . . These new words [including a new definition of multiple use in FLPMA] focus the definition more on environmental values and also require that all consideration must be given the long term, which includes future generations. Combining this definition with the above stated environmental goals shows new need to more carefully consider and utilize environmental values in all activities.[81]

In spite of recognition of these legal mandates to consider environmental and long-term values in the decisionmaking process in land management agencies, the momentum in BLM has been toward a great deal of energy leasing at the cost of preservation of environmental quality. Within the USDA Forest Service agency culture, pressure still exists to develop single use forest resource management.

As noted earlier, in 1993, the Clinton administration ordered all federal land management agencies, including both BLM and USDA Forest Service, to adopt ecosystem management principles in order to more effectively consider broader environmental considerations. According to its mission statement, USDA Forest Service managers are to use "an ecological approach to manage national forests and grasslands by blending the needs of people and environmental values in such a way that national forests and grasslands represent diverse, healthy, productive and sustainable ecosystems."[82] The Bureau of Land Management indicates that the principles of ecosystem management are consistent with the multiple-use mandates of the FLMPA. Accordingly BLM's managers are to sustain the productivity and diversity of ecological systems; use the best available scientific information when making resource allocation decisions; coordinate with federal and state governments and private landowners in planning; determine desired future ecosystem conditions based on historic, ecological, economic, and social considerations; minimize and repair land use impacts; use an interdisciplinary land management approach; reconnect isolated fragments of the landscape; and use adaptive management techniques.[83] Encouraging appropriate reciprocal relations between humans and ecosystems will be one of the most complex tasks these land management agencies will have to deal with in attempting to implement ecosystem management.[84]

Overall trends in natural resource management of public lands indicate increased demand for outdoor recreation opportunities in the United States. However, inadequate levels of funding at all levels constrain growth in provision of outdoor recreation. Only a few national parks, national wildlife refuges, and wilderness areas are established each year in the lower forty-eight states: insufficient to keep pace with population growth. Globally, only a few countries are expanding their national park programs. In less developed countries, attempts to maintain sensitive environmental areas are challenged daily by development considerations. Despite their many psychological benefits, outdoor recreation and wilderness experiences are increasingly becoming activities only for the more affluent.

The United States has achieved a remarkable record in the reduction of some pollutants, while others continue to be major environmental problems. The greatest success has been associated with what are called the conventional pollutants—those pollutants most easily treated, such as carbon monoxide in air or sediment in water. Toxic pollution problems continue to be increasingly burdensome and costly. For removal of even small quantities of toxic pollutants, pollution control costs are extremely high.

Outright prohibition of toxic releases into the air or discharges into bodies of water may never occur. The economic impact of such prohibitions would be considered unacceptable by business and industry. Yet more and more anomalies are appearing in animal populations around the world. Toxic pollutants are a suspected source of such anomalies. Pesticides and dioxin, for example, have been found to be endocrine disruptors. These are compounds that alter the production of reproductive hormones in both animals and humans.

As a result of strict air pollution control measures originating in the Clean Air Act Amendments of 1972, substantial progress has occurred with regard to the criteria air pollutants covered by the National Ambient Air Quality Standards (NAAQS): carbon monoxide, particulate matter, sulfur dioxide, ozone, lead, and nitrogen oxides. From 1970 to 1995 there was an overall decline in the emissions of five of the six criteria pollutants; the only air pollutant for which there was not a decline was nitrogen oxide.[1] Ozone also continues to be a problem: ozone standards have not been met in many urban areas as well. An effective air toxics program still remains to be developed, although the 1990 Clean Air Act Amendments required each of the states to develop an air toxics program to attain federal air toxics standards. It will be interesting to see how well the federal government and the states address the air toxics problem in the near future. From

1972 to 1990 only eight substances were ever regulated under the National Emission Standards for Hazardous Air Pollutants (NESHAPS).

The federal government has been less successful in achieving success in the cleanup of water pollution. The 1972 Clean Water Act (CWA) had a goal that all of the nation's water would be fishable and swimmable by 1985. The National Water Quality Inventory of 1996 indicated that 40 percent of the inventoried waters did not meet designated uses. Numerous shellfish beds are closed to harvesting as well. And the number of fish consumption advisories (warnings to anglers not to consume the fish that they catch) continues to increase.

The capacity of the United States to dispose of the waste that it generates has become overwhelmed over the past ten years. The 1984 amendments to the Resource Conservation and Recovery Act (RCRA) substantially tightened standards for landfills. Liners were required to minimize leaching of pollutants into groundwater supplies. Additional capping and intermediate cover requirements were added to reduce the risks of airborne pollution and pests in the vicinity of landfills. The result has been the closure of many landfills in the United States. At the same time the citizens of the United States have increased their production of solid waste, something on the order of five pounds per person per day. The largest landfill in the country, Freshkill, serving New York City is projected to close in the year 2000. New York's solid waste will have to go somewhere, along with New Jersey's, and Boston's, and so forth. Some states have attempted to bar the importation of solid waste from other states, but the Supreme Court has ruled that such prohibition violates the Interstate Commerce Clause. Furthermore, there are 30,000–40,000 abandoned hazardous waste dumps spread out across the United States. But the EPA has only designated about 1,300 of the waste dumps on the National Priority List (NPL) for which federal funds are provided for cleanup; this represents an increase of about 500 in the past ten years. The cost and complexity of pollution control will continue to challenge the brightest environmental administrators for many long years to come.

Pollution Control Policy

Pollution is the presence in the environment of matter or energy associated with human activities whose nature, location, or quantity produces environmental effects undesirable for humans and other life forms. As a process, pollution contaminates or alters the quality of some portion of the environment through the addition of harmful impurities. A pollutant is any extraneous material or form of energy whose rate of transfer between two components of the environment is changed so that the well-being of individual organisms or ecosystems is negatively affected. A pollutant may be any introduced gas, liquid, or solid adversely affecting human, plant, and/or animal life while causing a resource to become unfit for a specific human purpose. Thus, pollution may be considered the unfavorable alteration of the environment, wholly or largely as the result of human actions. The earth's life-support systems are threatened by numerous undesirable by-products of economic and industrial growth. These by-products, often characterized as hazardous or toxic substances, could affect virtually every earth ecosystem and resource base.[2]

The definition of "pollution" depends upon the public's decisions concerning proper use of the environment and determination of tolerable levels of pollution. Although scientists may define requirements for uses or describe harmful effects of particular substances, their thinking transcends the bounds of science when they try to prescribe levels of use of given substances in the environment. Clarence Davies points out that, "only by linking scientific knowledge with a concept of the public interest can one arrive at a working definition of pollution."[3] With the recognition that science cannot provide final answers or values in determining pollution policies, public and private forces compete to make these determinations through political and administrative processes. Thus, determination of appropriate pollution control mechanisms often is more political than technical in nature.

Pollution control does not necessarily entail recovery of resources in useful or available forms. Although rendered less noxious, pollutants may simply be displaced or dispersed more completely into the environment. For example, many air pollutants are converted into water pollutants or solid wastes. Carbon dioxide from fossil fuel consumption accumulates in the atmosphere and contributes to alterations in global temperatures. Similarly, sulfur oxides and nitrogen oxides from the same and other sources produce acid deposition through atmospheric chemical processes. Radioactive wastes accumulate in the atmosphere and invade, and thus contaminate, groundwater when they breach temporary storage sites and otherwise enter into the environment and become semipermanent sources of contamination.[4] In recognition of the importance of pollution as resource displacement, the Council on Environmental Quality cautions, "Although pollution may be the most prominent and immediately pressing environmental concern, it is only one facet of the many-sided environmental problem. It is a highly visible, sometimes dangerous sign of environmental deterioration. . . . Pollution threatens natural systems, human health, and aesthetic sensibilities: it often represents valuable resources out of place."[5]

Although the technical-scientific aspects of resource displacement are an important factor in pollution control, much of the work of government in this area involves translation of legislative intent concerning compliance standards and means of enforcement into specific policy actions. As a result of the limited technical knowledge of lawmakers as well as their desire to provide ample flexibility to implementing agencies, pollution control laws are often ambiguous, inconsistent, and unrealistic.[6] Some of the confusion results from legislative inability to reconcile conflicting values and interests without using vague language. John Baldwin notes that "the commonly used goal (of environmental law), 'to protect the public health and welfare,' is ambiguous because it does not define the term welfare, nor does it give any indication of the extent of control costs in decision making processes."[7]

Confusion also results from the fact that pollution control legislation involves complex technical and scientific processes. But environmental law cannot be based exclusively on technical-scientific considerations. Both generalists and technical-scientific specialists are needed in government to make environmental law more functional. Governmental personnel need—and must be capable of using—a great deal of discretion to decide how, where, and when to interpret and act upon pollution control legislation and policies.[8] Because political pressures exist that affect value judgments and decisions made in the pollution control policy process, caution must be exhibited in the degree of discretion granted. Further, adequate monitoring mechanisms are necessary to ensure appropriate policy implementation.

The need to involve many different agencies and levels of government in the decisionmaking process greatly adds to the complexity of political conflicts and processes. In the United States, federal, state, and local governments develop pollution control policies and procedures.[9] The states may propose standards and procedures that are given federal sanction. All three levels of government have some legal authority.[10] What will actually become pollution control policy may depend on the discretion exercised at each level of government.[11]

In order to facilitate a nationally consistent pollution control effort, some of the environmental mandates of the federal government establish minimum standards that must be maintained by the states, but they provide flexibility in implementation. This is accomplished through a process referred to as "acceptance of primacy," which is designed to give a state the opportunity to become the primary enforcement agent for federal policies. The Clean Air Act Amendments of 1970, the Clean Water Act Amendments of 1972, the Safe Drinking Water Act of 1974, and the Federal Insecticide, Fungicide, and Rodenticide Act (FIFRA) of 1972 are all examples of environmental pollution control laws that allow for primacy implementation. In the event that the state does not accept primacy, the federal government assumes the primary enforcement role. Therefore, if a state wants to keep control over its pollution control function, it must accept the burden of added implementation responsibilities dictated by the federal government. In order to encourage the states to accept primacy the federal government provides extensive cash subsidies initially but tends to reduce the amount of support over time. Nevertheless, as Patricia Crotty has noted, "Since states retain the right to accept primary enforcement responsibility under these laws and to rescind this acceptance, primacy offers the states a bargaining chip in dealing with preemptive mandates." [12] Because of the economic burdens associated with preemptive environmental mandates state and local government officials put significant pressure on Members of Congress for regulatory relief. In response, Congress passed the Unfunded Mandates Reform Act (UMRA) of 1995 (P.L. 104-4), which requires the federal government to provide funds to state and local governments for implementation of its mandates under certain circumstances.

Standards and procedures may be selectively ignored at one or more levels of government, adding to the general confusion. Baldwin notes that "because of agency uncertainty in enacting and enforcing standards and procedures, polluters are often quite successful in using the courts and administrative procedures to delay compliance." [13] The pressure to avoid private economic costs attending pollution control also influences the process by which pollution standards and procedures are formulated, inevitably weakening governmental efforts to exercise reasonable control.

Preservation of environmental quality and reduction of pollution levels necessitate reallocation of funds by private industry. Private industry reasons that such funds often could be better utilized as capital investment and profit for stockholders. Given the strong pressures to produce short-term profits, enterprises will keep long-term investments in pollution control programs and equipment to a minimum when they are given an opportunity by government.

Nevertheless, the pollution control industry, which employs millions of people, profits by protecting the environment. The Department of Commerce reports that the environmental industry generated $180 billion in revenue and employed 1.3 million people in 1995. [14] United States corporations spent about $92 billion in 1993 on pollution control related to air, water, toxic substances, and radiation. [15] Although pollution control is costly, it also stimulates new industrial technologies and increased productivity. [16]

When economists claim that pollution control reduces the nation's productivity, they fail to observe that a polluted environment imposes costs as well. In their analysis, Repetto and Rothman find that environmental regulations actually raise productivity by reducing the cost of a polluted environment.[17] And, after an extensive examination of studies on net exports, overall trade flows, and plant location decisions, highly regarded economists Adam Jaffe, Steven Peterson, Paul Portney, and Robert Stavins conclude that there is relatively little evidence to support the hypothesis that environmental regulations have had a large adverse impact on U.S. competitiveness.[18]

In the absence of critical self-assessment and under the rationale of profit and production quota ideologies, businesses add substantially to pollution problems in the United States. Individuals also dirty the environment in their desire for personal convenience. Viable long-range pollution control may require government intervention and regulation through (1) technological assessment, (2) planning for population growth and concentration, (3) assessment of social and environmental costs in order to limit causes of pollution, and (4) public participation and involvement in decisions on pollution control measures affecting society and the environment.

Some economists, however, believe that such a command-and-control approach is costly and inefficient and suggest use of tax incentives and market devices to reduce pollution levels.[19] John Cumberland suggests that the two approaches should be integrated. Corporations should be given economic incentives to reduce pollution, such as transferrable emission rights, which could be bought and sold. But at that point, when pollution levels are about to cause irreversible ecological damage, the polluter-pays option would cease and regulatory controls would be imposed.[20]

Richard Stewart also believes that the presumed conflict between environmental and economic development is false. He argues that "economically productive new investment is an important key to a better environment," because "it is easier and cheaper to build superior environmental performance into new products and processes than to modify old ones." Stewart concludes that it is preferable to build a better environment through investment than through closing down plants and unemployment.[21]

Both government and private pollution control policies are based on risk analysis and assessment procedures. Risk analysis involves assessing the risks of one action or alternative and comparing these to the risks of several alternative actions in order to analyze fully the consequences of proposed policies and actions. Although risk analysis is considered to be a valuable tool for policy decisions, it has many uncertainties in application, and it may be neither infallible nor very precise. Uncertainties arise from incomplete and inaccurate raw data, ignorance about causes and effects, and value judgments that are practically impossible to quantify.[22] In a brief prepared for Congress on risk management and environmental protection, Linda-Jo Schierow succinctly addresses both sides of this issue. Proponents of risk analysis want federal environmental protection programs to target the worst risks to health and the environment first. They believe this would reduce costs and increase flexibility. Opponents of risk analysis see it as excessive reliance of quantitative analysis for the evaluation of problems and solutions. They believe that such an approach ignores other important facets of policy decisions such as timeliness, fairness, effects on democratic rights and liberties, morality, reversibility, and aesthetic values, among others. The bottom line is that the quality of risk analysis depends upon the adequacy of data and the validity of the method employed.[23]

Solid and Hazardous Wastes

In industrialized societies, raw materials are extracted, refined, processed, and transformed into finished products. These processes produce large quantities of liquid and solid waste. Policy questions arise as to how these processes can be controlled in order to reduce and to dispose of the waste materials from production and consumption of goods. For example, residents of the United States produce roughly twice as much consumer solid waste as do people living in other nations with comparable economies and standards of living. Further questions involve the amount of waste that can be reused or recycled economically as well as the viable means for disposal of wastes, particularly toxic or hazardous wastes, in order to prevent or reduce damage to human health and the environment.

In the United States, billions of tons of virgin materials are removed annually from mines, forests, and croplands. In a 1984 report, the EPA records more than 6 billion tons of solid and hazardous waste. Over one-half of this total came from agriculture, but most of its waste was recycled to restore fertility to the soil. After agriculture (50.3 percent) and mining/milling (39 percent), the largest amount of waste came from manufacturing firms (6.4 percent). These companies disposed of more than 400 million tons of solid waste in a year. The finished products (materials and goods) and their packaging eventually also became wastes.[24] In 1994, the largest sources of municipal solid waste (MSW) in the United States (in millions of tons) were paper (81.3), yard waste (30.6), plastics (19.8), wood (14.6), glass (13.3), metals (12.7), textiles (6.6), and rubber and leather (6.4).[25]

Most solid wastes from manufacturing (sludge, slags, dust, and other inorganic or organic materials) are disposed of in landfills, open dumps, or impoundments, where pollutants can often leach into surface and groundwater. Even years after disposal, adverse environmental impacts and influences can occur. Approximately 10 percent of all industrial wastes (toxic chemicals, oils, acids, radioactive materials, and caustic substances) produced in the United States pose direct threats to human and other life-forms. Improper disposal of these wastes can result in tragic damage to life-forms and living systems.[26] Because of potential health threats from inadequate construction, the number of landfills in the United States was reduced from approximately 9,000 to 3,100 from 1984 to 1996.[27] Most of the closures were caused by the provisions of Subtitle D of the Resource Conservation and Recovery Act (RCRA) as amended, in 1984, by the Hazardous and Solid Waste Amendments Act (HSWA) (P.L.98-616). Subtitle D required all landfills to install liner systems, groundwater monitoring, and monitoring of soil surrounding landfills.[28] Massachusetts alone lost three-fourths of its landfills and over one-half of its disposal capacity by 1990. Given the approximate three to five pounds of waste each American sends to the dumps every day we are facing a "garbage crisis."[29]

The Resource Conservation and Recovery Act and its amendments address the most serious health and environmental problems that relate to solid wastes while reaffirming the goal of resource recovery under a moderate research and development program. "Resource conservation" is defined as the "reduction of the amounts of solid wastes that are generated, reduction of overall resource consumption, and utilization of recovered resources." Solid wastes include sludge and other solid, liquid, semisolid, and contained gaseous waste materials.[30]

Justifiable fears concerning disposal of solid and hazardous wastes continue to grow. In 1991 approximately 306 million tons of hazardous waste were generated in the United States. Furthermore, about 3.2 billion pounds of toxic chemicals were released into the environment.[31] Concerns about the amount of toxic materials and hazardous waste led to the creation of a "superfund" for cleaning up abandoned hazardous waste sites that threaten the health of nearby residents. The Comprehensive Environmental Response, Compensation, and Liability Act (CERCLA) of 1980 that created Superfund also included $1.6 billion for a five-year program to clean up thousands of leaking waste dumps.

Initial management of Superfund by the EPA was fraught with difficulties and corruption in the early years of the Reagan administration. However, several employees were dismissed, and eventually Administrator Anne Gorsuch resigned from office.[32] Partially in response to the many administrative problems associated with the original act, many changes in policy and management were incorporated into the Superfund Amendments and Reauthorization Act (SARA) of 1986, which authorized another $8.5 billion.

The new policies and procedures emphasize the importance of developing permanent solutions to hazardous waste problems and ensuring that all Superfund-related actions comply with applicable state and federal standards. Within a relatively short period of time after Gorsuch resigned, the pace of Superfund actions accelerated considerably, almost doubling in some categories.[33] Additional emphasis on hazardous waste management is still imperative in light of the tremendous amount of toxic and hazardous wastes produced in the United States that are disposed in environmentally unsound methods.

Polls indicate that the American people are very concerned about these problems. Many Americans feel that not enough is being done to clean up toxic waste sites. Furthermore, the majority of Americans often indicate that they would be willing to pay more taxes to fund cleanup programs.[34]

James Lester and Ann Bowman in *The Politics of Hazardous Waste Management,* Charles Davis and James Lester in *Dimensions of Hazardous Waste Politics and Policy,* and Charles Davis in *The Politics of Hazardous Waste*[35] clearly illustrate that hazardous waste management is one of the most perplexing problems of environmental administration, if not *the* most serious problem, as a result of its complexity and uncertainty. The detailed analyses and descriptive case studies of the authors in these volumes paint a vivid picture of the consequences of inadequate management of hazardous wastes, as well as the need for improved management policies and more stringent enforcement actions.

More and more American communities are discovering that they live near dumps that have been contaminated by chemical by-products such as dioxin, vinyl chloride, polybrominated biphenyls (PBBS), polychlorinated biphenyls (PCBS), and by other materials such as lead, mercury, and arsenic. The EPA has inventoried more than 39,000 uncontrolled hazardous waste sites across the United States, causing or having the potential to cause contamination. Since some of the sites are more serious than others, the National Priorities List (NPL) was established by the EPA for "remedial action." By 1995, the EPA had identified 1,374 of these hazardous waste sites for long-term remedial action and placed them on the NPL under Superfund.[36] However, the Office of Technology Assessment believes that there may be as many as 10,000 hazardous waste sites that pose serious threats to public health. Dangerous toxic waste dumps are located in almost every county of every state. These dumps present serious threats to water supplies, public health, and the economy.

The Office of Technology Assessment indicates that federal regulations may not effectively detect, prevent, or control the release of hazardous substances into the environment. Data inadequacies prevent comprehension of the scope and intensity of hazardous waste problems. Although the RCRA has provisions to prevent hazardous materials from being released into the environment, Steven Cohen notes that "the regulatory process it establishes is too cumbersome. . . . EPA's reaction to this problem has been to relax the program's requirements." Increased funding is necessary and the enforcement arm of Superfund must be strengthened. Cohen also recommends greater emphasis on research and development, observing that "we really do not know how to clean up a hazardous waste site."[37] The cost estimates for cleaning up a single abandoned waste site range from $1 million to $100 million, with the total cleanup cost ranging from $1 billion to $500 billion in the United States alone.[38]

In reference to solid and hazardous waste problems on a global level, the *International Register of Potentially Toxic Chemicals* of the UN Environment Programme (UNEP) indicates that "each year hundreds of millions of tons of hazardous industrial waste are produced, and much of this is discarded with little regard for, or knowledge of, its effects on human health and the environment. . . . The greatest enemy of our safe use and disposal of chemicals is ignorance." The *International Register* goes on to recommend that, in view of our ignorance, "there is a vital need for an international clearing house for scientific, technical, legal, and regulatory information for assessment and control of chemical hazards."[39]

A major issue in waste disposal policy is that social and environmental costs are not considered adequately. For example, the production process for 100 tons of steel from raw ore results in 280 tons of mine waste, 12 tons of air pollutants, and 97 tons of solid waste. Yet federal policies make such a process attractive economically through provision of depletion allowances, favorable freight rates, and preferential labeling, which, taken together, encourage the use of virgin materials rather than recycled materials. This occurs in spite of the fact that it is generally more economical and consumes less energy to use recycled metals than virgin materials.[40] For example, with recycled production and recycling almost 1,300 pounds less solid waste is produced than with production using virgin materials and incineration, and more than 2,800 pounds less solid waste than using virgin materials and landfilling, per ton of material processed.[41]

Valid questions can be asked about the real production costs and the real waste disposal costs of a manufacturing process. What environmental and social costs are borne by the public as the result of federal policy? Furthermore, the price of a given product may not incorporate the cost of its disposal or its recovery. The American system thus makes it very difficult to use material thriftily and to dispose of the resulting waste, when abundant and often inefficiently packaged goods do not reflect the costs of waste disposal.

Waste disposal, nevertheless, is subject to more stringent policies and regulations than in the past. Among its objectives, the 1984 amendments to RCRA proposed the complete elimination of open dumps and the upgrading of other waste disposal practices. It also offers federal assistance to states in their efforts to create waste management plans and to bring waste disposal systems up to federal standards. Faced with increasing amounts of waste and increasing costs and restrictions on disposing of wastes in sanitary landfills, many communities must also find suitable landfill sites. Land prices, environmental regulations, and public opposition often prevent acceptable sites from being used. As a result, many local governments and business officials are considering alternative disposal methods, including recovery and recycling of food, paper, and other materials, as well as burning wastes for energy. With the predicted increases in wastes, more efforts are being made to eliminate wastes at all levels of extraction, manufacturing, transportation, and consumption.[42]

Given all of the environmental and health considerations associated with the siting of proposed landfills, the acronyms NIMBY and LULU are becoming more widely recognized (NIMBY stands for "not in my back yard" and LULU represents "locally unwanted land use"). Congressional debates over the EPA's landfill ban regulations heighten an understanding of why such terms are popular.

Congressman John Spratt of South Carolina testified about the development of what is now the largest licensed hazardous waste landfill in the country—GSX Corporation in Pinewood, South Carolina. Under the guise of developing a fullers' earth mining operation, an undisclosed principal acquired the land and GSX began operation of the landfill. Between 1978 and 1996 the GSX landfill received more than one billion tons of highly toxic wastes from locations all over the country, nearly two-thirds of which was produced outside of South Carolina. Given that the volume of dangerous emissions from hazardous waste facilities varies directly with the size of the landfill, it is understandable that the citizens of nearby Sumter are concerned about the GSX site. Consequently, when the EPA proposed to assess adverse health and environmental effects of landfill releases 500 feet from disposal sites under the RCRA and not at the boundary of such sites Congressman Spratt protested vigorously. At the hearing he pointed out that according to language in the 1984 Hazardous and Solid Waste Amendments to RCRA "no migration of hazardous constituents from the disposal unit for as long as the wastes remain hazardous" is permissible. Allowing for a buffer of 500 feet before enforcement actions would take place, would not be satisfactory.[43] In the absence of such enforcement guarantees by government citizens will continue to clamor, "NIMBY!"

Until landfills are replaced through technological advances, the urgent need for high-quality hazardous waste containment sites will continue. Further, such sites will need to be selected within a reasonable period of time while acknowledging citizen concerns. A selection process should be used that gains and maintains the confidence of a highly concerned public. Such a system should

1. Establish clear, predictable standards that eliminate from consideration any unsafe sites and unsafe disposal methods
2. Objectively and honestly determine the facts about each potential site— *in a way that permits everyone affected to be heard*
3. Put the decision in the hands of a constituency that is, of necessity, broader than just the neighbors of the site
4. Establish timetables that are known to everyone in advance
5. When the site is selected, compensate neighbors for the burdens that they are bearing for the common good. This compensation may come in direct dollar payments, or perhaps indirectly in the form of community improvements.

Although this proposal may not be the answer to all of the problems, it is far more palatable than additional overflowing landfills, more groundwater contamination, and other abandoned hazardous waste sites like Love Canal in New York.[44]

Another goal of the RCRA is the development of comprehensive solid waste management programs for each state. This includes improved control over solid waste disposal practices and better resource recovery as well as utilization of conservation techniques whenever possible. The CEQ notes that the federal role in such activities is to "(1) establish guidelines for state solid waste plans; (2) designate criteria for classifying land disposal facilities; and (3) publish an inventory of unacceptable disposal facilities . . . in terms of effects on surface and ground water, air quality, and public safety, as well as cover material. . . . Such facilities will have to be phased out through state control efforts."[45]

The 1984 HSWA to the RCRA established waste minimization provisions. "Waste minimization" is a phrase that appears in RCRA that combines the concepts of waste reduction, recycling, and treating wastes before disposal. The goal is prevention of waste or reduction of the toxicity of wastes, before they are sent to landfills, in order to protect human health and the environment. As a result of the 1984 amendments, the EPA issued three waste minimization regulations that encourage companies to reduce waste:

1. Companies that generate hazardous wastes and ship them off the site of generation must certify on each manifest accompanying the shipments that the company has a waste minimization program in place.
2. Companies that manage the waste of others, or that generate waste and manage it themselves on the site of generation, must annually certify that they have a waste minimization program in place. This certification is placed in the permit file that is maintained on company property.
3. All hazardous waste generators are subject to RCRA biennial reporting requirements. Companies that generate hazardous wastes and ship them off the site of generation are subject to a special waste minimization section in the biennial reporting requirements. They must describe waste minimization efforts undertaken during the year and report any changes actually achieved in the volume and toxicity of wastes.[46]

There are several reports that address the problem of waste reduction but approach the problem from slightly different perspectives. The Office of Technology Assessment's report *Serious Reduction of Hazardous Waste* defines "waste reduction" as "in-plant practices that reduce, avoid, or eliminate the generation of hazardous waste so as to reduce the risks to health and the environment." INFORM's report *Cutting Chemical Wastes: What 29 Organic Chemical Plants Are Doing to Reduce Hazardous Wastes* (1985) defines "waste reduction at source" as occurring when "the source of the waste is altered in some way so that the amount generated is reduced or eliminated altogether." The Environmental Defense Fund's report *Approaches to Source Reduction* (1986) defines "source reduction" as "any technique by which the amount of hazardous substances imposed on society's waste handling capacities is reduced." Finally, the EPA's *Report to Congress: Minimization of Hazardous Wastes* (1986) defines "waste minimization" as "the reduction, to the extent feasible, of hazardous waste that is generated or subsequently treated, stored, or disposed of."[47] Despite the obvious benefits in reducing hazardous and solid waste at the source before they become waste management problems, the EPA does not support mandatory waste reduction regulations. Of the voluntary compliance efforts those by the 3M Corporation appears to have the most successful record. From 1975, when it initiated a Pollution Prevention Pays program, to 1986, 3M supposedly achieved a 50 percent reduction in all wastes leading to a savings of $300 million.[48]

Through the Solid Waste Disposal Act and other legislation, the federal government has assumed a supportive role in the establishment and implementation of local and regional programs for solid waste problems. Both recycling and research and development initiatives are encouraged. Various laws now provide for (1) research, development, and demonstration of new techniques; (2) university training programs; (3) technical assistance; and (4) grants-in-aid and other cooperative activities.

As public appreciation of the extent of the economic and social costs of solid waste has grown, the concept of solid waste management has evolved. The assumption behind the concept is that humans can devise social-technological systems that will wisely control the quantity and characteristics of wastes.[49]

Although less significant ecologically than the problems already discussed, littering is another major waste disposal problem. Extensive accumulations of litter can disrupt the natural environment even to the point of endangering ecosystems. Billions of dollars are spent annually by national, state, and local governments in efforts to clean up the vast array of discarded materials that have been carelessly disposed of by the general public. Nonreturnable bottles and aluminum cans have added considerably to the general litter. In the United States public and private lands continue to suffer from the impacts of littering, especially those used for recreational purposes. More effective control efforts would require a change in public attitudes toward the problem; intensive and extensive research into the causes for littering; and implementation of effective and enforceable government antilittering policies. Few littering offenses actually result in fines or other forms of punishment.

Problems associated with the disposal of toxic or hazardous wastes have resulted in more governmental involvement than problems associated with littering since entry of toxic and hazardous substances into the environment can lead to serious human health problems symptomatic of accompanying environmental degradation.

It should be noted that "toxic" and "hazardous" are relative terms, so that it is not particularly useful to classify all chemical substances as toxic or nontoxic. Toxic effects depend upon the composition and basic properties of a given substance, upon the dosage, route, and condition of exposure, upon the susceptibility of organisms, and upon other factors. Toxic substances are chemicals, or mixtures of synthetic and natural chemicals considered to be poisonous to humans, other animals, or to plants. Pesticides, some industrial chemicals, drugs, hazardous wastes, and radioactive materials are generally classified as toxic.[50]

In sufficiently large quantities, even common substances may become toxic. Paracelsus noted in the sixteenth century that "all substances are poisons; there is none which is not a poison. The right dose differentiates a poison and a remedy." A common measure of toxicity is the amount required to produce death in 50 percent of the animals dosed with the chemical. This is referred to as LD50, the lethal dose to 50 percent of exposed organisms. For example, common household salt (sodium chloride) results in LD50 for humans at 4,000 mg/kg. The insecticide DDT results in LD50 at 100 mg/kg. But dioxin (TCDD), which is a growing contaminant pollution problem, produces LD50 at 0.001 mg/kg.[51]

Before the Toxic Substance Control Act (TSCA) of 1974 and RCRA were passed, federal control over toxic substance regulations was handled poorly through fragments of previously passed pollution control legislation. The Toxic Substance Control Act deals primarily with premarket testing and regulation of toxic substances. RCRA addresses much broader issues, ranging from recycling of waste to control of hazardous waste disposal. Together, these laws provide the legislative mandate for federal regulation of the generation, transportation, treatment, storage, and disposal of hazardous wastes. Through the Toxic Substance Control Act, EPA maintains a system designed to control hazardous wastes from the time the waste is generated through its ultimate disposal and is empowered to impose significant civil and criminal penalties for failure to comply with these regulations.

Future legislation could address hazardous waste issues with an array of programs mandating (1) waste reduction at the source, (2) waste separation and concentration, (3) hazardous waste exchange, (4) recovery of basic materials, (5) destruction by incineration with attending energy recovery, (6) waste detoxification and neutralization, and (7) waste-volume reduction.[52]

Unfortunately, many industrialized countries, the United States included, "solve" waste problems simply by transferring toxic wastes to less developed countries, which are less able to prevent or mitigate environmental and health impacts.

As a result of stricter regulations in the United States, corporations seek out locations overseas for disposal of unwanted hazardous wastes. Without stricter rules governing the export of toxic wastes from industrialized to less developed countries, human health and ecosystems are likely to suffer in the less developed areas. Without adequate implementation of the Basel Convention on Transboundary Shipment of Hazardous Waste, corporations can export thousands of Love Canals to nations that are ill equipped to withstand the negative impacts of hazardous wastes. Pressure in the industrialized nations to ship overseas occurs because of lower overseas disposal costs and lack of adequate treatment and disposal facilities in their countries. The 1989 Basel Convention on Transboundary Shipment of Hazardous Waste explicitly prohibits the shipment of hazardous wastes to less developed countries unless the country is notified beforehand and it is established that adequate treatment facilities exist in the receiving country.

Another strategy of American and multinational corporations to avoid strict toxic substance control and hazardous waste management regulations is the manufacture of hazardous substances outside the national boundaries where such regulations are imposed. In the less developed countries where these products are being produced, there is little or no capacity to control by-products of manufacturing processes.

The Bhopal, India, incident where more than 2,000 people were killed as a result of an explosion in an American-owned chemical manufacturing plant provides a grisly example of how multinational corporations exploit less developed countries. With regard to such problems the CEQ and the Department of State recommend (1) development of procedures for regulating export of hazardous wastes; (2) scientific study to determine problems related to storage of hazardous wastes in developing countries; (3) improvement of means to handle hazardous wastes domestically in areas using products producing the wastes; (4) development of programs identifying and utilizing science and technology to facilitate recycling and recovery of useful products from hazardous wastes; and (5) development of international agreements to coordinate control of hazardous substances and wastes. There is a need to approach problems associated with toxic wastes on a global basis because these problems do not respect national boundaries. The Basel Convention partially addresses these concerns.

Some of the best approaches to hazardous waste management are used in West European countries. Countries such as Denmark, West Germany, and Sweden are substantially ahead of the United States in developing regulatory procedures for controlling hazardous wastes. Denmark has developed a nationwide hazardous waste management control system with nationally uniform standards that are implemented by local governments. The system includes a major fiscal subsidy and an extensive network of collection sites combined with a centrally located processing facility.[53] In an assessment of the missing links in hazardous waste controls in the United States, Bruce Piasecki and Jerry Gravander observe that several West European countries developed regulatory and managerial systems for hazardous waste problems fifteen years before such problems gained public recognition in the United States.[54]

The German approach is characterized by a combination of federal guidelines and local initiatives that shape hazardous waste control policies. The key elements in the German system are the federal Waste Disposal

Law of 1977, a public-private cooperative management effort, and government subsidies to encourage private companies to process their own waste when possible. In a 1987 study of hazardous waste minimization in Europe, Alan Williams reported that the Germans also required hazardous waste minimization as a condition for issuance of environmental permits and had developed comprehensive technical standards for hazardous waste avoidance, treatment, and disposal alternatives.[55]

Disposition of radioactive wastes poses two unique problems, both nationally and globally. First, radioactive wastes have some potential for use in the manufacture of nuclear weapons. Thus, control of these materials must be much more stringent than is the case with less potentially dangerous hazardous substances. Second, radioactive wastes are different in kind from other hazardous substances. Since the fission process produces large quantities of radioactive materials that are capable of destroying life, there is a need to have them permanently prevented from entering the environment. Further, handling of radioactive waste materials requires a level of perfection never previously imagined: "Unlike the disposal of any other type of waste, the hazard related to radioactive wastes is so great that no element of doubt can be allowed to exist regarding safety. . . . In general, the complex behavior of radioactive materials in the often subtle interrelationships among the various life forms and their physical environment makes prediction of the harm caused by releases of radioactive wastes highly uncertain." [56]

No nation has yet conducted a demonstration program for satisfactory disposal of radioactive wastes. Current possibilities include deep sea dumping, deposition of wastes in underground repositories (carved out of rock or salt formations), and deep space dumping. Existing international agreements allow low-level radioactive waste materials to be disposed of in oceans.

Unfortunately, radioactive waste storage problems are much more complex than they first might appear. This is in part due to the lack of knowledge of the health impacts of low-level radiation. If human health is jeopardized with exposure to low-level as well as high-level radiation, then safe disposition of large amounts of low-level radioactive wastes may prove as difficult as disposition of more obviously toxic high-level radioactive materials.

The amount of waste material is rapidly increasing, a pattern that is likely to continue if existing nuclear plants continue in operation and if planned plants are indeed brought on line in the future. The typical nuclear reactor produces about 20–30 metric tons of spent fuel per year; this amounts to 2,000 metric tons annually for the U.S. nuclear plants.[57]

Some of the by-products from the nuclear plants have half-lives approximately five times as long as the period of recorded history.[58] Depending on who is making the judgment, it is estimated that plutonium remains toxic to humans for a period of 250,000 to 500,000 years. It appears highly unlikely that any radioactive waste storage site could be guaranteed to remain stable for that long. However, the Nuclear Waste Policy Act of 1982 (NWPA) specifies that the high-level radioactive waste depository only be geologically stable for 10,000 years. Not counting the extensive amount of high-level reprocessed waste produced by the military, the United States is projected to have accumulated approximately 60,000 metric tons of spent fuel by the year 2010, the earliest projected date by which the Yucca Mountain repository is to be open.[59] The original date for opening Yucca Mountain repository was 1998. Furthermore, the cost of selecting and establishing the permanent repository for radioactive hazardous waste under the NWPA is projected at $30 to $40 billion. Original projections were $1 billion to $2 billion.[60] Bills were passed in both houses in 1997 to require the Department of Energy to build an interim site near Yucca Mountain, but were not signed by the president.

Siting a permanent radioactive hazardous waste repository is fraught with political controversy.[61] Through a multiattribute utility analysis of the sites nominated for the first waste repository in 1986, the Department of Energy determined that the best site would be Yucca Mountain in Nevada, which is adjacent to the Nevada Nuclear Test Site. However, the Yucca Mountain site is potentially subject to earthquakes and is in the midst of a former volcanic area. A group of California Institute of Technology scientists reported in the March 27, 1998, issue of *Science* that they detected movement in the Yucca Mountain area ten times greater than previously estimated, indicating potential earthquake development.[62]

Along the political dimension it is worth noting that a 1987 University of Nevada poll indicated that three-fourths of the citizens of Nevada did not want the repository in their state.[63] Although Nevada is already a nuclear testing location, the citizens resort to NIMBYism when it comes to storage of nuclear waste. People simply do not want nuclear waste in their backyards. Perhaps the irony in the debate over the location of a permanent radioactive waste storage site is the much greater threat that an active nuclear power facility represents. The loss of cooling capacity at a nuclear power plant could lead to a meltdown of the core and a nuclear explosion that could result in the deaths of millions of people. In spite of such concerns, the Diablo Canyon nuclear power plant in California is located along the highly unstable San Andreas fault and probably represents a far greater threat than Yucca Mountain. Given the thousands of years of half-life of radioactive waste and the geologic transformations that are likely to occur over such a span of time, there is considerable doubt whether any area is perfectly suitable for nuclear waste or possibly any extremely hazardous waste.

In the United States and elsewhere, the nuclear power industry has experienced a substantial slow-down. Various factors, including escalating start-up costs for plant construction, lower energy demands than the nuclear industry originally expected, public opposition to nuclear power, short-term availability of fossil fuels, and the utter failure of industry and government to solve radioactive waste storage problems, have created this situation. Nevertheless, the United States and other industrialized nations will continue to face serious radioactive waste storage problems from existing sources.

The nuclear power industry may still look forward to marginal expansion, especially in the less developed countries, where questionable or demonstrably unsafe nuclear waste storage methods (for example, unsanctioned ocean disposal or shallow land burial) are likely to be employed. Possible recommendations include (1) encouragement of development of nonnuclear energy alternatives; (2) promotion of internationally accepted facilities for safe storage and disposal of nuclear wastes; (3) encouragement of cooperative study by various nations to determine policies, methods, and locations for radioactive waste disposal; and (4) adoption of international arrangements and agreements to ensure that ocean commons are protected.[64]

Air Pollution

The United States has achieved remarkable progress in reducing air pollution since the Clean Air Act was established in 1970. According to the EPA, the Clean Air Act saved from $5.6 trillion to $49.4 trillion in health, property, and other damages between its passage in 1970 to 1990. This represents a worst-to-best scenario with a central estimate of $22.2 trillion in benefits. During the same time frame, direct compliance costs equal $0.5 trillion.[65] This amounts to net benefits of $5.1 trillion to $48.9 trillion.

In the early 1980s, there were mounting pressures from industries and utilities to relax air quality standards. However, efforts to relax air quality standards met with stiff opposition from environmental groups. Acrimonious debates over air pollution control reform lasted for more than a decade. Finally, Congress passed a far more restrictive clean air law in 1990. Congressman Henry Waxman (D-California) led the fight for thirteen years, unrelenting in his pursuit of a stronger law to more effectively protect the air. The 1990 Clean Air Act Amendments addressed many of the most glaring weaknesses in the act, such as air toxics, ground level ozone, and acid rain.

The EPA reports that most areas of the country are now in compliance with air quality standards for conventional pollutants such as particulate matter, sulfur dioxide, and carbon monoxide. Still, more than 80 million people lived in counties that exceeded one or more of the NAAQS (National Ambient Air Quality Standards) in 1995. Ground-level ozone was the single greatest problem, based on population and number of areas not meeting the standards.[66] Ground-level ozone causes respiratory problems and is a major component of smog. *The U.S. EPA's 25th Anniversary Report: 1970–1995* indicates that between 1970 and 1995 total emissions for all of the criteria pollutants declined, except nitrogen oxide. The total emissions of lead declined 98 percent, ozone declined 23 percent, sulfur dioxide declined 32 percent, carbon monoxide decreased 23 percent, and particulate matter declined 78 percent, but nitrogen oxide emissions increased 14 percent. These successes are even more impressive considering that from 1970 to 1994, the U.S. population increased 27 percent, the economy grew by 90 percent, and the vehicle miles traveled increased by 111 percent.[67]

Generally, there has been strong public support for air pollution control policies since the passage of the Clean Air Act. According to a 1982 Lou Harris poll, 93 percent of respondents were in favor of retaining or strengthening the Clean Air Act, while not a single segment of the public polled wanted a relaxation of environmental laws. Nationwide, 39 percent of voters polled indicated they would vote against candidates who favored weakening clean air standards.[68] In a Roper Organization poll conducted in June 1992, 72 percent of the respondents believed that air pollution control laws still had not gone far enough, another 18 percent felt that the air laws struck an appropriate balance. Only 5 percent of those polled believed that air pollution control had gone too far.[69]

Air pollution is associated with concentrated populations, industrial growth, and high motor vehicle use characteristic of urban areas. These combined sources often exceed the capacity of the atmosphere surrounding urban areas to dilute pollutants. Major pollutants such as sulfur dioxide, carbon monoxide, nitrogen oxides, and particulate matter occur naturally in the atmosphere. However, the air is considered to be polluted when levels of these materials become harmful to life, cause damage to materials and structures, or impede visibility. Excessive or harmful levels of air pollution are generally caused by chemicals, smoke, or toxic substances that are by-products of human activities. Wind, natural chemical processes, and weather distribute and disperse these air pollutants.[70]

Air pollution results from the addition of various substances and energy forms to the atmosphere. Other considerations include physical, chemical, biological, and psychological impacts that are determined by humans to be undesirable or detrimental. Air pollution can occur in indoor locations as well as in the outdoor (ambient) atmosphere.

All of the following aspects are impacted by air pollution: (1) human health (increases in death and illness from cancer, heart, and lung diseases linked to air pollution); (2) plant and animal life (chronic injury inflicted on living organisms and ecosystems and on agriculture, forests, and ornamental vegetation by increasing quantities and varieties of air pollution); (3) materials and buildings (sulfur dioxide contributing to early aging and damaging of a variety of materials, ozone, and particulate matter, resulting in costly damage and soiling of paints and other materials); (4) climate, acid rain, and other global effects (air pollution altering the natural climate and atmosphere unpredictably with resulting negative effects on the global environment); and (5) visibility (air pollution dimming visibility, obscuring scenic and aesthetic views, and interfering with safe transportation).[71] The negative effects of air pollution are compounded by combinations or mixtures of pollutants.

Air and water pollution control policies formulated in the early 1970s established goals, standards, and procedures that went well beyond proven technical and scientific capabilities and abatement experience. Rather than the standard incremental policy approach with limited change, this new approach pushed professionals to develop methods to attain and maintain air quality levels required under the law. Both the 1970 air and the 1972 water pollution control laws provided the basis for a larger federal role in pollution control. Rigorous and credible compliance procedures were stipulated under these laws. Thus, policies in both air and water pollution control went beyond typical incrementalism into speculative and forward-looking policy. Although experts attested that the goals and procedures were theoretically attainable, both have engendered a great deal of controversy in efforts to transform speculative public policy into actual accomplishments.[72] In a comparison of the air pollution control policies of the United States and Sweden Lennart Lundqvist notes, "With only slight exaggeration, one could characterize the U.S. approach as one of going beyond available means to establish a seemingly absolute objective of protecting public health while the Swedish approach has been one of adjusting objectives to available means."[73]

The "forced gap" between technical-scientific capacities and legal objectives indicates an interesting reversal of the general rule in which laws and policies follow technological advances after negative effects have already occurred. This situation also dramatizes increasing needs for technological assessments to lay the groundwork for developing future scenarios. Innovations as well as potential preventive measures can be formulated for more effective pollution control in the future. The remarkable reductions in the air pollution emissions from 1970 to 1996 attest to the validity of the action-forcing speculative policy approach of the Clean Air Act.

Air quality standards are the levels of air pollution prescribed by law or regulation that cannot be exceeded during a specified time period in a defined area. Ambient air quality standards refer to the maximum allowable levels of specific pollutants permitted under the law. The EPA is required to identify the highest concentration levels of air pollutants that will not endanger public health and to establish air quality standards at or below these levels. Initial regulatory efforts for pollution control under the Clean Air Act centered around six "criteria" air pollutants and several hazardous air pollutants. The 1990 Clean Air Act Amendments expanded control efforts toward air toxics, acid rain precursors, and ground-level ozone control strategies.

The criteria air pollutants for which National Ambient Air Quality Standards have been established are particulate matter, sulfur dioxide, carbon monoxide, nitrogen oxides, lead, and ozone. Each state is required to develop a state implementation plan (SIP) for the attainment of these standards. For areas where the pollution levels are so great that they cannot attain the NAAQS, additional control plans are required such as transportation control plans (TCPs). A large number of urban areas nationwide fail to meet the Clean Air Act standards for one or more criteria pollutants. These nonattainment areas are subject to transportation control planning requirements involving such strategies as mandatory vehicle inspection and maintenance programs and ways to reduce vehicle miles traveled in these urban areas. Examples of the latter are restricted travel lanes on expressways and access to bridges and tunnels for buses and van pools during rush hour traffic periods. However, implementation of such control strategies are costly and some states have been reluctant to initiate programs. Of the states required to develop vehicle inspection and maintenance programs to meet carbon monoxide and ozone air quality standards, one-half of them failed to meet the EPA's deadline.

In the early 1970s in addition to the criteria pollutants, the EPA established the National Emission Standards for Hazardous Air Pollutants (NESHAPS) program. However, the complexity of developing control strategies for toxic air pollutants caused major delays in developing standards for them. By 1985 EPA had only developed eight standards and identified an additional twenty other toxic air pollutants for potential designation under Section 112 of the Clean Air Act. Substances included under NESHAPS are asbestos, benzene, beryllium, coke oven emissions, inorganic arsenic, mercury, radionuclides, and vinyl chloride. Standards were also drafted for methylene chloride, cadmium, ethylene dichloride, butadiene, carbon tetrachloride, and toluene.[74] The standard-setting process for identification and designation of toxic substances as hazardous air pollutants is time-consuming, costly, and fraught with heated debate over the potential health effects of these pollutants. Richard Liroff reports that a six-month study conducted by the EPA indicates that routine outdoor emissions of toxic air pollutants produce approximately 1,300 to 1,700 cases of cancer each year.[75] Another important study of the EPA Total Exposure Assessment Methodology Team reports that the indoor levels of airborne toxic organic chemicals are often much greater than outdoor levels in both rural and urban areas.[76] Liroff cautions that too much research and not enough regulation may amount to "paralysis by analysis" and that the EPA needs to increase its regulatory action under Section 112. As a partial result of the inaction of the federal government in this area, several state and local governments have gone ahead and established their own toxic air pollutant management programs.[77] The 1990 Clean Air Act Amendments address toxics by requiring the EPA to establish standards with maximum limits for 189 toxic compounds. By 1995, the EPA had issued standards for hazardous air emissions for eighteen industries and claimed that these standards reduced the amount of toxic air emissions by 900,000 tons annually.[78]

In addition to routine exposures to airborne toxic pollutants, there is the constant threat of accidental chemical releases, such as the one that occurred in Bhopal, India. In response to concerns over such incidents, the U.S. Congress created Title III: Emergency Planning and Community Right-to-Know provisions, along with the Superfund Amendments and Reauthorization Act of 1986 (SARA). These provisions mandate "a comprehensive planning effort for hazardous chemical releases, involving coordination between local, state, and federal governments and industry." In her analysis of how communities and emergency management personnel handled four actual airborne toxic releases, Susan Cutter identifies three obstacles that hamper contingency planning for such releases: lack of adequate data and information about toxic hazards in the community; inexperience in dealing with airborne toxic releases; and insufficient knowledge of how the residents and emergency managers are likely to act during such incidents.[79] In a study of the national pattern of airborne toxic releases, Susan Cutter and William Solecki identified 571 *acute* toxic airborne releases in the U.S. over a four-year period. Fifty-two percent of the incidents occurred in seven states: California, Illinois, Indiana, Louisiana, Ohio, Texas, and West Virginia. Illinois had the highest number of incidents.[80]

Policy approaches involving complex standards and regulations are complicated by political and economic pressures on agency personnel to utilize discretionary authority to lower standards or relax enforcement of regulations. For example, the power industry, which is the greatest source of sulfur dioxide in the United States, continually pressures the EPA to lower standards. The revised ozone and particulate matter standards proposed by the EPA in 1997 met stiff resistance from the power industry and many other industries who question the adequacy of the EPA's scientific methods. The industries do not want to pay higher costs for the installation of more efficient pollution control equipment because the added costs would reduce their competitiveness.[81] Although a great deal of the responsibilities for implementation of pollution control regulations is delegated to state and local governments, the national government—through the EPA—has an oversight role to assure adherence to national air quality standards. But under difficult economic conditions, many state governments may be susceptible to pressures to relax enforcement of standards in order to attract or keep industries.

Even if air pollution problems were addressed adequately within the United States by the government in cooperation with the private sector, air pollution would continue to affect both American and populations in other countries. Air pollution does not respect national boundaries any more than it respects state boundaries. Tracking of the global transportation of air pollutants reveals worldwide increases in pollution levels, these in part due to growing populations and to expansion of industrial activities in less developed countries. The World Bank reports that 70 percent of the world's population breathes unsafe air. Because of the polluted air, 300,000 to 700,000 people die prematurely and many of these people are in developing countries.[82]

The facts about the global impacts of air pollution underscore the need for global cooperation in maintenance and improvement of air quality worldwide. A program sponsored by the World Meteorological Organization and the World Health Organization (WHO) is attempting to (1) monitor pollutants on a global basis, comparing levels and trends in human activity and environmental impact; (2) assemble information relevant to assessment and improvement of air quality; and (3) collaborate with member states in efforts to develop the necessary skills and facilities to deal with air pollution questions of national and international consequence.[83] Although it is not possible to make quantitative projections of air quality around the world with much precision, it is certain that, given present policies and practices, air pollution problems will be compounded as the result of failure to regulate air quality adequately in both less developed and industrialized countries. Current observations of scattered cities in less developed countries show high levels of hazardous pollutants at levels far above those recommended by the WHO. Air pollution has increased and will continue to do so in various industrialized countries due to accelerating conversion of fossil fuels, especially coal, into usable energy.[84]

Some global air pollution problems are potentially devastating. The build-up of human-produced carbon dioxide (resulting primarily from combustion of fossil fuels) has been linked to the greenhouse effect. Such a build-up would reflect some of the earth's heat back to the earth's surface, creating a warming effect. Relative to the problem of carbon dioxide build-up in the atmosphere the National Academy of Sciences estimates that there has been a definite increase in the earth's temperature over the last 100 years. By the year 2050 a two- or three-degree increase is expected. This would result in average worldwide temperatures higher than any experienced for hundreds of thousands of years.[85] Although it is difficult to predict exactly how temperature changes might affect weather patterns and the environment, the result, according to one future scenario, might be the melting of the polar ice caps, a process that, in turn, might create coastal floods and change weather patterns causing inland deserts. Further impacts on plants could exacerbate growing food shortages worldwide.

Probably the best means available to reduce and control carbon dioxide emissions would be to lessen amounts of this material generated by the burning of fossil fuels. On June 12, 1992, the United States and 153 other industrialized and less developed countries signed the Framework Convention on Climate Change. Upon ratification, which occurred March 24, 1994, the signatory governments committed themselves to voluntary reduction of greenhouse gases or other actions that would increase the absorption of greenhouse gases.[86]

The Intergovernmental Panel on Climate Change (IPCC), composed of scientists from many different countries, has been formed to investigate the nature and extent of climate change the global warming is likely to generate. The IPCC has issued two major reports that clearly demonstrate that climate change is occurring and that major terrestrial, ecological, and social implications will result.[87]

Another means would be to reduce levels of cutting and burning of tropical rain forests. Assuming that a certain level of energy production is considered necessary to human survival at more than a bare subsistence level, reduction in fossil fuel consumption would have to be accompanied by production of energy derived from other sources. Globally, reduction of carbon dioxide levels could be achieved through reforestation and implementation of new cooperative programs intended to explore and develop alternative energy forms concentrating particularly on renewable energy resources.

Some scientists and many corporations disagree on the extent of global warming, and some argue that the global climate is actually cooling.[88] These dissenters argue that the many different climate change models are flawed and are based on inadequate data and that there is no conclusive one-to-one correlation that global warming is indeed a result of the burning of fossil fuels. The neutrality of these dissenters is doubtful. David Helvarg reports that "the Administration is under heavy pressure from industry to delay or weaken tougher E.P.A. air pollution standards" and that "in the United States, big oil is conducting a successful multimillion-dollar disinformation campaign on global warming."[89]

Another global air quality problem is the depletion of the stratospheric ozone layer surrounding the earth, which filters ultraviolet radiation. Stratospheric ozone serves to reduce the amount of ultraviolet radiation that reaches the earth's surface. Thus, ozone protects human populations from skin cancer while also inhibiting declines in agricultural and marine productivity.[90]

Oxone depletion is reportedly caused predominantly by the release into the environment of chlorofluorocarbons (CFCs) which were used as refrigerants, as propellants in most aerosol products, and in many industrial processes. Between 1979 and 1986 assessments indicated that the concentration of ozone in the stratosphere declined by an average of 5 percent per year and over the poles by 30 to 40 percent and up to 60 percent during the seasonal ozone hole over Antarctica.[91] The amount of CFCs released worldwide has dropped. But some countries continue to use aerosols, while many others make extensive use of CFCs in industrial products such as cooling fluids and insulation. In some countries the uses of these pollutants may still be rising.

Additional domestic and international research is needed to determine further the effects of ozone depletion on human beings, plants, animals, and ecosystems. Increased effective international regulation and integration of research efforts are also necessary. One such research effort is an interagency cooperative scientific program, begun in 1987, to investigate the Antarctic ozone hole. The National Aeronautic and Space Administration, National Oceanic and Atmospheric Administration, National Science Foundation, and the Chemical Manufacturers Association are collaborating with scientists from several American and other countries' universities in this research effort. Evidence from this study, the National Ozone Expedition (NOZE1), indicates that the ozone hole is not caused by changes in solar activity. What was found was a highly disturbed chemistry, consistent with the human-caused chemical change theory associated with CFCS.[92]

On March 22, 1985, the Vienna Convention for the Protection of the Ozone Layer was established. Through this convention, several countries agreed to cooperate in the conduct of scientific programs to assess risks to the ozone layer and to control activities that "have or are likely to have adverse effects" on the ozone layer. Subsequently, on September 16, 1987, the United States was the first of forty-seven nations to sign the Montreal Protocol on Substances that Deplete the Ozone Layer, pursuant to the Vienna Convention. The protocol froze production and use of the most dangerous chlorofluorocarbons (CFC-11, CFC-12, and CFC-113) at 1986 levels two years after the protocol went into effect. Another 20 percent reduction was to occur two years later. Still another 30 percent reduction in production was to occur six years later, if a majority of the signing countries agreed. Supposedly, U.S. and European trade associations representing CFC users and producers supported limits on CFCs. But the ability of individual countries to establish strict regulatory policies to ensure that reductions occurred on schedule in the face of likely opposition from such producer and user groups affected their implementation, despite limits established in the Montreal Protocol.[93] Subsequent research indicated that the ozone layer was deteriorating at a much faster rate than previously assumed, so the planned phase-out of CFC-11 and CFC-12 was accelerated to the end of 1995. Currently, over 120 countries have signed the Montreal Protocol.[94]

Future research will also have to address the problem of the cumulative effects upon the biosphere of various types of air pollution (carbon dioxide in the atmosphere, acid rain, ozone depletion, and other types) taken together. How does overall degradation of air quality affect climatic and atmospheric processes? In the United States, although limited scientific research on cumulative atmospheric effects has aided policymakers attempting to deal with air pollution problems, there often is insufficient basic knowledge to provide an adequate basis for decisionmaking.

Without intensive fundamental research on the effects of global pollution upon the earth's climate and atmosphere, science will be unable to adequately portray alternative future scenarios relative to the effects of probable increases in atmospheric pollutants. The data simply will not be available. And, without either the data or the scenarios, policymakers will be in the absurd position of having to make blind guesses as to which courses of action are the more likely to maintain or improve air quality throughout the world. International cooperation is needed to establish overall intergovernmental coordination of data-gathering and planning activities focused on problems of air pollution. Industrialized nations will have to pool data with that gathered by the World Climate Program, which was initiated to address air quality problems encountered in the less developed countries. (The World Climate Program is intended to take full advantage of present knowledge of the world's climate in order to attempt to identify and prevent potential adverse human-caused changes in climate.)[95] The Intergovernmental Panel on Climate Change has also made tremendous contributions toward a greater understanding of the relationships between pollution and climate change, but more information is necessary to make appropriate decisions.

The need for more research was clearly demonstrated at the 1997 meeting of the Framework Convention on Climate Control signatories in Kyoto, Japan. The Kyoto Protocol establishes how much each country is to voluntarily reduce greenhouse gas emissions. Disputes occurred about the scientific methods used to estimate emissions and the assignment of higher burdens on industrialized nations. The United States contributes about 25 percent of all greenhouses and U.S. representatives agreed to a 7 percent reduction in emissions from 1990 levels by 2008–2012. But China also contributes about 25 percent, and since it was treated as a developing nation it did not have to agree to make substantial reductions. Many of the chemical and petroleum producers put pressure on Congress to not ratify the Kyoto Protocol because of the lack of responsibility assigned to developing nations for the greenhouse gas problem.

Pesticides

A pesticide is any chemical compound, substance, or mixture of substances—including plant regulators and defoliants—used to destroy, discourage, or mitigate the negative impacts of "pests" (insects, rodents, or weeds, for example) considered harmful to humans or potentially productive animal and plant communities. Herbicides are intended to kill weeds, insecticides to kill insects, and fungicides to kill fungi. In the United States farmers use more pesticides—chiefly herbicides—than any other group. However, many different pesticides are used in homes, institutions, and industries. Most pesticide products are made from one or more of approximately 600 active ingredients. There are about 23,000 pesticides registered with the EPA.[96] One billion pounds of pesticides are utilized annually in the United States. In 1996, the estimated volume of pesticide business worldwide amounted to over $30 billion.[97]

Use of synthetic organic chemicals to control various pests expanded dramatically after World War II. Congressional Quarterly reports: "The use of agricultural chemicals began in the mid-nineteenth century when the trend toward intensive farming and specialized crop production (monocultures) created imbalances in nature that frequently led to drastic increases in pest populations. As agriculture became a commercial production industry, pesticides began to play an increasingly central role."[98] In her national best-seller, *Silent Spring* (1962), Rachel Carson warned people about the possible hazards of widespread use of pesticides. She argued that many pesticides used in agriculture and forestry had "unknown and cumulative toxic effects that could be gauged only after years of tests." Because so little was known about these pesticides, she recommended that "their use should be curtailed." She also criticized the U.S. Department of Agriculture's endorsement of the growing use of pesticides. Farmers were censured for using pesticides in amounts exceeding limits imposed by government regulation.[99]

The eloquent warning of *Silent Spring,* which helped intensify public interest in environmental affairs, was that humans were introducing dangerous substances into the environment without first knowing what their impacts might be:

> As crude a weapon as the cave man's club, the chemical barrage has been hurled against the fabric of life—a fabric on the one hand delicate and destructible, on the other miraculously rough and resilient and capable of striking back in many ways. These extraordinary capabilities of life have been ignored by the practitioners of chemical control who have brought to their task no "high-minded orientation," no humility before the vast forces with which they tamper.[100]

Serious research into the environmental impacts of pesticides began in the late 1960s. This research revealed harmful chemical build-ups of persistent chemicals like DDT, for example, in eagles, peregrine falcons, and other predator birds as the result of pesticide use. Research also showed that, by a process called "biological magnification," certain pesticides are stored in organisms low in the food chain and are transferred in ever increasing amounts as they move up the food chain.[101] Recent research has made an alarming association between pesticides and their potential impacts on animal hormones that affect reproductive systems. Chemicals like pesticides and dioxin appear to affect the ability of endocrine glands to produce hormones in the proper amount. Endocrine disruptor research has shown that pesticides can mimic, amplify, or block natural sex hormones like estrogen and testosterone resulting in sexual anomalies. This endocrine problem occurs in the most polluted areas and is linked to animals and humans worldwide.[102]

In fact, the EPA believes that the long-range airborne transport of toxic substances such as organochlorine-based pesticides is an emerging global environmental issue. Specific concerns are associated with a class of chemicals referred to as persistent organic pollutants (POPs). Some of the reasons for the increased concern are that these substances are toxic to both humans and animals; they do not degrade readily in the environment; they accumulate in the body fat of humans and animals; and they often change from a solid to a gaseous form that can be transported into the atmosphere. A substantial amount of scientific evidence indicates that relationships exist between POPs and genetic, reproductive, and behavioral abnormalities in wildlife and humans. These persistent organic pollutants also are associated with increased amounts of cancer and neurological deficits in humans.[103]

In response to the perceived severity of this issue, representatives from more than one hundred countries met in Washington, D.C., in 1995, at a conference sponsored by UNEP and the United States. At this meeting the participants agreed to formulate a legally binding agreement to more effectively control the use of selected persistent organic pollutants.

The negative environmental impacts of DDT are symptomatic of the enormity of the pesticide problem. Its use having been prohibited in the United States in 1972, DDT (an extremely persistent pesticide), still remains in the American environment today, in large part because of its continuing use in other countries, particularly developing countries. Concentrations of DDT continue to appear, and it may still be found in a variety of organisms, including the penguins of Antarctica.[104] The pesticide is dangerous to human as well as to other animal organisms. Many individuals have traces of DDT and other pesticides in their systems. The National Cancer Institute has demonstrated that some pesticides can induce tumors in laboratory animals.[105] Americans are estimated to have twelve parts per million of pesticide residues in their systems. This is almost twice the level permitted for most foods under interstate commerce laws.

Highlighting concern for mitigating the impacts of pesticides on human beings, the CEQ points out that, "pesticides reach humans through the food chain or by direct contact. . . . Long term and chronic health effects occur as the chemicals are ingested and inhaled (also through food). Bioaccumulation, the buildup of toxic materials in tissues, is evident in fish and birds as well as in humans."[106]

In dealing with "pests," insecticides and herbicides are the major pesticides used. As the problems associated with the use of persistent insecticides became better understood, the production of these chemicals declined, especially after 1975. However, more toxic substitutes that were less persistent were used, such as carbamates and organophosphates. It should be noted that the endocrine disruptors mentioned above appear to be linked specifically to organophosphates. In contrast, the production of herbicides such as atrazine and 2,4-D grew tremendously from 1960 to 1980 (by 700 percent) to replace mechanical means of removing weeds, primarily in forests.[107] After the pesticides are applied—usually more than once—they disperse throughout the environment, accumulate in soil, and wash into streams, rivers, and lakes, where they affect aquatic life, birds, and other wildlife. Over two-thirds of the insecticides used in agriculture are released from aircraft. Less than one-half this amount reaches the crops that are supposed to be protected, and very little actually ever comes into contact with insects. Unnecessarily large quantities of insecticides thus are released and dispersed into the atmosphere and, subsequently, into a wide variety of ecosystems.[108]

If pesticides were applied just to pests, only 1 percent of the amount currently used would be necessary. The effects of pesticides are far-reaching and complicated by other environmental factors. David and Marcia Pimentel note that "the number of nontarget species may decline because of pesticide contamination or because of severe weather conditions. The ways in which insecticides and herbicides travel through the environment and the dosages that eventually reach beneficial species cannot be easily traced."[109]

Furthermore, repeated applications of pesticides to given pest populations result in the "natural selection" of individual members of that population that can tolerate doses of pesticides stronger than the doses normally required to kill the majority of the population. Resistant members can breed and produce resistant strains, thus nullifying the effects of the pesticide applied.[110] Warfarin, for example, a popular rodenticide for rats, is rapidly losing its effectiveness because rats are building a genetic resistance to it.

Herbicides, insecticides, rodenticides, and predicides (chemicals like the compound 1080 used to kill predators such as coyotes) are employed by various governmental agencies in the United States. The Bureau of Land Management, for example, carries out extensive herbicide programs to remove sagebrush on public lands for the purpose of increasing the amount of forage available for livestock. However, wild species, such as antelope who feed specifically on sagebrush, and sage grouse, who rely on sagebrush for cover, suffer accordingly. Insecticides are used in many USDA Forest Service programs to control forest insects, particularly beetles. The

USDA Forest Service uses extensive amounts of herbicides for silvicultural purposes, such as clearing of undergrowth to promote growth of targeted species. Application of poisons to control rodents, predators, and other pests is a regular function of the USDA under animal damage control programs.

When animal damage control cannot be justified on the grounds of improving the economy or protecting human health, it sometimes becomes an end in itself. Some sheep ranchers, suffering economically, blame coyotes and other predators for their economic woes. These individuals sometimes advocate indiscriminate poisoning in a kind of stubborn unwillingness to look for other causes for their problems. Indiscriminate poisoning—as opposed to utilization of limited individual controls for specific predators—can lead to unnecessary poisoning of nontarget species. Poisoning of nontarget species can also occur when animals eat the carcasses of other animals. Sometimes indiscriminate poisoning can even threaten the survival of rare and endangered species as in the instance of the black-footed ferret, whose existence was jeopardized as the result of consumption of rodenticides intended to eradicate prairie dogs. Similarly, the golden eagle and the timber wolf suffer indirectly from indiscriminate employment of pesticides.

In the United States, responsibility for regulating the use of pesticides resides with the EPA, which has had this responsibility since 1970, when it was transferred from the USDA. Although responsibility for pesticide regulation was delegated to the EPA, there is still some competition among agencies and between federal and state regulatory bodies where pesticides are concerned. Not only the USDA but various state agriculture agencies as well are engaged in pesticide studies and programs under authority delegated to them from the EPA under the Federal Insecticide, Fungicide, and Rodenticide Act (FIFRA), originally passed by Congress in 1947.

Although pesticide control legislation has existed in the United States since 1947, regulation of pesticides has been an important environmental concern only since the 1960s. The first major legislation to result was the 1972 amendments to FIFRA, referred to as the Federal Environmental Pesticide Control Act of 1972, which gave the EPA enhanced regulatory and enforcement powers. This act stipulated that all pesticides have to be registered with the EPA, which has the power to control the manufacture, distribution, and use of pesticides. The act also gave the EPA power to ban use and require proper disposal of hazardous pesticides. The EPA can also impose penalties for improper use of pesticides. Under the 1972 law, pesticides were divided into two categories, "general use" and "restricted use." The restricted use category is to be used only by qualified applicators because of the potential hazards involved. Under this authority EPA banned DDT in 1973 and has banned several other dangerous pesticides since then.[111]

The 1978 Federal Pesticide Act introduced changes in pesticide management to simplify the regulation of pesticides. Besides streamlining the regulatory process, the 1978 act also provided for public access to data pertaining to the effectiveness of various pesticides as well as to impacts of pesticides on human health and the environment.[112] Legally, all pesticides produced in the United States must be registered and used only according to specific instructions. However, restricted use of a given unregistered pesticide may be granted to "qualified" applicators under specific conditions. Such restrictions are at times ignored. Unregistered pesticides are often illegally used. The banned compound 1080, a dangerous predicide that, when used enters the food chain and reaches nontarget species, was privately used to control rodents and coyotes. Further, lifting of the executive order banning use of 1080 led to the granting of EPA permits to use 1080 for rodent control.

On August 3, 1996, President Clinton signed the Food Quality Protection Act (P.L. 104-170) which amends FIFRA and the Federal Food, Drug, and Cosmetic Act (FFDCA). The primary purpose of the Food Quality Protection Act was to remove pesticides from the purview of the FFDCA's "Delaney Clause," which banned the presence of any cancer-causing substance in any amount in food products. The new law also is to expedite pesticide registration under FIFRA for minor uses and to improve data collection with regard to the effects of pesticides in children's diets.[113]

The most comprehensive study ever conducted on pesticide policy, Christopher Bosso's *Pesticides and Politics,* informs us that the world of pesticide politics changed enormously between 1947 and 1972. In 1947 when the FIFRA was first enacted, the purpose of the act was to assure product quality. But in 1972, when major revisions were adopted to amend FIFRA, the focus had changed to prevention of harmful effects. Participation in the policymaking process also changed from a narrow group of industry and farm user groups to a broad collection of industry and farm user groups, as well as environmentalists and state governments. The low visibility of the proceedings associated with the 1947 act permitted the development of a policy favorable to user interests. The broad political participation associated with the 1972 amendments produced a new policy with an emphasis on balancing product quality considerations with environmental risks associated with pesticide use. Bosso suggests that the increase in oversight and participation in pesticide policy was a result of the post-Watergate reform movement.[114]

Many hazardous materials, like pesticides that are banned for use in the United States by the EPA, are exported to other countries. This practice poses many difficulties. By their very nature most pesticides are sufficiently toxic to cause serious health and environmental hazards. This is especially evident when pesticides are applied inappropriately and in excessive quantities. Further complications result from the variance in regulations pertaining to the use of pesticides from one country to the next.[115] The Federal Pesticide Act of 1978 and amendments strengthened safeguard and notification procedures for substances intended for export. Pesticides destined for export are now subject to the same labeling requirements as are applied to pesticides intended for domestic use.

With respect to the export of pesticides and other hazardous substances, it is entirely appropriate that "the principle of environmental responsibility," a principle first formulated in the 1972 Declaration of the U.N. Conference on the Human Environment, should apply. As Thomas Schoenbaum suggests, "At the threshold, the international community must insist that states take seriously their responsibilities, as reflected in principle 31 of the Stockholm Declaration, to protect the environment and to ensure that activities within their jurisdiction of control do not cause damage to the environment of other states or to areas beyond the limits of national jurisdiction."[116]

Noting an increase among developed and less developed countries of the use of monocultures vulnerable to plagues of pests, the CEQ and the Department of State warn us about the potential seriousness of the environmental consequences of expanded chemical applications. Even though their use is designed to raise crop yields, the chemicals may also adversely affect production. Furthermore, pest-predator populations are often destroyed by the chemicals and, as the pests develop resistance to the pesticides, they can become an even greater threat. Despite a significant rise in the use of pesticides, food losses to pests remain enormous. It is quite likely that increased pest resistance to pesticides will only decrease the effectiveness and increase the cost of pesticide applications.[117] The United Nations Environment Program elaborates on the problems attending increased use of pesticides:

> The continued large-scale use of pesticides has led to the appearance and proliferation of resistant strains of pests, as the result of the operation of natural selection. Increasing the dosage of pesticides merely delays the evolution of resistant races. The application of different types of pesticides has led to the evolution of pests that are immune to a wide array of chemicals. . . . FAO reported . . . that 392 species of arthropods (insects, mites, and cattle ticks) had become resistant

to pesticides. About 50 species of plant pathogens had so far been reported resistant to fungicides and bactericides, and 5 weed species resistant to herbicides.[118]

With the recognition that traditional methods generally serve to exacerbate rather than solve pest problems and that the world will continue to demand increased food supplies, more sophisticated pest control approaches are necessary.

One alternative approach is integrated pest management, which the CEQ defines as "a system for controlling pests through a combination of techniques, including natural predators and biological controls, use of resistant (crop) varieties, modifying environmental conditions, close monitoring of pest levels, and use of chemical pesticides only when needed as part of an overall strategy."[119] Integrated pest management entails the integrated use of multiple tactics for maintaining pest populations at tolerable levels. This method should help preserve a productive agricultural economy while reducing health and environmental costs. Promotion of integrated pest management, particularly in the early stages of planning and project design, requires international cooperation, especially in the less developed countries where extreme misuse of pesticides is prevalent.[120]

In 1979, in considering the international scandal associated with pesticide exports, Anthony DeCosta pointed out:

> In recent years, the lack of restrictions controlling international trade of hazardous chemicals has had dire consequences. The WHO (World Health Organization) estimates that 500,000 people throughout the world are poisoned by pesticides each year, 5,000 fatally, although no detailed statistics are available. . . . Most (third world) peasant and health care workers continue to mix chemical solutions with their hands and carry bucketfuls of DDT and other pesticides on their heads. Taken together, the results indicate a scandal of global proportions.[121]

David Pimentel estimates that pesticides now kill about 220,000 people per year.[122] The export of pesticides banned in the United States to less developed countries may have a boomerang effect for the United States. The exported materials threaten the health of the foreign consumers while contaminating foods and food products produced in less developed countries and imported back into the United States.[123] A similar pattern occurs with other industrialized countries. The United States imports annually 600 different food commodities valued at greater than $13 billion from 150 nations, many of which allow unrestricted use of dangerous pesticides.[124]

Noise Pollution

Noise is a major environmental hazard that can adversely affect human health and the quality of people's lives. Conceptually, noise is sound of an unwanted, obnoxious, and disturbing nature that has no redeeming value. Medical research reveals that noise presents serious health hazards (deafness, mental and physical illness, and even death) to human beings. The CEQ states: "Nearly half of the U.S. population is regularly exposed to levels of noise that interfere with normal activities such as speaking, listening, and sleeping."[125]

One of the more serious consequences of exposure to excessive noise is its adverse affect on hearing. It is generally agreed that steady exposure to approximately eighty-five or more decibels can cause permanent hearing loss. Traffic and other urban noises such as riveting machines, jet airplanes, and high-pitched music far exceed this limit. It is estimated that a great many people (10 percent of the population in the United States) are exposed to noises of duration and intensity sufficient to cause permanent hearing reduction. Hearing loss may occur gradually, but the damage is irreversible, due to destruction of highly specialized cells that cannot regenerate. And hearing loss has profound effects on people who, in addition to having trouble identifying and understanding sound, may find their own speech distorted or otherwise interfered with. It is common for persons with hearing loss to feel isolated and depressed.[126]

Increasing evidence points to other physical and mental illnesses—particularly associated with stress—resulting from noise pollution. Although most noises do not signal danger, persons still react to noises as if they represented threats. Adrenaline is released, blood pressure rises, muscles tense. There is some evidence of a link between noise and cardiovascular problems, particularly hypertension. Excessive noise can also constrict blood vessels, increasing the possibility of high blood pressure and peptic ulcers. It is speculated that high-level noise may disturb and cause permanent harm to unborn children. Further, excessive noise can affect the entire autonomic nervous system, producing distress, neuroses, and possibly mental illness. Excessive noise can also interrupt normal processes of sleep and relaxation, both essential to good health and peace of mind. Finally, in addition to human physiological and mental problems caused by excessive noise, there is some evidence that noise shockwaves can actually cause physical and structural damage to buildings and homes.[127]

Some of the more intangible aspects of environmental quality can be seriously impaired by unwanted, excessive noise. The relatively recent advent of recreation requiring motorized equipment—motorcycles, snowmobiles, scramblers—interferes with other forms of recreation on public lands. Some forms of recreation—hiking, nature study, fishing, hunting—require quiet, undisturbed contact with nature. Many users of public land have made strenuous efforts to seek out quiet open space, away from urban and other human-generated noises that intrudes upon enjoyment. Motorized equipment noise can also adversely affect wildlife. A quiet setting is a very important aspect of recreational experience in the outdoors, particularly in natural areas of wilderness. The interruption of quiet by noise destroys or degrades the wilderness experience for most people. More recently ORVs have been joined by helicopters. It is almost impossible to enjoy a wilderness experience at the Grand Canyon anymore. The National Park Service authorized helicopter rides over the Canyon several years ago. An airport opened near the South Rim about 1970. From 1970 to 1980 flights increased tenfold and, today, as many as one hundred helicopter takeoffs and landings occur per hour. As they cruise along the Grand Canyon, the noise reverberates off the walls in a deafening manner. With the growth in the number of ORV users and helicopters on public lands it appears that there is insufficient regulation of noise to ensure preservation of recreational quality.[128]

The main source of urban noise is surface transportation. The ancient Romans had similar problems; chariot traffic was banned from Rome at night. Still, today there are relatively few controls over surface transportation noise. Although noise regulation traditionally has been the responsibility of state and local governments, these governments have experienced great difficulty establishing and enforcing regulations for reducing avoidable and damaging noise. Thomas Schoenbaum points out that "many communities have enacted ordinances which establish a maximum noise level measured at the property line of the offending source. Noise abatement planning law should also be incorporated into building codes, site planning, and zoning requirements." Nevertheless, few cities have comprehensive noise abatement programs.[129]

The Noise Control Act was passed into law in 1972. This was the first comprehensive federal program ever devised to curb noise harmful to the human environment. The act gave the national government authority to set standards that would eliminate certain commercial sources of noise.[130] The EPA was directed "to promote an environment for all Americans free from noise that jeopardizes their health and welfare" and was given authority to regulate new products in commerce that were "major sources of noise." It was also directed to establish noise-labeling requirements both for noisy products and for products intended to reduce noise. In 1978 Congress amended the 1972 law by passing the Quiet Communities Act, which encouraged development of noise control programs at the community and state levels. This act established a necessary linkage between federal regulatory programs and local noise control efforts.

In its analysis for rule making, the EPA concentrated on transportation and construction noise because they were of primary concern to most local communities.[131] But, as was the case with other environmental health and safety programs, the federal noise abatement program was gutted as the result of drastic budget cuts that went into effect during the Reagan administration and has never recovered adequately.[132] In 1992 the Administrative Conference of the United States (ACUS), which advises federal agencies on procedural matters, recommended that Congress repeal the Noise Control Act or delegate specific noise-control responsibilities to the EPA and provide funds for them. ACUS made the recommendation because the federal authority of the act still preempted state and local actions but was not being implemented by the EPA.[133]

Before the EPA noise abatement program was curtailed, the agency was charged with the responsibility of lowering noise emissions from products identified as major sources of noise. The major noise sources identified by EPA were medium and heavy trucks, motorcycles, buses, garbage trucks, wheel and crawler tractors (used in construction), portable air compressors, pavement breakers (jack hammers, rock drills), power lawnmowers, and truck refrigeration units. The EPA also issued noise control regulations for locomotive and rail cars. Through the Federal Aviation Administration (FAA), the EPA issued noise control regulations for aircraft as well.[134]

One of the more serious noise pollution problems is associated with aircraft noise specifically around airports. Low-flying jets may expose persons living or working in areas surrounding airports to noise levels in excess of 115 decibels. More than one-half of the major airports in the United States have become embroiled in noise-related litigation, complaints, or political disputes. Complaints have ranged from alleged hearing losses to possible property damage.[135] In an article describing problems associated with noise emanating from aircraft in the Los Angeles vicinity, *Newsweek* reported:

> Cracked windows, rattling crockery, drowned out conversation, people who live near airports constantly complain about such troubles. But these may be the least bothersome [impacts] of jet noise. Constant exposure to the roar and whine of aircraft, report two acoustical experts, may represent serious hazards to health. . . . Stress-related diseases, particularly strokes and cirrhosis of the liver (alcoholism), were associated with large increases in deaths by the airports as compared to deaths in another part of LA.[136]

Progress has been made in controlling some aircraft noises through the efforts of the EPA and the FAA. Yet the efforts of both agencies have encountered severe resistance. Similar resistance is encountered in other areas of noise control as well. The most adamant opposition to noise control measures generally comes from corporate sources that would have to bear much of the additional cost entailed with modification of production methods and transportation systems in order to reduce noise. But existing and anticipated technological controls offer great potential for achieving noise reduction in a cost-effective manner. Typically such new technologies emerge well in advance of regulatory measures.

Although underfunded, most states have enacted noise control legislation regulating motor and recreational vehicles, particularly snowmobiles. Despite their limited budgets, the states are in a unique position to perform leadership roles in the provision of technical support to communities. Through the interstate commerce clause of the U.S. Constitution and in cooperation with the U.S. Department of Transportation, the EPA has authority to regulate noise-related aspects of highway transportation. The Occupational Safety and Health Administration also is empowered to protect workers on the job from excessive noise. However, this authority to regulate does not necessarily mean that noise control measures have been effective. Since noise pollution occurs primarily within state and local jurisdictions, the ability of the federal government to act is limited. Through the interstate commerce clause the federal government can affect the production and use of equipment that results in the crossing of state boundaries, but the actual on-site control of noise is the responsibility of local governments under their respective police powers. As a result of such constraints, the EPA's authority to act with regard to noise pollution exists in the form of technical assistance and guidance to local governments as to appropriate regulations.

Based on surveys of the international scene, noise from surface transportation is perceived to be the most serious and widespread noise-related problem in Europe and Japan. In the early 1990s, the world's motor vehicle population, consisting of private cars and commercial vehicles, was approximately 300 million. There were only 100 million vehicles in 1960. The Organization for Economic Cooperation and Development (OECD) reports that more than one-third of the 400 million inhabitants of OECD countries are exposed to an unacceptable level of noise daily, in excess of 65 decibels.[137] Worldwide there is a pressing need to reduce human exposure to traffic noise. An integrated strategy combining legislation, urban planning, road design, soundproofing of buildings, traffic control, and other measures might have some chance of at least mitigating noise pollution problems.

Many countries have adopted regulations to control maximum permissible noise levels for different types of motor vehicles. Switzerland, France, Australia, and Japan, for example, all have noise abatement strategies and promote the development of communities with reduced levels of industrial, traffic, and domestic noise.[138]

These countries regulate noise emissions from the source for a greater number of products than the United States. They also have attempted to implement a greater variety of noise abatement approaches. Other countries, including some of the former Soviet nations, have also implemented noise regulations. On the international level, efforts have been initiated to standardize noise emission limits so that new products will not produce international trade problems.[139] According to the UNEP, noise abatement efforts have been in effect globally for some time:

> Provisions for preventing and compensating for occupational deafness have been made in many countries. Some countries have established permissible noise exposure levels for workers; others have also adopted regulations to control maximum permissible noise levels for workers; others have also adopted regulations to control maximum permissible noise levels for the different categories of motor vehicles.

With regard to aviation, UNEP further reports that under the umbrella of the International Civil Aviation Organization

> standards and recommended practices have been developed for noise certification of subsonic turbojet aeroplanes and for helicopters. Work is in progress on extending noise certification requirements to cover the remaining types of aeroplanes not yet covered by the international standards. ICAO has also developed guidance material on operating procedures for aircraft noise abatement, with special focus on safety aspects.[140]

Water Pollution

Water pollution occurs when any substance or energy form associated with human activities alters the natural or desired quality of a body of water. "Water pollution" describes the discharge of liquids, solids, and other substances likely to create a nuisance or to cause the water to be injurious, harmful, or objectionable to humans and other life-forms.

Water pollution may result either from point or nonpoint sources. Nonpoint pollution sources such as sedimentation, pesticides, organic agricultural wastes, and fertilizers are difficult to measure and control. Water pollution results from both human and natural activities. The label "pollution" is employed when wastes from human and natural activities flow into a water system in such quantities that the natural capacity of the system to cleanse itself is either reduced or destroyed.

Normally, naturally self-cleansing water systems use oxygen to break down organic pollutants into harmless or inoffensive forms. When too much waste enters a water system, however, the natural purification processes break down. The consequence is polluted water unsuitable for a variety of functions. Waterborne sewage consists mainly of wastes, heavy metals, and toxic compounds such as pesticides and chemical fertilizers. The major pollution sources are municipalities, industries, and agriculture, especially hog farms.

Water quality is determined by the extent to which bodies of water can support aquatic life while also meeting standards for the protection of human health and the preservation of human life. In order to determine the extent to which a water system may be polluted, baseline standards are developed and the relative quality of each individual system is measured against these standards. Efforts to maintain or restore water quality are based on control of the amounts and kinds of material that are dumped into waterways, deposited on land, or released into the air. The water quality of streams, rivers, lakes and oceans depends on the capacity of each of these water systems to render specific pollutants relatively harmless. This cleansing capacity varies with the types and amounts of the pollutants as well as with water temperatures, rate of water flow, degree of sedimentation, and mineral content.[141]

Preserving drinking water of sufficiently high quality that human health will not be jeopardized is the top priority in water pollution control. Industrialized societies with their multitudes of toxic synthetic chemicals, heavy metals, and other hazardous pollutants are growing more concerned than in the past about preserving the quality of drinking water sources.[142] Particularly in urban areas of less developed countries where sewage problems have not been addressed adequately, water is too often unsanitary. In the United States experts predict a severe water crisis unless improvements are initiated by Congress and other public entities and agencies at all levels of government in the management of water supplies. The crisis will not be caused by poor supply management but by pollution, waste, and excessive demand in areas already short of water. The crisis would appear to be as much institutional and systemic as it is environmental.

There is a general lack of recognition of the intimate relationship between water quality and water quantity in a given region or river basin. Agencies responsible for water management seemingly fail to recognize their interdependence. And even when specific pieces of legislation express an understanding of the need to address relationships between water quality and water quantity, this understanding may not be incorporated in a fully comprehensive legislative program. Water supply projects are frequently managed separately from water quality projects. Also, the numerous governmental entities and agencies charged with the responsibility of water management have overlapping and often conflicting functions.[143]

In practice, water pollution control efforts can lack definition, direction, and effectiveness. A 1970 study of the factors pertaining to water quality in the metropolitan area of Albuquerque, New Mexico, revealed that multiple federal agencies, state agencies, regional agencies, one county agency, one joint city-county agency, as well as several city agencies and military bases had responsibility for ensuring an adequate supply of water and maintaining the quality of water. Additionally, at least thirteen federal and four state agencies provided funding and technical assistance for solving water quality problems.[144] This situation is still quite typical. At the federal level, there are still many different programs dealing with water supply and water pollution.

The situation is so confusing, in fact, that the Office of Management and Budget has difficulty determining how much is spent annually on water programs. This absence of direction and understanding ceases to be amusing when it is realized that in almost every congressional district in the United States there is some type of serious water-related problem. Various problems include leaky municipal water drains, locally polluted groundwater, disappearing groundwater, water shortages, drought conditions, accelerating annual flood damages, and poisoning of lakes and other water systems through acid rain.[145]

Moreover, although additional sewage treatment plants are created each year, the amount of water requiring treatment is also on the increase, so the sewage-related water pollution levels continue to remain relatively the same.

Pollution from uncontrolled and nonpoint sources such as agricultural runoff, mine wastes, drainage from toxic dumps, and storm-sewer overflows continues to be a problem. While water quality is improving in some areas, in others it is deteriorating. In 1996 U.S. rivers, lakes, and coastal waters were cleaner than they were in 1970, but almost 40 percent of the waters surveyed are incapable of supporting designated uses, and 97 percent of the Great Lakes surveyed were impaired.[146] Also, anglers were warned to not consume fish or were banned from fishing in 1,700 water bodies in 1995 because the fish were either chemically contaminated or diseased. One-fourth of all shellfishing beds were closed for harvest in 1990 because of pollution.[147]

The 1987 Clean Water Act contains a provision that has improved the collection of water quality data. In accordance with Section 305(b) each state is to conduct an assessment biennially that will provide an identification and definition of the causes of water quality problems (both point and nonpoint) within their respective jurisdictions. On the basis of this assessment each state is required to target water quality control action priorities based on such indicators as environment and public health risks, value of affected aquatic habitat, and benefits to be realized. Then the states are expected to establish management programs for the targeted areas. The EPA recommends that the states conduct workshops to establish the targets and gather appropriate data from knowledgeable public and private sources.[148]

In 1984 the less visible problem of groundwater pollution from toxic substances led health officials in dozens of states to issue health warnings to their citizens. The National Wildlife Federation reported in 1985 that "concern about water quality was focused underground, as evidence mounted of contamination of aquifers that supply drinking water to half of all Americans." Nevertheless, there was little funding available to the EPA for monitoring and enforcement of prohibitions against groundwater contamination caused by hundreds of potentially dangerous pollutants.[149] A serious EPA effort to develop a long-term strategy for protecting the quality of groundwater was blocked initially at the cabinet level by the Reagan administration because it was considered to be just another environmental "raid on the treasury," constituting a federal cooption of state prerogatives.[150] But evidence of a growing number of groundwater contamination problems continued to mount, and the EPA was pressured to release its groundwater protection strategy in August 1984.[151]

The EPA groundwater protection strategy clearly acknowledges that "states, with local governments, have the principal role in groundwater protection and management." The four major components of the strategy that address critical needs are

1. Short-term buildup of institutions at the state level
2. Assessing the problems that may exist from unaddressed sources of contamination—in particular, leaking storage tanks, surface impoundments, and landfills
3. Issuing guidelines for EPA decisions affecting groundwater protection and cleanup
4. Strengthening EPA's organization for groundwater management at the headquarters and regional levels, and strengthening EPA's cooperation with federal and state agencies.[152]

Partially in response to EPA initiatives on groundwater, by 1991 forty-four states had developed or were in the process of adopting and implementing groundwater protection strategies. Seventeen states had EPA-approved wellhead protection programs and twenty-nine other states had programs under development.[153]

The Safe Drinking Water Act Amendments of 1986 (SDWA) expanded the EPA's authority to tackle difficult problems associated with groundwater pollution. Specifically, the new rules required the EPA to take a more aggressive stance on enforcement with regard to violations of the act, provide new monitoring requirements for disinfection, and more stringent underground injection control. Although these strategies are laudable, the extensive requirements for information required for effective implementation of the SDWA amendments may have been somewhat unrealistic, given the absence historically of effective baseline monitoring programs.[154] The 1994-95 CEQ environmental quality report indicates that even though the U.S. drinking supply is one of the safest in the world, one-fifth of Americans get water from a facility that is violating at least one national health standard.[155]

Many water problems have resulted from lack of coordination among programs at all levels of governments. The U.S. Congress identifies four areas of policy failure that have exacerbated water problems:

1. Lack of coordination between federal and state programs dealing with water quantity and quality
2. The failure of water agencies to recognize the interdependence between groundwater and surface water
3. Government policies that discourage conservation
4. The proliferation of cost-sharing arrangements for federal grants that have led cities to solve their least pressing water problems first.[156]

Proper coordination along with comprehensive policies clearly identifying priorities are thus clearly mandated for ensuring effective water supply, water quality, and water conservation programs. The significance of such mandates becomes far more apparent when one considers that the amount of groundwater pumped in the United States has grown from roughly 35 billion gallons per day in 1950 to about 81 billion gallons per day in 1990 according to the U.S. Geological Survey.[157]

In general, administrative coordination among government agencies has often been lacking. Each agency tends to operate in relation to its own policy, budget, values, and vested interest. Often, only token cooperative effort occurs as a result from these individual orientations and concerns. Consequently, in the area of water pollution control there has been not so much a policy as a hodgepodge of specific agency missions and policies. The intrusive participation of state and local water agencies, each with its own vested interest, has further complicated matters. Add to this mix various contributions from the private sector and the sources of overall policy failure are more apparent. Where there is a plethora of specific missions and policies, efforts at coordination are doomed practically before they begin. Pressing water quality issues may be ignored.

Conflicting water quality standards may be set at various levels of government. Economic priorities can come into conflict with environmental and public health priorities. Agencies at all levels compete over water jurisdictions, especially in the West. Many different governmental entities may control segments of the same water system and conflicts occur.

Political and geographical boundaries that are effective for management of other aspects of environmental activity lose their meaning where the flow and spread of polluted water (surface or ground) are concerned. Helen Ingram and Dean Mann observe that "many experts have argued that a river basin approach giving regional authorities the power to levy effluent fees would be an improvement." But they also note that in the political climate of the 1980s environmental legislation legitimizing this approach did not exist.[158]

In the 1990s, however, a new watershed approach to managing water resources was established to help overcome many of the jurisdictional cooperation problems. The watershed approach constitutes a new way of thinking that integrates a comprehensive ecosystem management approach into water quality management. The new approach is designed to address many different water quality problems that occur throughout a watershed including nonpoint and point sources of pollution as well as habitat degradation.[159] The common elements of a statewide watershed management approach are management units, management cycles, stakeholder involvement, strategic monitoring, assessment, prioritization and targeting, development of management strategies, management plans, and implementation of the plans.

Why watersheds? The EPA believes that managing water resources by watershed will achieve better environmental results. Because watersheds are defined naturally by hydrology, they provide more logical units for management of water resources. The resource becomes the focal point, and managers can get a better picture of the problems by understanding overall conditions in an area and the stressors that affect those conditions. The EPA points out that the watershed approach will reduce reporting requirements, simplify wetlands permitting, and provide greater flexibility. Under reporting, the biennial Section 305 (b) Clean Water Act reports will become five-year reports, but states will submit electronically information collected according to the state's schedule.[160] The increased stakeholder participation requirements will produce greater coordination between government agencies and affected parties during the development of comprehensive management strategies leading to more effective resolution of water quality problems.

Water quality standards for pollution control stipulate the characteristics and degrees of water quality. These characteristics and degrees may vary depending on the sources of the water and the uses to which the water will be put. A management plan for regulating water quality by a given agency typically considers water uses, water quality criteria that will protect those uses, methods for implementation and enforcement of water treatment plans, methods for protecting existing high quality water, and water quality standards for specific bodies of water. Normally water quality standards are based on three elements: determination of uses of each stretch of interstate or coastal water, determination of amounts of various kinds of pollutants allowed in these waters, and formation of implementation plans documenting step-by-step those remedial measures needed to control or prevent pollution. The process of determining appropriate water quality standards is enormously complicated. Authority for establishing such standards is contained within various pieces of federal and state legislation. Actual responsibility for setting water quality standards, however, rests with both federal and state agencies. The EPA establishes discharge limitations for effluent from industrial facilities and municipalities through National Pollutant Discharge Elimination System (NPDES) permits for discharges into navigable bodies of water, while the states set ambient water quality standards for waters within their jurisdiction.

Some of the major problems encountered in the establishment and implementation of water quality standards are:

1. Unclear definition of administrative authority and responsibility
2. Insufficient incentive provided to both public and private institutions to take into account the consequences of their actions that affect water quality
3. Separation of planning from operating responsibilities, with consequent lack of follow-up
4. Inappropriate agency jurisdictional authority, responsibility, and resources
5. Excessive emphasis on actions by state and local governments
6. Weak mechanisms for enforcement of water quality standards
7. Lack of integration of water quality standards into comprehensive environmental programs.

Federal water pollution control legislation follows the same philosophical approach as federal air pollution control legislation. Both categories of legislation propose strong federal leadership intended to encourage the states to set water and air quality standards. Along with the establishment of quality standards, the states are expected to set target dates for achievement of water quality goals. In some instances new technologies would be needed in order to achieve these goals. Thus, legislative approaches to solving air and water quality problems tend to be "speculative" and futuristic rather than incremental and current.[161]

Agencies that attempt to achieve speculative goals also encounter resource shortages and technical-scientific limitations. The very fact that such goals are speculative allows agency decisionmakers greater discretion in interpretation and application of the laws and regulations. Discretionary maneuvering room also increases possibilities for internal and external pressures on agencies to influence if not control outcomes.[162]

For enforcement of regulations to be consistent and have impact, the regulations themselves must be clear and precise. For enforcement of regulations to be meaningful as well, agency personnel must have strong ethical commitments to act in the public interest, regardless of legal ambiguities and private pressures. Preservation of human health and protection of the environment, which are obviously not mutually exclusive goals, should dictate decisions, especially where discretionary actions are possible. Individuals sincerely committed to such values would be likely to avoid taking unnecessary risks with either human health or the environment. Within these parameters, however, professional agency personnel should be allowed enough freedom to address their responsibilities creatively and imaginatively. For example, in one EPA decision to revise new source performance standards, the decisionmakers failed to articulate coherent environmental policy. Commenting on this case, Bruce Ackerman and William Hassler stated, "The people . . . who shared the . . . decision are remarkable for their high intelligence and conscientiousness. Their failure to make sensible policy is a symptom of organizational, not personal breakdown—a failure to give decisionmakers incentives to ask the hard questions raised by any serious effort to control the environment."[163]

In another case of administrative discretion, the EPA granted repeated extensions for nonattainment of carbon monoxide and ozone standards. In 1987 Representative John Dingell, chair of the EPA oversight committee, questioned this abuse in a letter to the EPA Administrator, Lee Thomas. Dingell wrote, "EPA's different interpretations of the Clean Air Act, in our view, reflect different EPA emphases on public health and economic well-being. The current EPA policy delays the sanctions, which were designed primarily as additional measures to protect public health, while also delaying any economic consequence of the sanctions." Dingell went on to indicate that the EPA supposedly imposed construction bans for one or more pollutants in more than 400 nonattainment areas because they failed to have an approved SIP in effect by the deadline. But six years later seventy-five areas in seventeen states still had unenforced construction bans as a result of continuing SIP problems. Dingell expressed his concern over this abuse in administrative discretion in his concluding comments:

> The GAO report raises questions about EPA's willingness to enforce, by construction ban sanctions, the law for violations of the Clean Air Act. Enforcement is spotty and not evenhanded. The report also shows that when EPA started on a course of enforcement 'many of the affected states and many legislators believed EPA was too strict by proposing the construction moratorium as quickly as possible.' EPA responded quickly by changing its interpretations of the law and adopted a new policy to appease those objecting to strict law enforcement. GAO says this relaxation 'is inconsistent with the Clean Air Act.' Yet it appears that even under a 'strict' enforcement regime, these objectors, including Congress, had little to fear because this sanction (even when imposed), without vigorous EPA monitoring and enforcement, is weak or toothless. But other sanctions are available and they are not being used. It appears that when it comes to stationary sources, EPA has little 'stomach' for enforcement through sanctions in the case of nonattainment deadlines. That, in my judgment, is wrong. That does not mean that I favor construction bans. I do not. Like Illinois, I think it is counterproductive from a pollution standpoint. But I believe the law must be enforced. I am not satisfied that this is the case under the present EPA policy.[164]

In 1988 the EPA extended the carbon monoxide and ozone nonattainment deadlines to 1998 amid numerous complaints from environmentalists and state officials.[165]

In the United States, the federal government began its activities in the area of water pollution control with the passage of the Rivers and Harbors Act of 1899. The act prohibited discharge of wastes other than sewage into the navigable waters of the United States unless permits to discharge other materials were granted by the U.S. Army Corps of Engineers.[166] This was followed by several modest pieces of legislation culminating in the Federal Water Pollution Control Act of 1972. This monumental act established the National Pollutant Discharge Elimination System, which gave the EPA the authority to control point sources of pollution. It also provided the basis for one of the most expensive public works programs in history through the Section 201 grants to state and local governments for wastewater treatment plants.[167] The federal government provided nearly $60 billion and state and local governments contributed another $20 billion for this program.[168] This municipal construction grant program was replaced in 1990 by a state revolving fund program that provides loans to state and local governments for construction of sewage treatment facilities that have to be repaid to the federal government.

The 1972 act and its amendments fundamentally changed the American approach to water pollution control. Effluent limitations rather than ambient water quality standards were imposed. The ultimate goal was "zero discharge" of pollutants into the navigable waters of the United States by 1985. The 1977 amendments provided mid-course corrections for programs associated with the 1972 act. New authority for controlling toxic pollutants was also added.[169] Further corrections were included in the 1987 Clean Water Act related to an integration of point and nonpoint source control efforts. Such action was prompted by the 1982 National Water Quality Inventory and the 1984 National Water Quality Inventory. The 1982 report indicated that six out of ten EPA regions reported nonpoint sources as the major cause of water quality problems. The 1984 report noted that nonpoint pollution was a major problem in twenty-four states and "the primary cause of use impairments (for example, degradation of a drinking water supply) of rivers and streams in three of the western regions." The 1987 act also authorized $400 million in grants to the states for implementing nonpoint source pollution management programs.[170] After twenty-five years of enhanced water pollution control following the imposition of national pollutant discharge standards, EPA reports that industrial and municipal discharges have been substantially reduced, and that the leading cause of water pollution in the U.S. remains nonpoint source pollution.[171]

While control of water pollution from point sources has been and still can be relatively effective, there has been little progress in controlling water pollution from nonpoint sources—crop production, rangeland, pastureland, feedlots, animal operations, and animal holding areas, as well as runoff from cities, roads, and mining. In many situations pollution from nonpoint sources prevents waters from being swimmable or fishable. To address the complexity and massiveness of water pollution from nonpoint sources would require integrated and comprehensive regional land use planning and management. Such an approach has yet to become a reality. But the new emphasis on watershed management may make major improvements in this area.

There also has been little progress in controlling groundwater pollution that comes largely from toxic contamination and is very difficult to control. Groundwater can be defined as subsurface water occurring underneath a water table in rocks, soils, or geological formations that are fully saturated. Permeable subsurface formations—aquifers—yield significant amounts of water to wells and springs. The volume of groundwater is approximately fifty times the annual flow of surface water. Approximately 25 percent of the fresh water used for all purposes in the United States is supplied by groundwater. Fresh water and groundwater are, however, intimately related. Groundwater provides the base flow for streams and rivers. Rainfall percolates down and fills the volume above groundwater thereby providing additional surface water. Although natural filtration processes cleanse most pollutants from groundwater, it is still very vulnerable to pollutants, particularly to those that do not break down.[172] This vulnerability becomes particularly significant when it is realized that over 40 million rural Americans depend on their own wells for drinking water. Treatment for contamination of well water before use is both ineffective and prohibitively expensive.[173] Prevention of groundwater pollution also depends heavily upon control of nonpoint pollution sources.

Water resources are threatened globally. The CEQ and the Department of State report *Global 2000* summarizes the difficulties:

> Water pollution from heavy application of (persistent) pesticides will cause increasing difficulties (for both industrialized and less developed countries). Water pollution in LDC's is likely to worsen as the urban population soars and industry expands. Already the waters below many LDC cities are heavily polluted with sewage and wastes from pulp and paper mills, tanneries, slaughterhouses, oil refineries, and chemical plants. . . . Virtually all of *Global 2000* study's projections point to increasing destruction or pollution of coastal ecosystems, a resource on which the commercially important fisheries of the world depend heavily. It is estimated that 60–80 percent of commercially valuable marine fishery species use estuaries, salt marshes, or mangrove swamps for habitat at some point in their life cycle.[174]

Both contamination of water and demand for reliable water will greatly increase. If contamination is to be contained and demand met, industrialized and less developed countries will need to develop comprehensive water management schemes based on hard data and the knowledge of how to improve water quality and exploit groundwater potential without exhausting groundwater resources.

Andrew Gross provides a highly illuminating picture of the state of water quality management worldwide:

> In Canada, there is good cooperation now between public authorities and the private sector, though legislation is viewed as weak and so is enforcement. The pulp and paper industry decreased its pollutant

level to that of 25 years ago, despite doubling production at its mills. Latin America, Brazil, Mexico, and Venezuela are constructing major wastewater treatment facilities, training inspectors, setting effluent guidelines, and allowing tax breaks on pollution control equipment. But fast population growth and urbanization create enormous pressures; in rural areas, facilities are inadequate too. Enforcement of laws is weak.

In Western Europe, effluent guidelines and river basin management proved successful in parts of France and Germany. Countries along the Rhine are making definite efforts to clean up "the sewer of Europe." Sweden leads the world in relative environmental spending, but still has problems with acid rain polluting its lakes (originating from the Benelux nations, France, and Germany). Some companies, like Hoechst, spend 15 percent of their capital budget on pollution control equipment. In Eastern Europe, the former Soviet states are finally becoming aware of the cost of polluting, but the record of water/wastewater treatment is "spotty." Hungary and Czechoslovakia have had an effluent charge system since the early 1960s and have reduced waste loadings in their rivers. In the Mediterranean Basin, 18 nations signed an agreement to decrease chemical discharges, but there is much to be done to clean up this inland sea.

In Africa—Middle East there is concern with lack of capital, lack of trained manpower, absence of river basin arrangement, adequate water supplies, and a lack of sewage treatment. Several Central African nations have brought some pressure on mining companies to reduce pollution. Most Asian countries have very few monitoring stations for either surface or groundwater. Iraq, India, and Pakistan show damage from overirrigation. Both African and Asian countries plead for more assistance from the developed nations and from multinational agencies. Japan is the leader, along with Sweden, in relative spending for cleanup; while much has been accomplished, much remains to be done because of concentration of population and heavy industries. Australia did some early planning along the southwest crescent, but here too treatment facilities could be improved.[175]

The World Health Organization estimates that substantial numbers of people—some 60 percent of the population—in less developed countries lack adequate water supplies because of deteriorating water quality or lack of access to potable drinking water. The result is hundreds of thousands, perhaps even millions, of parasitic infections and illnesses each year. Approximately one-fourth of the people living in urban centers lack safe and reliable drinking water supplies. With the explosive expansion of human settlement in and around urban centers by the year 2000, urban water problems can only become more critical.

About two-thirds of those living in less developed countries lack access to suitable waste disposal systems. At the time of the *Global 2000* report, some 90 percent of rural populations and 25 percent of urban populations had inadequate sanitation facilities.[176] Efforts are being made by the UNEP, in cooperation with the World Health Organization, in the areas of water quality and sanitation. They include (1) a project intended to test and apply methods for surveillance of drinking water quality in rural areas, and (2) a project evaluating health hazards of wastewater reuse intended to provide less developed countries with guidelines enabling them to minimize health hazards.[177]

The thousands of abandoned carcinogenic, mutagenic, and teratogenic hazardous waste sites across the United States pose further threats to human health. But the Superfund Amendment and Reauthorization Act of 1986 provided $8.5 billion for the development of permanent remedies to reduce the health threats of hazardous waste. The ultimate crux of the matter in all pollution control matters is the extent to which citizens are willing to pay or alter their lifestyles in order to reduce the risks.

There are now more people living in cities than in rural areas in the United States. This is increasingly a worldwide phenomenon. Rural areas in many less developed countries are incapable of meeting the food and energy needs of burgeoning populations. The inmigration of poorly skilled, impoverished people is, however, overwhelming the capacity of urban ecosystems' capacity to assimilate them. Over the next ten years, three-fourths of the increase in world population will be in less developed countries, and most of this increase will be in cities already bursting at the seams. Shantytowns are forming around virtually every major city in the developing world. There is no infrastructure in these shantytowns to accommodate the waste or to provide for the health needs of the "inmigrants." The inmigration problem is also becoming an environmental refugee problem for many industrialized countries in the North, as can be seen in the United States with the influx of Latin American immigrants into California, Texas, and Florida.

Concentrations of people in inner-city areas have been shown to be subject to stress: stress on one another and stress on ecosystems. In addition to psychological stress, increasing evidence demonstrates that environmental pollution in urban areas is damaging to the health of residents. A disproportionate share of U.S. asthma deaths and illnesses is concentrated in poor, urban, African American children. Increased incidence of asthma is also associated with cockroaches and dust mites in poorly maintained inner city dwellings. There has been an increased incidence of hospital visits associated with air pollution incidents in cities around the world, especially of the elderly. Older inner-city areas also are plagued by the presence of lead-based paint and water lines with lead solder. Numerous inner-city children have learning disabilities associated with consumption of lead-based paint chips or breathing air contaminated with lead particles. With greater population there is also a greater amount of noise, which affects sleeping patterns and learning among urban children. This is especially true near airports and major traffic arteries. Living in cities may be exciting and challenging, but it appears to be dangerous to your health as well.

Urban and Regional Policy

Urban areas have a high concentration of people in a comparatively small geographic space, a substantial diversity of related activities exhibiting a high degree of interaction, and physical form characterized by a combination of a variety of built-up and vacant spaces in close proximity.[1] The urban environment results from the interrelationships and interactions among human population concentrations, the built environment, and the biophysical environment. Urban areas cannot be treated simply as extensions of human beings. Environmental considerations necessitate recognition of the underlying bases of regional ecology and of the effects of concentrated human populations on the land and other living systems. Not only political but also growing environmental pressures require that government approach urban policy on a regional basis.

Regional environmental approaches to urban problems are very difficult to undertake because (1) individual environmental functions and problems cannot be considered in isolation from one another; (2) urban problems are often subject to expedient political solutions, regardless of the environmental consequences; (3) community development goals frequently conflict with each other; (4) databases are inadequate; and (5) there is little commitment to setting priorities and addressing environmental and other urban problems.[2] The new Community-Based Environment Protection (CBEP) program initiated in the mid-nineties by the EPA is designed to address several of these problems with an ecosystem focus that extends beyond municipal borders.[3] Throughout this chapter reference is made to CBEP to suggest how urban environmental management can be improved.

Environmental problems are magnified in urban areas as a result of the concentration of people. The population of the United States lives on a small percentage of the available land, predominately along natural and man-made transportation corridors such as coastlines, river banks, and interstate highways. The transportation linkages foster greater interaction, leading to commercial activities. As commerce grows, more and more people are attracted to urban centers. But the increased density leads to many environmental problems.

The natural cleansing ability of ecosystems to purify air and water is overwhelmed by the numbers of people, and extensive pollution concentrations result. Partially as a result of the increased pollution and economic decline of cities, the American public has grown disenchanted with city life and large segments have moved to the suburban and exurban areas. Massive energy inputs are required to sustain urban life, and the associated entropy is rising to such levels in the urban environment that the continued existence of urbanization has been called into question.[4]

The dominant pattern of settlement in the United States is multicentered metropolitan regions. This represents a shift from the urban-rural patterns of the past. The shift has been facilitated by the creation of major networks of highways and communications systems and by development of new housing projects in suburban and exurban areas, just beyond the suburbs inhabited chiefly by "well-to-do" families. Many formerly rural areas have been entirely transformed into suburban and exurban housing settlements or into areas of mixed use. These areas are dotted with industrial facilities, private residences, and minifarms, and crisscrossed by transportation systems, communication networks, and power grids. Especially where population pressures are pronounced, older ways of life embodied in rural areas have been altered drastically. As urban areas sprawl out into the countryside, new "edge cities" are being formed, as the jobs and entertainment move to the suburbs where the professionals live, irrevocably changing lifestyles.[5]

Since 1955 more than two out of three new houses have been built in suburban or exurban areas in the United States. Many people want to occupy new housing in areas of low density. A recent trend toward a gentrification of some rundown urban centers has developed, however, as upper-middle-class families displace the poor. Despite this trend, opinion surveys show that people generally prefer to live in smaller communities with low-density housing, open space, clean air, relative quiet, and other amenities. Similar desires are apparent in other industrialized nations, such as Japan and the countries of Western Europe.[6]

From 1650 to 1850, the world's population doubled from 0.5 billion to 1.13 billion. Over the next one hundred years the world population doubled again to 2.5 billion. By 1997 there were almost 6 billion people living on the Earth. But the International Institute for Applied Systems Analysis in Laxenburg, Austria, estimates that the world population may reach 10.6 billion over the next eighty years and then to stabilize.[7] Less developed countries (LDCs) have experienced and will continue to experience substantial shifts in population from deteriorating rural areas to cities. Given present trends, Mexico City is expected to have about 30 million people by the year 2000, while greater Bombay, greater Cairo, Jakarta, and Seoul are projected to have 15–20 million people each. In all, more than four hundred cities in less developed countries are expected to exceed the one million population mark by 2000.[8] Approximately 3 billion people will be living in urban areas, with three-quarters of them in developing nations. Before 2010, for the first time in human history, more people will be living in urban areas than rural areas. The World Bank indicates that this trend will be accompanied by a substantial spread of urban poverty which "will become the most significant and politically explosive problem in the next century."[9]

These rapid population increases will exacerbate already critical pollution problems, both directly and indirectly:

> Rapid urban growth will put extreme pressures on sanitation, water supplies, health care, food, shelter, and jobs. . . . The majority of people in large LDC cities are likely to live in 'uncontrolled settlement'—slums and shantytowns where sanitation and other public services are minimal at best. In many large cities—for example, Bombay, Calcutta, Mexico City, Rio de Janeiro, Seoul, Taipei—a quarter or more of the population already lives in uncontrolled settlements and the trend is sharply upward.[10]

In an essay on the inexorable growth of megacities around the world Eugene Linden poses an intriguing question: does this growth portend global epidemics and pollution or will it stir feelings of self-reliance that point the cities in the direction of salvation?[11]

In this age of metropolitan spread urban life in the United States has broken out of the symbolic boundaries of the classic city. Government has become deeply involved in the evolution of urban areas insofar as the evolutionary process affects social affairs, economic growth, and the state of the environment. It is becoming more obvious that national urban policy should have both an environmental and a geographical foundation. But coordination of federal, state, and local government has proven to be extremely difficult.[12] With the present complex array of governmental entities at all levels, implementation of unifying environmental policy has not yet been possible. The forces behind the development of national urban policy have waned, leaving concern for urban affairs mostly with the local sector.

There is growing recognition of the need to address urban policy on a regional scale. Already in 1971 Walter Scheiber observed that the councils of government (COG) format of regional planning for urban areas had spread across the United States. This formalized government coordination process grew out of an awareness that urban problems extend beyond jurisdictional boundaries.[13] Although many of these multipurpose regional policymaking entities have no legal power to compel adherence to their policies, they have, nevertheless, come to be recognized as necessary adjuncts to government. Increased support from the federal government to urban areas has also stimulated the need for greater planning coordination at the broader metropolitan level in order to achieve economies of scale.[14] Most of these quasi-governmental bodies are voluntary in nature, depending financially on local government as well as federal and state grants.[15] The councils of government regional planning process provides the means to insert greater rationality into individual urban government economic development decisions, taking the entire urban region into consideration. Accordingly COGs have the capacity to mesh both environmental and developmental goals into regional policies.

In some ways people in our postindustrial society—suburbanites employed in information-based, service occupations—are restricted from perceiving the need for regional environmental planning. Such people tend to see larger environmental needs in abstract and symbolic terms, since they are physically and psychologically distanced from the more immediate and severe urban environmental dilemmas. They are likely to feel directly only short-term local impacts, and to identify with short-term, politically expedient decisions having little or nothing to do with the major environmental concerns such as sanitation, land use planning, and pollution control for the whole metropolitan area.

In the United States the physical and aesthetic deterioration of the urban environment is readily apparent. So also is the failure to develop sound regional environmental policies. Yet opportunities exist for the implementation of responsible policies for urban areas. Environmental objectives and alternatives need to be integrated and related to various sectors such as housing and economic development. This will both result in efficient administration and avoid creation of adverse environmental and other impacts. However, areawide coordination and planning are necessary for the integration of sectors of various local environmental systems.[16]

Urban areas are integrated living systems of people in their environment. They are fragmented, severely stressed ecosystems that, with all their human, technical-scientific and industrial resources, operate according to the principle of *environmental unity*. In spite of systemic fragmentation, all elements of the urban environment are interrelated and interdependent. A change in one element will lead to changes in all of the others.[17] One outstanding feature of urban or metropolitan societies is that they are overwhelmed by rapid changes in population, in social and biological needs and wants, and in technology. In environmental terms urban systems exhibit erratic behavior. At the decision-making level much of the change experienced in urban areas operates without control. Crisis-spawned policy is formulated and administered in response to symptoms felt in fragmented ways, but seldom considered comprehensively. Short-term considerations—economic development, progress, and growth—tend to determine policy.

Given the sheer weight of growing populations in urban areas of the United States, a plethora of federal, state, and local agencies has appeared in response to present and anticipated problem areas. The sheer number of such agencies has served to guarantee development or myopic, disjointed, and ineffective urban policy responses, especially in environmental areas. The incompatible mix of policymaking agencies ranging from the U.S. Department of Housing and Urban Development to various local agencies has produced little in the way of developing actual, operational, and realistic urban environmental policy on a regional basis. The pace of growth has increased by several magnitudes since 1967, when Dennis McElrath wrote that "the political culture (including agencies) has not kept pace with the dramatic changes which are occurring in society, nor with their environmental correlates."[18] McElrath pointed out then the growing interdependence between society and the physical environment, which goes beyond political boundaries, activities, and policies. He identified two major obstacles to the development of effective urban environmental policy: the absence of an integrated concept of public responsibility for environmental concerns, and the existence of a political structure that impedes comprehension of policy issues in meaningful terms.[19]

Due to policies initiated as a result of the National Environmental Policy Act, environmental impact statements are now prepared regularly for many urban projects, especially those that receive federal funding. Interagency reviews as well as input from involved federal, state, and local agencies are mandatory. The review process forces at least some serious consideration of environmental impacts. Public hearings that are required as part of the process can also focus decisionmakers on environmental issues and problems. Although public participation in the environmental impact assessment process can have a positive influence on the decision-making process, such participation does not guarantee environmentally responsible decisions. Politically powerful individuals and groups often succeed in dictating priorities with little attention given to values in the environmental public interest. Further, bias in the lead agency responsible for preparation of environmental impact statements can result in misrepresentation of public input in an effort to promote agency interests, regardless of the environmental costs.

The American urban culture is characterized by a high level of mobility.

As a result of this mobility, many urban Americans are not familiar with and lack the necessary concern for issues that goes hand in hand with long experience in a given urban area. Urban policies, then, are often formulated to serve the needs of a transient society with inadequate consideration for future values.

Transient citizens often feel alienated from and apathetic toward local government. Lacking a sense of permanence or stability, such individuals typically lack a long-range commitment to enhancement of the urban or regional environment. Despite public opinion polls to the contrary, this transient population shares with their more stable fellow citizens a habitual tolerance of environmental abuses and problems. Such problems are treated simply as everyday realities that are unchangeable. Because a great majority of apathetic urban dwellers fail to participate in governmental activities, agency administrators need to expend greater efforts to obtain and ascertain citizen inputs.

Administration of surveys concerned with environmental and human values might be one means of assessing the opinions of individuals who otherwise remain silent. Serious efforts to educate persons affected by environmental problems would facilitate greater public participation in urban environmental affairs. Regardless, residents of urban areas are the most obvious contributors to and victims of environmental deterioration. Since the majority of Americans will spend most of their lives in urban areas, they have both the right and the obligation to devise the means for living in environments free from avoidable pollution and other forms of degradation. Environmental legislation of the 1970s and 1980s encouraged and sometimes mandated the cleanup of urban areas. In the 1990s, the Clinton administration initiated the Community-Based Environmental Protection program and the Brownfields Economic Redevelopment Initiative. The Brownfields program is designed to help cities clean up abandoned properties that posed safety and health risks for residents and to turn the properties into revenue generating entities.[20]

However, the right to live in an environment free from avoidable pollution cannot be exercised unless those affected have access to accurate information concerning the nature and extent of the danger from pollutants. There is a close correspondence between the right to live in a clean environment and the right to know about hazardous chemicals stored, manufactured, used, or disposed of in their vicinity. Community right-to-know efforts are closely tied to efforts by organized labor to obtain information about existing and potential workplace hazards.

The right-to-know movements sprang from a common concern that supposedly harmless chemicals to which workers and the general public had been exposed might actually cause acute or chronic diseases. After almost a decade of controversy related to worker health issues, unions began to push vigorously for chemical labeling and wider disclosure of information concerning hazards. Frustrated by long delays in getting the federal Occupational Safety and Health Administration to issue uniform labeling standards, unions turned to state and local governments for relief. Philadelphia passed the first right-to-know ordinance in 1981, which contained strong community disclosure provisions. At the state level the community right-to-know provision was first considered in the 1981 New Jersey legislative session. After three legislative sessions, the New Jersey legislature enacted the nation's most comprehensive labeling law addressing both worker and community disclosure issues.[21]

Another milestone in the history of community right-to-know provisions was the publication of California's Model Local Ordinance in 1982. Developed to provide the basis for discussion in a series of workshops held around the state, the model ordinance was hailed widely by citizen groups as a prototype for the nation. Designed to complement California's worker right-to-know law, the Model Local Ordinance was established for use by county government. However, California officials stressed that it could easily be modified for use by a municipality. Extension of the right-to-know concept beyond the workplace received another boost late in 1982, when the International Fire Fighters Association passed a resolution at its annual meeting to seek disclosure of information on chemicals to which its members might be exposed in the course of their duties. Entry of the fire fighters into the debate gave the appearance of taking the issue of potential danger due to exposure to hazardous chemicals from a theoretical to a more practical level. But public concern had already escalated because of earlier PCB, vinyl chloride, and asbestos scares.[22]

By 1986 Elder Witt was able to report that "31 states had right-to-know laws, and in hundreds of communities, industry had begun to join hands with community leaders in the Chemical Awareness and Emergency Response Plan developed by the Chemical Manufacturers Association."[23]

The federal government responded to the problem of community right-to-know when on October 17, 1986, the Superfund Amendments and Reauthorization Act of 1986 (SARA) was enacted into law. In the same year, Title III of Public Law 99-499 established the Emergency Planning and Community Right-to-Know Act of 1986. This act stipulates specific requirements for federal, state, and local governments and industry regarding emergency planning and community right-to-know reporting on hazardous and toxic chemicals. Title III has four major parts: sections 301 to 303—emergency planning; section 304—emergency notification; sections 311 and 312—community right-to-know reporting requirements; and section 313—toxic chemical release reporting missions inventory.[24] Implementations of the provisions of Title III enhance the ability of local administrators to deal with toxic and other hazardous chemical releases through better coordination and planning with federal and state agencies.

Although local governments may have the actual responsibility to plan for and handle chemical disasters, the state governments coordinate their efforts under Title III. Governors appoint state emergency-response commissions to guide and monitor Title III efforts in their states. All states had such commissions by April 17, 1987. The state commissions divide their respective states into local planning districts and appoint a local emergency-planning committee for each district. Thirty-five states chose counties as planning districts, five chose regions, and five elected to designate the entire state as a planning district. Administratively, some states complicated the coordination process by dividing responsibilities for implementing Title III between the state environment and/or health agency and the agency responsible for dealing with emergencies.[25]

Managers are cautioned about the potential legal liability associated with failure to maintain proper reporting procedures under the Emergency Planning and Community Right-to-Know Act of 1986. The act contains specific criminal sanctions for knowingly and willfully failing to comply with certain reporting requirements.[26]

The Environmental Protection Agency has further responded to the need to provide citizens access to environmental information through the establishment of regulatory information and numerous databases on the internet. The EPA Center for Environmental Information and Statistics provides citizens with a convenient online source of reliable and comprehensive information about environmental quality in their community. Simply by typing in their zip codes citizens can gain access to information for that area on air quality, drinking water, surface water quality, hazardous waste, and toxic releases.[27]

Despite the recent slowing down of the rate of environmental deterioration in urban areas as a result of NEPA, other federal environmental legislation, state environmental legislation, and local environmental ordinances, all major urban areas in the United States continue to face major environmental problems. Consequently, there remains an urgent need to develop a sound national urban policy that provides direction for environmentally sensitive urban development in the United States. Similar needs still exist at the urban level internationally. At a 1986 meeting of ministers of urban affairs from Organization for Economic Cooperation and Development (OECD) member countries, the ministers expressed total agreement about the need to protect and upgrade the urban environment.[28]

With present urban growth trends and accompanying environmental and related problems, powerful pressures may force a nationally administered urban regional policy to be forged. Should such an eventuality occur, new American lifestyles would almost assuredly result. Regardless of whatever happens formally, substantial changes and innovations in national urban and regional policy are inevitable in the urbanization process.

The Inner City

The Council on Environmental Quality has argued that special attention should be given to the inner city where many of the most severe environmental problems are juxtaposed against social and economic problems. The inner city constitutes the decaying, aging urban areas interwoven within the central city as well as with surrounding sections of the city sharing the same environmental characteristics. Those living in the concentrated inner city experience a wide range of environmental problems, some apparent and some hidden, but most of them characterized by severity. Some of these problems—air pollution, congestion, noise, litter and garbage, sanitation, deteriorating living spaces, rat and other pest infestations, and absence of open space, among others—are more pressing in the inner city than elsewhere, though all of them affect other urban sectors.[29]

Due to the complexity of inner-city problems and the absence of simple solutions to them, the CEQ avers that environmental problems in urban settings cannot be differentiated sharply from nonenvironmental problems. Moreover, many traditional environmental objectives are articulated in new forms by the inner-city poor whose immediate focus mainly is upon economic and social considerations. The primary concern of these residents is poverty and attendant ills such as inadequate housing, poor health, unsanitary conditions, and inadequate education and recreation, all of which lower the quality of life. Consequently, the urban poor's environmental concerns are highly intertwined with their inner-city experience. For them, the concept of environment "embraces not only more parks but better housing; not only cleaner air and water, but rat extermination."[30]

Urban areas, including inner cities, function as ecosystems, but fragmented and incomplete ones. Although a given city may constitute only a partial ecosystem, that city, nevertheless, is part of still larger systems dependent for its survival upon energy and materials supplied from outside its immediate system. Neither the regional ecosystem nor the city stands alone. Each affects the other. The city, for example, disturbs the region through the introduction of pollution and other forms of degradation into the regional system. The regional system helps to prevent the city from experiencing environmental collapse.[31]

Stewart Marquis attributes much of the destructive power of cities to governmental and private-sector management that is deluded by a belief in the possibilities of mechanical-force technology while failing to comprehend the possibilities for combining natural with human controls of the urban environment. He recommends social planning and engineering of complete ecosystems incorporating humans and their activities as critical parts "to deal with complex interdependencies between lesser systems, the social life, artifactual, the natural, as they function in the larger ecosystem."[32]

Inner-city problems cannot be treated in isolation from a larger environmental context. Narrowly targeted approaches neglect essential environmental considerations. Underlying causes and effective solutions to environmental problems are often disregarded or ignored. Urban and federal managers addressing inner-city problems not only tend to ignore the wider environmental context but follow a wide variety of specific approaches and missions that often conflict with one another. Such conflicts work directly against the evolution of integrated social and environmental programs that may be necessary for the survival of cities.

The development of solutions for inner-city environmental problems will require effective and comprehensive approaches that address the variety and complexity of inner-city problems. However, such efforts should be tailored to the unique needs and conditions of each locality. The Community Based Environmental Protection and Brownfields programs, initiated by the EPA in the 1990s, acknowledge these concerns and are being used to deal with them in a more effective manner.

Most of the resources necessary for coping with broad environmental problems can be located within the respective urban area. Federal assistance is appropriate only after local resources have been utilized in a systematic manner that identifies locally determined priorities. Federal assistance, then, should be utilized to aid specific localities meet their recognized needs.

The CEQ reported in 1971 that fragmented approaches to inner-city problems are "generally inadequate to deal with the interdependent nature of the inner city environment. Programs have also failed when they failed to involve the residents themselves. And the way financial assistance was provided impeded the effort."[33] On the one hand, there is an obvious need for more and better community participation by inner-city residents in the decisionmaking process. Consideration of the values and social structures of these residents is imperative for the development of adequate solutions that will result in an improvement in the quality of life and environment of the inner city. But too often such considerations are ignored. On the other hand, the public participation process can be rife with political conflict and subject to manipulation through the influences of pressure groups. Under such circumstances environmental concerns can become almost hopelessly confused and entangled in debate. The CEQ cautions that the opportunity for public participation "should not obscure the fact that many of the forces that shape the intimate environment of the inner city resident are often beyond his knowledge or control."[34]

Major Environmental Problems of the Inner City

Air pollution. Business districts and nearby industrial areas bear the heaviest air pollution loads, and large concentrations of the urban poor live in and around business and industrial districts. Nonurban residents are exposed to far lower concentrations of particulate matter in the air than inner-city residents. In fact, the incidence of disease and death associated with asthma is disproportionately high among African American children who live in urban centers and who have a lower socioeconomic status. Exposure to excess amounts of indoor air pollution and allergens is suspected.[35] A study of hospital admissions in European cities showed a significant increase in daily admissions for respiratory illnesses with higher levels of ozone, one of the most troublesome air pollutants in urban areas.[36]

Water pollution. Water pipes in inner-city housing are frequently old and inadequately maintained, often containing lead-based pipe or joint cementing compounds. Consequently, inner-city dwellers are exposed to much higher than average amounts of lead in their drinking water. Toxic chemicals in urban drinking water can be consumed directly in the water but also can be released into the air through showers, baths, dishwashers, and cooking. Researchers in Europe have found significant levels of the toxics trichloroethylene and tetrachloroethylene in urban populations in Zagreb, Yugoslavia, Milan, Italy, and Galicia, Spain.[37] Furthermore, inner-city residents have very limited access to clean, water-based recreation; they are forced to utilize nearby water sources, such as harbors and rivers and streams that wind through downtown areas. These sources often contain dangerously high levels of bacteria and other pollutants.[38]

In spite of the availability of programs to address water pollution problems, economic rather than human health factors tend to dominate urban decisionmaking. John Adamson observes:

There are a myriad of governmental programs available to deal with water pollution problems in the urban environment, including the inner city. There are federal grants available for wastewater line and treatment plant construction, clean up of receiving streams and rivers, and the acquisition and preservation of urban parklands by water bodies. However, the cost in time and commitment on the part of local officials is sometimes too great. . . . [The] more pressing problems of creating (or salvaging) a tax base have recently taken precedence over the attempts to recreate a decent living environment for the inner city's residents who tend to be one of the least powerful political groups in most cities.[39]

Solid waste. Junk, litter, and garbage are abundant in poverty areas. These attributes have the psychological effect of discouraging individual and community efforts to create cleaner and more sanitary neighborhoods. Unattractive and odorous garbage strewn across the inner city also invites disease-carrying rodents. An estimated 60 to 80 percent of rat bites occur in inner city areas.[40]

Neighborhood deterioration. Overcrowding of individual and multiple dwelling units, poor building maintenance, widespread dilapidation of structures and abandonment of buildings by landlords are common inner-city problems. Lead is present in paint, plaster, and caulking of older buildings. Particles of these materials are sometimes eaten by children. Prolonged or recurrent ingestion of lead produces dangerously high levels of lead in the bloodstream. More than 400,000 children have been afflicted with lead poisoning that causes lethargy, convulsions, mental retardation, and even death.[41] More recent estimates suggest that a much higher number of children are affected by mild lead poisoning, a number now declining because of restrictions on lead in gasoline. Undue lead absorption among children in high-risk exposure categories has changed from more than 40 percent in some cities in the 1960s to about 5 percent in 1981, according to community programs for the prevention of childhood lead poisoning.[42] In recent years the number has been reduced considerably as lead has been replaced in paints and copper and plastic are used for plumbing.

Lack of open space and recreation. Very little of the public recreation area in the United States lies within forty miles of the center of metropolitan areas containing populations of more than 500,000. Inner cities continue to experience a decline in available open space and parklands. Transportation corridors and industries along waterways often cut off the access of inner-city residents to suitable recreation areas.[43] However, the President's Commission on Americans Outdoors (PCAO) recommended that greenways be established as "fingers of green that reach out from and around and through communities all across America, created by local action." This is being accomplished through the use of abandoned railroad rights-of-way, public and semipublic property, and riversides.[44]

Transportation problems. Constant highway construction through inner cities costs many residents their homes. Existing traffic on both highways and streets increases air pollution and noise.

Noise. Inner cities have intensive noise pollution problems. The causes for excessive noise range from transportation and construction sounds to the sounds of crowds of people engaged in everyday activities. Excessive noise negatively affects the mental health and the quality of life of inner city residents.

Given these and other environmental problems of the inner cities, serious questions can be raised concerning the overall quality of life in inner cities. Obviously, the environment molds the individual, both physically and mentally. Paul and Anne Ehrlich note that the dehumanizing effects of life in slums, ghettos, and other crowded, noisy, and smoggy environments aggravate the problems people already have, producing symptoms of mental disturbance or emotional stress.[45]

Local government administrators could substantially improve the quality of life in inner cities. First, public participation of inner-city residents in government efforts to address environmental problems that affect them directly is essential. Second, government initiatives for the solution of environmental problems need to be comprehensive and broad-based. Third, community revitalization should be encouraged through provision of adequate funding and planning assistance. Finally, attention should be focused on cities as places for people to live, work, invest, and recreate as harmoniously as possible. Typical governmental approaches focusing on symptoms and short-term solutions are inadequate, given the magnitude of the urban environmental crisis. To some extent, inner-city processes and problems may go beyond government control or influence. Governmental activities alone will not be decisive. Inner-city decline or improvement will be governed by human tastes, economic factors, and general demographic processes of regional rather than local origin.[46]

Suburbia

Suburban areas that were once compact and attractive are now too frequently sprawling and disorderly. These symptoms of decline are in large part attributable to the American romance with the automobile and the massive scale of its use. One land developer and mortgage banker, testifying before a congressional committee, described the steps leading to suburban sprawl:

> A farm is sold and begins raising houses instead of potatoes—then another farm. Forests are cut, valleys are filled, streams are buried in storm sewers. Traffic grows, roads are widened, service stations . . . and hamburger stands pockmark the highways, traffic strangles. An expressway is cut through and brings clover leafs which bring shopping centers. Relentlessly, the bits and pieces of a city are splattered across the landscape. By this irrational process noncommunities are born—formless, without order, beauty or reason—with no visible respect for people or the land. Thousands of small, separate decisions— made with little or no relationship to one another nor to their composite impact—produce a major decision about the future of our cities and our civilization, a decision we have come to label "sprawl."[47]

There is little visible governmental control of sprawl that each year consumes hundreds of thousands of acres in its largely uncontrolled outward push from areas of relatively dense population. This form of land consumption is known as the "tyranny of small decisions," each of which is made for individual gain or profit. Each of these individual decisions that contribute to the general decay of the land is usually one that will be most profitable to a real estate developer.[48]

Nationally, land use policy is a mess—incoherent and ineffective. Much of the urban policy is land use policy. In this context of potential land use calamities, growing numbers of municipal governments in the United States have adopted restrictive land use controls. Variations include limitations on the number of building permits per year (Petaluma, California), requirements for specific types of urban infrastructure such as roads and sewers (Ramapo, New York), moratoria on sewer construction (Montgomery County, Maryland), large lot requirements (portions of Vermont affected particularly by second home development). Collectively, these modest efforts at control have been seen as a "quiet revolution in land use control." Economists have explained this revolution as a response to increases in the negative externalization of costs of land development upon existing home owners. Sociologists explain the same revolution as resulting from diffusion of policy innovations.[49]

On the political level, people in the suburbs tend to dominate the metropolitan areas in the United States. One explanation lies in the greater number of people in the suburbs surrounding central cities. Population density in central cities declined from 7,517 persons per square mile in 1950 to approximately 4,167 in 1980. For the same time period the population density of suburbs increased from 175 persons per square mile to 223, while population density for the whole United States increased from 51 persons per square mile to approximately 61. The pattern of urban sprawl makes it difficult to consider urban and rural areas as distinct from each other. Typical metropolitan areas are a patchwork of urbanized and nonurbanized land. On the average only 10 percent of the land in metropolitan areas is actually urbanized. The remaining areas exist as croplands, forests, pastures, natural areas, or other nonurbanized lands. This patchwork settlement pattern results in (1) integration of land uses that incorporate a great variety of types of employment, recreation, and lifestyle; (2) intense competition for available land; and (3) further dependence on the automobile.[50]

Suburban settlement patterns may be correlated with the relatively new phenomenon of counterurbanization in which small towns and nonmetropolitan areas grow in the face of the stagnation of large cities. The population of metropolitan areas has been growing more slowly than the national average.[51] Glenn Fuguitt, Tim Heaton, and Daniel Lichter report that between 1950 and 1980 "the metropolitan share of the U.S. population increased from 56.5 percent to 76 percent. By the 1970 to 1980 period metropolitan growth slowed considerably (down to 2.7 percent from more than 8 percent each in the two previous decades). During this time period internal growth was negative (1 percent) while emergent new metropolitan areas grew by 3 percent."[52] These trends have continued in the 1990s.

Reduced rates in urban growth in many areas have resulted in declines in industry, employment, and tax base; especially in the snowbelt states. There is also greater external control of existing resources. Although many urban areas are still gaining in population, counterurbanization represents a distinct reversal of settlement patterns through the 1960s.[53] Furthermore, the movement from central cities to small towns and nonmetropolitan areas is likely to continue because light manufacturing operations and high technology businesses locate in such places because of substantially lower costs. Michael Conzen describes this phenomenon as "the proliferation of exurbia—scattered residential developments set in agricultural districts without urban amenities, and thus dependent for economic and social services upon distant villages and towns."[54] Such development places tremendous burdens on small town administrators, who generally lack the resources necessary to meet sudden increases in demands for services.

In 1965 less than one-half of all metropolitan residents lived in the central cities of the United States. With the exodus of many affluent citizens from central cities to suburban areas, there was a dramatic increase in the number of poor and disadvantaged citizens who replaced them in the central cities. A majority of the new inner-city residents were the rural poor, who came from Appalachia, the South, and from rural Central America and the Caribbean. These new residents were poorly equipped to survive in and deal with urban environmental problems. Thus, central cities experience a loss of leadership and a declining tax base while the surrounding suburban and exurban areas continue to grow.

Most major metropolitan regions need more structured and comprehensive planning and zoning. Relatively compact urban centers have been transformed into disorganized and congested megalopolitan areas with little real attention given to environmental quality considerations. Yet opportunities still exist for many large urban regions to achieve a balance between development and open space. One such opportunity is through the establishment of greenbelts—open areas surrounding areas of relatively high population density. Judith Kunofsky and Larry Orman suggest that greenbelts could provide a permanent means of protecting open areas among cities in a metropolitan region. This approach would be a more formalized version of the concept of greenways suggested earlier for linking countryside and city. While most of the land in greenbelt zones would remain in private ownership, governments could secure the future of greenbelt areas through a combination of means such as land use regulation, land acquisition, and creative applications of city and county zoning authority.[55] Conservation easements would be most appropriate, since the land could be dedicated in perpetuity for conservation purposes such as maintaining open space.

Greenbelt alternatives have been attempted in such American cities as Seattle, Boston, Cincinnati, and Boulder. Such approaches pose attractive alternatives to suburban development that often reflects little sensitivity to natural features and the quality of the environment. Environmentally sensitive areas such as ridges, slopes, and forests are affected negatively by urban sprawl. Often their beauty or utility as watersheds or habitat for wildlife is diminished. Inadequate government control is exercised over development in floodplains, on excessively steep slopes, or near aquifers (geological formations that serve as groundwater recharge areas). Data and expertise are generally available on ecological features such as subsoil composition, aquifers, ground cover, and wetlands. Such information is not usually sought by planning and zoning boards.[56] Development often appears to operate apart from and without consideration of nature. Expediency, rather than environmental sensitivity in both private and public spheres, characterizes urban sprawl.

In some rapidly developing suburban communities, city and county officials express concern over protection of the environment from the negative impacts of urban sprawl. In such areas developers are required to be sensitive to environmental values through concepts such as "impact zoning." This environmentally aware approach can serve as the basis for zoning ordinances and land use regulations. It would help to counteract the practices of granting waivers, extensions, and amendments by city and county governments that reduce the effectiveness of zoning and land use regulations.

Environmentally sound, local land use regulations often are adopted by small, well-governed, upper-middle-class suburbs. In these areas there appears to be a conscious desire to associate land use decisions with the capabilities of the environment to tolerate the results of such decisions. In contrast, regional and county governments generally do not adopt environmentally sound regulations, probably because of the less homogeneous population of these larger units of government reflecting more disparate interests. Uncontrolled land development, according to Ian McHarg, bears no relationship to natural or ecological processes, values, or intrinsic suitability. If anything, areas of high quality with prominent natural features become targets for expensive suburban developments.[57]

Consequently, in vacant areas often administered at the regional or county level urban sprawl usually takes place. County governments commonly are extremely ill-equipped to plan for or control growth and development in an orderly manner so that natural geological features are preserved. The organizational structure of county governments facilitates urban sprawl. Historically, county governments in America were organized as creatures of state governments, intended to disperse rather than consolidate power and authority. Dispersion of power has indeed occurred. But, when county governments were originally created, no one was able to foresee the roles to be played by automobiles and subdivisions in the creation of urban sprawl. Furthermore, since county governments often do not have a specific person or office in charge of planning, growth, or finances, it is difficult for any official or group of officials to assume responsibility for and control of land use in suburban areas.

County governments are the predominant units of government in most growing metropolitan areas. A county differs from a city to the extent that it is not created at the behest of its citizens. Counties are established by the state "to serve as a kind of political outpost of state government, apply-ing state laws and administering state business at the local level."[58] Often lacking strong central administration, county governments are subject to political domination by developers and pressure groups. The weakness of county government in some areas, operating without organizational leadership and lacking direction, makes them particularly vulnerable to pressure groups interested in development opportunities, regardless of the environmental and social costs. In most instances county government is incapable of formulating and implementing environmentally sound policy decisions with regard to suburban growth. A consequence of mul-tiple autonomously elected officials at the county level is decentralized and often uncoordinated policymaking, with unincorporated suburban areas left basically ungoverned and vulnerable.

Much environmental damage brought on through urban sprawl is ir-reversible. Concrete and asphalt cover over and permanently destroy topsoil. Furthermore, the removal of topsoil from natural processes and agricultural uses and its replacement with artificial covers reduces the penetration of rainwater into the soil and causes rapid water runoff with attendant detrimental effects on water tables.[59] Environmentally sensitive floodplains and ridges will never recover fully from the influence of sub-urban development. Certain plant and animal populations are similarly affected. For example, because of a lack of planning controls more than 4,000 homes and 60 businesses are located in the floodplain of the San Luis Rey River in southern California. In order to protect the homeowners, habitat for the least Bell's vireo songbird was inundated through a water control project.[60] Uncontrolled surburban growth led to the destruction of habitat for the California condor as well, resulting in its extinction in the wild. Uncontrolled suburban development degrades the environment and the quality of life for present and future generations.

Suburbia: Future Possibilities

Despite all the difficulties, some efforts have been made to increase the potential for future development of metropolitan areas without concur-rent destruction of the environment. Ian McHarg, for example, formu-lated ecologically sound development plans for the Baltimore region: the land was examined for intrinsic opportunities or constraints upon urban development and a concept he calls "physiographic determinism" was used, on the grounds that optimum development should respond to the operation of natural processes. By taking advantage of natural opportu-nities and constraints, the region would gain millions of dollars along with numerous environmental benefits, rather than the negative economic projections and environmental consequences associated with uncontrolled growth. Beginning with the basic assumption that a given area "is beautiful and vulnerable," McHarg suggests that cooperative planning for orderly suburban development can operate with the following assumptions, which may be taken sequentially:

> Development is inevitable and must be accommodated.
> Uncontrolled growth is inevitably destructive.
> Development must conform to regional goals.
> Observance of conservation principles can avert destruction and ensure enhancement.
> The area can absorb all prospective growth without depreciation.
> Planned growth is more desirable than uncontrolled growth and more profitable.
> Public and private powers can be joined in partnership in a process to realize the plan.[61]

In addition to blending economics with ecology and private with public powers to control suburban growth, other more overtly political forces are at work to resist uncontrolled development of natural areas outside the more heavily populated central cities. Resistance is being marshaled through political, economic, and litigation activities undertaken by organized and ad hoc pressure groups. Where the potential environmental consequences of proposed development can be demonstrated to be significant, public participation in developmental planning is mandated by law. In some locations and under certain circumstances urban developers cannot accomplish everything that they want to do. Group pressures can have positive impacts. The Nature Conservancy, for example, has national and regional programs for acquiring natural areas for conservation before such areas are subjected to private development and urbanization, often subsequently transferring ownership to public bodies like a state parks system. The Conservancy organizes the defense of natural areas on a local basis, providing legal and financial assistance to others interested in acquisition of natural areas for conservation.

⟨In the relative absence of policy and legal mandates for the orderly and environmentally sensitive development of suburbs, a certain amount of improvisation may be necessary.⟩The CEQ identifies one positive innovation in suburban development planning that could be used to retard the negative effects of urban sprawl:

> The combination of open space with cluster zoning or planned unit development lowers the initial community service costs because of

smaller networks of roads and utilities, and it makes for a more livable environment for the long term. These developments can vary from new satellite communities that blend in with natural landscapes to small, sensitively designed clusters of townhouses around a common green. The challenge is to break traditional ways of building and development and to break down the economic, political, and institutional barriers that obstruct widespread use of these innovations.[62]

However, some problems have developed since these innovations were first characterized. For example, some regulatory pressure will have to be brought to bear on developers who misapply the concepts of cluster zoning and planned urban residential development, especially where the use of these concepts encourages increased population densities in areas already experiencing negative environmental side effects from crowding. But such efforts still have tremendous value in that innovative solutions to the problems of development in suburbia are necessary.

Urbanization contributes to rural cultural decline in the form of loss of traditional social and cultural values and institutions. People move from rural to urban areas, and urban influences spread to hasten decline in rural cultures. Furthermore, the displaced rural poor, when subject to urban environments, in effect often become cultureless. A sense of anomie was common for those rural poor who migrated from Appalachia and the South in the 1950s and 1960s to northern cities in search of jobs and the glamour of city life.

A similar phenomenon exists now in many developing countries in Africa and Latin America, where inmigrant encampments spring up on the fringes of major cities where there are virtually no development controls. Totally lacking urban infrastructure, these communities experience tremendous hardships, ranging from a lack of basic amenities to disease and death as a result of the absence of environmental health precautions. These squatter settlements or shantytowns specifically lack safe water and waste disposal facilities. Overcrowding in flimsy, temporary structures is a common feature of daily life. In the less developed countries, it is increasingly recognized that human settlements are affected by numerous and complex political, social, and environmental factors that cannot be dealt with except through comprehensive and coherent planning and administration.[63]

In summarizing the major urban (suburban and inner-city) issues for developed or industrialized countries throughout the world, the United Nations Environment Program points to the "deterioration of inner city areas, the need for energy conservation as the result of rapidly increasing fuel costs, and public participation in improving the quality of life." Other areas of concern identified include decreasing growth rates, emergence of environmental concerns, more turbulent social conditions, and strained economic circumstances.[64]

Urban issues of this magnitude cross jurisdictional lines and are of concern to entire regions. Therefore, resolution of these issues requires cross-jurisdictional solutions. Environmental policy issues are magnified at the local level for a variety of reasons. Local governments are at the bottom of the governmental hierarchy and many of the rules and regulations created by higher units of government apply to local governments. Although national governments may set minimum standards for air and water quality, regional, state, and local agencies also have environmental responsibilities.[65] Nothing less than a full integration of efforts of all levels of government will be required if the urban environment is to be protected. The EPA Community-Based Environmental Protection initiative emphasizes an integrated effort involving multiple jurisdictions working together in a harmonious relationship.

Community-Based Environmental Protection

In recognition of the fact that environmental problems seldom are isolated to a single urban area, the Environmental Protection Agency initiated the Community-Based Environmental Protection (CBEP) program. Through CBEP the EPA provides leadership and technical assistance to state and local governments to assist them address regional environmental problems. Such assistance helps local governments develop ecosystem management initiatives that transcend political jurisdictions and protect larger ecosystems. The program is designed to help people cooperate in the development of ecosystem-based solutions to localized environmental problems that have economic, social, and environmental effects in the larger region. The first step in the protection of ecosystems is for local governments and citizens to identify how localized environmental problems affect the health of the total environment beyond municipal boundaries and, then, to cooperate in the development of solutions.

After the problems have been identified, the community must identify the stakeholders who have a common interest in protecting the environment and quality of the area. Essentially, these are the people who live, work, and have businesses in the community. Once the stakeholders are identified, they must work together to develop ecosystem protection plans. Such plans take into consideration economic, social, and ecological conditions under the broad umbrella of community values. That is, how does the community want to change existing economic, social, and ecological conditions and in what direction? The cooperative nature of the goal-setting process creates long-term support for and a sense of pride in the solutions that are developed.

The basic steps of the CBEP process are to (1) assess environmental conditions; (2) involve stakeholders; (3) identify a vision of an ideal community through a consensus of stakeholders; (4) set goals for sustainable ecosystems, quality of life, and a sustainable local economy; (5) gather information on ecosystem health, local economy, and local quality of life; (6) develop strategies to address problems in order to meet goals; (7) periodically assess progress; (8) reevaluate goals; and (9) repeat the process on a continual basis in order to sustain the ecosystems on which the health of the community depends.[66]

The EPA has been a leader or partner in community-based efforts throughout the country. Its role varies from one community to another. Where communities cross state boundaries or in places that are nationally significant, the EPA may lead the CBEP effort or be an active partner in the design and implementation of solutions. For example, the drainage basin of the Long Island Sound involves several states. Through a stakeholder-based National Estuary Plan, the EPA coordinated the development of a Comprehensive Conservation Management Plan subsequently approved by the EPA and the states of New York and Connecticut. The development of the watershed management plan involved the New York Department of Environmental Protection, Westchester County, the New York town of Lewisboro and the Connecticut communities of Norwalk, New Canaan, Redding, Weston, and Ridgefield.

The EPA also assumed a leadership role in the development of a complex management strategy to restore the South Florida Everglades ecosystem. After three years of negotiations, the Governor's Commission for a Sustainable South Florida, the Federal Everglades Restoration Task Force, Florida state agencies, numerous local governments, and private sector and public interests came to a consensus agreement on how to restore the Everglades ecosystem. The Central and South Florida Project Comprehensive Review Study agreement addresses a sixteen-county region. Among a host of actions, the plan calls for "modified water deliveries" and redirection of the South Florida urban sprawl back toward communities east of the Everglades.[67]

The EPA also has worked closely with local governments in areas of exceptional risk, such as the Biologically Integrated Orchard Systems (BIOS) Project in Central California, and in environmental justice projects, such as in the South Bronx. The BIOS project is a broad-based collaboration of the EPA, the multimillion-dollar almond industry, nonprofit organizations, farmers, private organizations, and California state agencies. This collaborative effort is focused on toxicity from organophosate pesticides along the San Joaquin River valley. In the South Bronx, the EPA and state and local agencies are using a community- and partnership-based approach to address citizen concerns about environmental problems in the industrialized residential area of South Bronx. Environmental problems addressed in this collaborative effort range from the siting of waste treatment facilities to air pollution. Asthma is a major concern for the area. As mentioned earlier in this chapter, a disproportionate percentage of the children that suffer and die from asthma live in American cities.[68]

The EPA assists other communities around the country where the agency is not directly involved by sharing data, equipment, and lessons learned. The agency has created the CBEP Internet Home Page to facilitate the transfer of such information. The information can be accessed at this EPA website: <http://www.epa.gov/ecocommunity>.

Land Use Planning

Land use planning concepts apply to both urban and nonurban areas. In recent times, land use planning at all levels has become more responsive to environmental considerations. Effective land use planning calls for environmental awareness and concern in the determination of the location, nature, and extent of development. John Baldwin asserts that in addition to knowledge and sensitivity, planning requires "the natural and human history of the locality, the ecosystem structure and function, and a sense of time, space, line, proportion, pattern, and symbolism."[69] Land use decisions must also take into account the capacity of the land itself to withstand developmental impacts. Planning efforts must be made to avoid adverse environmental consequences.

Whether urban or nonurban, land use planning is a political process in which values compete and power conflicts arise. Thomas Dye points out that in large cities "the planning staff may be the only agency that has a really comprehensive view of community development. Although they may not have the 'power' to decide about public policy, they can 'initiate' policy discussion through their plans, proposals, and recommendations."[70] Environmental considerations may or may not be incorporated into the planning process. Regional intergovernmental relations further complicate the process. Each agency, whether federal, regional, state, or local, has its own policies and interests that affect planning processes. Moreover, no governmental agency has full responsibility and authority for the environment of an entire region. Consequently, planning agencies must interact politically with a variety of other agencies having some responsibilities in relation to a given area and land use plan. Brokerage politics, bargaining, and manipulation become integral parts of the planning process. In this process environmental value considerations often are ignored or given only tacit acknowledgment.

Out of necessity, land use plans must take potential economic development and population growth into account. Such projections often are the principal impetus behind decisions that encourage overdevelopment and overpopulation that normally result in detrimental environmental effects. For example, in the southwestern United States economic development and population growth are heavily dependent upon the availability of water. Hence, local water engineers and planners may plan water developments to accommodate future needs of the metropolitan area or region. However, these water plans can also foster further development and growth over and above projected levels, thus creating demands for water that may exceed available supplies. In such planning processes insufficient thought often is given to determination of appropriate levels of economic development and population growth compatible with the preservation of environmental quality. An apparent assumption made by many planners is that substantial future growth is good and must be encouraged and accommodated, regardless of environmental and other consequences.

The CEQ questions whether planning can actually evaluate ecological factors adequately without comprehensive, long-range considerations. Master plans sometimes become collections of complementary, functional goals that emphasize growth and efficiency with no real concern shown for long-term environmental impacts.[71] In fact, public officials at various levels have at their command sophisticated land use tools for planning and directing development that achieves a balance between economic and public interests. But inadequate institutional arrangements often result in little effective use of these tools.[72]

Planning is little more than analysis of past trends and present processes with the intent of portraying possible future scenarios for a given region. The soundness of the planning process depends on the comprehensiveness of the approach, particularly when environmental factors are taken into consideration. When important variables are excluded from consideration, the negative effects on society and the environment can be quite substantial.

Planning generally means change. Goals are set with respect to economic development, population growth, and desirable levels of environmental quality. A comprehensive and long-range planning approach relates social and environmental goals to developmental goals in order to facilitate all of the goals in as harmonious a fashion as possible. The process should be continuous, particularly in view of the fact that the mere presentation of possible future alternatives does not guarantee that any of these alternatives will result.

Land use planning is an interdisciplinary process of evaluating, organizing, and controlling the present and future uses of land and its resources in terms of their suitability for the perpetuation of the long-term stability of a given environment. This includes an overall ecological evaluation with respect to the specific kinds of uses as well as evaluations of the social, economic, and physical contexts of the land concerned. Land use planning is valuable not only for nonurbanized land but also for urban areas in which development is spread over wide areas involving numerous, conflicting land uses. Planning should anticipate problems and propose solutions— assuming that there are mechanisms available for prompt and vigorous change in public policy—before environmental problems have become unmanageable.

Land use planning requires solid support from all levels of government. Areawide or regional organizations with legally entitled planning functions like metropolitan planning commissions often have little real authority. Such organizations depend on other governmental entities for funding and have to rely heavily on voluntary compliance with planning goals. Typically, city and county planning boards must rely on funding from separate budgets in both cities and counties. City councils and county commissioners can overrule or simply fail to implement land use plans and planning advice. Furthermore, governmental entities often consider that their responsibilities and plans should terminate at their political boundaries. Therefore, individuals and groups wishing to initiate activities that might not conform with planning objectives of one jurisdiction can take advantage of jurisdictional limitations by locating just outside its boundaries. Jurisdictional confusion near and along boundary lines can also result in undesirable development patterns in the disputed areas. Unplanned and uncontrolled growth also becomes possible simply because controlling governmental agencies do not exist or cannot be identified.

Across the United States planning agencies have little power to either reject or regulate subdivisions. Governmental approval is usually assured for subdivision developers who can get what they want through manipulation of zoning and other requirements. Developments often take place without significant governmental constraint on nonurbanized lands. In general, faulty land use decisions grow out of the reliance on local, short-term planning objectives. John W. Gardner warns that:

There must be an inducement so strong for state and local governments to do comprehensive planning on an appropriate geographic
scale and to conform with national goals and objectives that it is
politically and economically unpalatable for them to do otherwise. . . .
Participation on the part of local government in any regional environmental program should be as great as possible, but it must be
recognized that environmental protection problems will have to be
solved on the metropolitan or regional scale.[73]

As things stand at the present time, however, metropolitan or regional
planning perspectives are the exception rather than the rule. Planners are
forced to rely on the tools available to them. Zoning is one tool that can
serve as a powerful instrument for ensuring protection of the environment while promoting a higher quality of life.

In many industrialized nations, zoning is an accepted environmental
planning instrument utilized at various levels of government. This is particularly apparent in such densely populated countries as the United Kingdom, Germany, and the Scandinavian countries, where zoning restrictions
are well established and strongly enforced. In efforts to protect the natural
integrity of countryside, open spaces, greenbelts, agricultural lands, and
forests, land is zoned and protected from urban sprawl and other forms
of land development. The Netherlands, which is one of the most densely
populated countries of the world, has a nationally directed urban planning process with specific environmental protection objectives. The process is implemented by local governments but it is so carefully controlled
that it has produced a ring of cities around the country with a green core
of open space over time.[74]

In the United States, zoning to protect the integrity of the environment
is increasingly undertaken. Some cities like Davis, California, have extremely stringent environmentally oriented zoning regulations. For most
of the country, however, failure to establish and enforce protective zoning
restrictions has led to environmental degradation through the removal of
agricultural lands and open space around urban areas. Because of negative
consequences such as these the CEQ concluded, close to thirty years ago:

The traditional local zoning system is ill suited to protect broader
regional, state, and national values [let alone local values]. Local governments have a limited perspective on, and little incentive to, protect scenic or ecologically vital areas located partially or even entirely
within their borders. Economic pressures often spur development to
the detriment of the environment because of local government dependence on property taxes.[75]

Zoning is a form of police power that can, if used, serve to protect the environment, whether it is occupied by humans or not. Zoning authority is delegated by the state to local governmental units to use within their boundaries for the protection of the health and welfare of their citizens. The courts have established the right of communities to regulate land use through zoning and other forms of restriction in situations where it is clear that the public interest is served. The courts have also allowed local governments to restrict private property rights in order to allow preservation of open space.[76] Although zoning implies limits on private property rights, zoning restrictions have proven very easy to change through local political and economic pressures.

Zoning and land use planning could achieve more if they were conceived in terms of areawide, state, regional, or even national applications. However, application of zoning authority to larger areas is not enough to guarantee that the environment will receive sufficient protection. The planners themselves must regard environmental values as important. McHarg notes: "Each year I confront a new generation of graduate students, secure in their excellence, incipient or confirmed professionals in one or another of the planning or design fields. My most important objectives in this first encounter are to challenge professional myopia, exclusively man-centered views, to initiate consideration of basic values and to focus particularly on the place of nature in man's world—the place of man in nature."[77] Broader perspectives toward environmental values are needed by decisionmakers as well as by planners. This is particularly evident because of the need to secure administrative support for the planning process so that planning recommendations can be acted upon. Still, there are few or no mechanisms available to really plan or control land use.

With the lack of effective land use planning and control, urban sprawl and other forms of uncontrolled development have dominated the American scene. As it is true of other industrialized countries, the United States is using land more extensively and intensively than ever before. Certain kinds of land resources are becoming scarce, such as coastal wetlands and barrier islands. Millions of acres also are being taken out of productive use.

A significant proportion of the increases in nonmetropolitan growth has occurred in the least populated counties of the United States. The Conservation Foundation's 1982 report noted that forty-six states had registered higher growth rates in nonmetropolitan counties than in metropolitan ones. With this trend continuing, many sparsely settled areas are now undergoing intensive development with significant impacts on the environment.[78] Commenting on these trends, the Conservation Foundation observed that the development was occurring in lands that were previously inhospitable, lands that were too hot (Florida and the southwest); without water (the arid and semiarid southwest); inaccessible (barrier islands); too cold (Alaska); or subject to high risk of natural disasters (floodplains and hurricane and earthquake zones). The inhospitability of these lands was a form of protection from human development. With the aid of advanced technology, humans have now overcome natural forces and have changed the quality and extent of these fragile lands, altered their natural biota, and brought pollution to areas that were relatively untouched. In many places, new development is substantially affecting traditional patterns of life.[79]

Nonmetropolitan growth is occurring in new and fragile lands both in the United States and around the world. This situation calls for wise long-range planning and management that operate on a regional basis that takes full cognizance of environmental factors. Considered in relation to changing populations, changing technologies, and changing economies, planning will have to be flexible, innovative, and continuous. Planning will also have to be anticipatory, so that undesirable human settlement in fragile, undeveloped areas may be avoided. John Baldwin notes, "The environmental planning and management field is soberingly complex and still in its early stages of development." He believes that, even at the local level, community planning "requires a basic understanding of the interrelationships of ecological and social systems, the 'rules of the game,' and decisionmaking processes."[80]

Sharing similar concerns, the executive director of the UNEP points out that "the difficult problems faced by urban managers in view of the rapid growth of cities, particularly in the developing countries, [are] the lack of adequate and environmentally sound waste management facilities and the problems inherent in providing basic services such as clean water and adequate shelter." The delicacy, complexity, and urgency of wise planning are readily apparent: "The balance to be struck between population, resources, environment, and development is increasingly elusive. Large numbers of people still drift to the cities as the prospects of survival in the rural areas become ever more dim, due to drought, loss of agricultural potential and heavy population pressure on land."[81] The simple fact is that unless humans around the world develop a capacity for effective and comprehensive long-range planning, patterns of environmental degradation may reach a point where humans in any significant numbers will cease to have a recognizable role in the complex of living systems commonly known as the biosphere.

Over the past ten years numerous international meetings have been convened to address environmental and developmental issues. Foremost among those meetings are the 1992 United Nations Conference on Environment and Development (UNCED) in Rio de Janeiro, also known as the "Earth Summit"; the 1994 United Nations International Conference on Population and Development in Cairo; and the 1996 United Nations Conference on Human Settlement II (Habitat II) in Istanbul.

Several major treaties have also been signed to promote greater harmony among nations while dealing with critical environmental problems. The 1989 Basel Convention on Transboundary Movements of Hazardous Waste and Their Disposal, the 1992 United Nations Framework Convention on Climate Change, the 1992 Convention on Biological Diversity, and the 1993 Environmental Side Agreement to the North American Free Trade Agreement (NAFTA) all display an increasing willingness among the nations of the world to address environmental problems.

In her description of an environmental agenda for the United Nations, Hilary French points out that the countries of the world have now made hundreds of agreements, declarations, action plans and international treaties on the protection of the environment. These now number over 800 (if one includes less binding types of agreements), with more than 170 being environmental treaties.[1]

Yet the availability of freshwater supplies in many countries is critically low, the world's oceans are overfished, tropical forests are disappearing, cities are seriously overcrowded, and greenhouse gases continue to warm the planet and threaten human existence. Poor management practices and differences in cultural values compound the environmental issues. No amount of negotiations among diverse individuals or the establishment of innovative management strategies can address the world's environmental problems if the world's population continues to grow exponentially. According to the United Nations, the earth now has approximately six billion people and is projected to peak at about 10.4 billion in the year 2050 and then level off, assuming current rates of growth.

International Environmental Administration Issues

Environmental administration issues must become matters of national concern before they can become matters of international concern. Concern, and the beliefs on which it is founded, rather than scientific findings shape issues. For although scientific, technical, and economic factors are relevant to environmental administration, political psychology also influences environmental issues as they evolve. In fact, the intrinsic significance of an issue—its ultimate importance for human welfare or survival—seldom determines its political urgency.[2] While local psychological factors have a major bearing on how issues develop, the effects of many environmental problems extend beyond political boundaries and must be addressed as issues of international environmental administration.

Among the many pressing environmental issues of international significance, the following require immediate action, according to Lynton Caldwell:

1. Genetic loss (threatened extinction of presently endangered species)
2. Ecosystem disruption and destruction (massive loss of habitat, genetic material, quality of life, and regenerative capabilities—marine as well as terrestrial)
3. Deforestation (many of the above effects as well as destruction of forest-dwelling peoples, soil deterioration, flooding, siltation, and possible reduction of atmospheric oxygen)
4. Desertification (caused or exacerbated by human activities, reducing food and fiber productivity and simultaneously causing wind erosion of topsoil and impairment of atmospheric clarity by dust)
5. Contamination of the environment—air, water, soil, and biota (by industrial toxicants including radioactive materials). Degrading and depletion of fresh water (many of the above effects, eutrophication or acidification of lakes and streams, and exhaustion of groundwater aquifers).[3]

Despite the fact that categorization and prioritization of environmental issues may be recognized as arbitrary, often reflecting the values associated with an individual, organization, or nation, many issues still have international dimensions and must be addressed from international perspectives.

Various environmental problems such as carbon dioxide buildup, atmospheric release of radioactive materials, and ozone depletion—although often regarded as regional or national issues—may have international or even global repercussions.[4]

The U.S. government typically attempts to deal with three types of international environmental issues: (1) pressing domestic issues such as acid rain, solutions to which require international cooperation; (2) issues affecting areas under international jurisdiction, such as the world's oceans or Antarctica; and (3) issues affecting less developed countries that are in need of U.S. technical or material assistance (with an understanding that ecological instability can induce political instability).[5]

By their very nature international environmental issues are affected by rapid, unpredictable, technical-scientific developments. The magnitude of these developments often places severe demands on human societies and ecosystems. Exemplifying such difficulties is biotechnology—the use of knowledge drawn from the life sciences to manipulate living processes and organisms. Biotechnology has the capability not only to change the characteristics and life span of almost any existing life-form but also to create new life-forms. Impacts that are as potentially intrusive as these also need to be addressed within a broader framework of international environmental policy and administration.

Certain major environmental problems that confront the United States, among other countries, illustrate some of the difficulties encountered in international environmental administration. Among these key problems are (1) chlorofluorocarbon emissions and the depletion of the stratospheric ozone layer (a global pollution problem); (2) environmental transboundary problems such as acid rain (environmental difficulties shared between two countries with a common border such as the United States and Canada); and (3) tropical rain forest elimination (a global or regional resource problem).

The magnitude of these environmental problems requires multilateral approaches for problem solving. No single nation could possibly accomplish the actions necessary to reduce substantially or eliminate such wide-ranging problems. As early as 1916, with the establishment of migratory bird treaties, it was recognized that individual country or bilateral actions were inadequate to deal with transnational, regional, and global environmental problems.[6]

Acid Rain: A Transboundary Problem
of Global Proportions

Acid rain, first recorded in Manchester, England, in 1872 by a British chemist, is caused chiefly by the burning of coal and petroleum. A product of the industrial revolution, it has come to be a pernicious environmental problem of global proportions.[7] In *Acid Earth: the Global Threat of Acid Pollution,* John McCormick argues that acid rain is not exclusively a North American or West European phenomenon affecting lakes and forests but a worldwide problem whose effects are evident wherever there is industry and an abundance of vehicular road traffic. Evidence of the negative impacts of acid rain may be found in such unexpected places as Zambia, South Africa, Malaysia, Venezuela, and even the Arctic.[8] At the United Nations Conference on the Human Environment in 1972, Sweden first raised the acid rain issue as an international problem. Since that time an abundance of information has been developed, suggesting that acid rain causes environmental damage on an international scale.[9] Acid rain may also appear as snow, fog, mist, or clouds of gas. In whatever form, atmospheric acid pollution contributes to diminished tree growth and disease in trees, sterilizations of soils, pollution of rivers and lakes, crop reductions, and corrosion of buildings.[10]

In characterizing the nature and causes of acid rain problems, Anne LaBastille notes: "The problem of acid rain starts, most experts agree, with the worldwide burning of coal, oil, and natural gas. Despite general adherence to existing environmental controls, the smokestacks of electrical generating plants, industrial boilers, and smelters release sulfur dioxide and nitrogen oxides, the chief precursors of acid rain."[11] Nitrogen oxides are also emitted from automobile exhausts and from chemical fertilizers. Prevailing winds and high smokestacks, along with increased emissions of sulfur dioxide and nitrogen oxides, cause acid rain to fall over wide geographical areas. Acid fog or smog also appears in many urban areas, causing severe human health problems and contributing to the deterioration of building surfaces and landmarks. The interaction of air inversions and nitrogen oxides from heavy vehicular traffic frequently causes acid smog conditions in Los Angeles and San Francisco.

It would be a serious mistake to underestimate the present and potential negative impacts of acid pollution. The negative effects are numerous:

> All told, several thousand lakes (in the U.S.A. alone with thousands more in other countries) have been damaged. A number are lifeless, perhaps irreversibly so. Acid precipitation may cause damage to forests, soils, crops, nitrogen-fixing plants, drinking water, and building

materials as well. It is known that over a period of three to five years simulated acid rain begins to acidify soils and release metals. Most food crops do not grow well in acid soils and the increased uptake of metals in crops could affect human health in certain cases. The extent to which acid rain will affect food production is still unknown but it could be significant. [12]

Many scientists believe that these and other problems, such as leaching of chemicals by groundwater, pollution of water sources, and fish kills, are only the beginning. Other serious long-term effects are developing. In Scandinavia 20,000 of Sweden's lakes are acidified, while 80 percent of the lakes and streams in southern Norway are either "dead" or "critical." Fish populations have been wiped out in fully 4,680 square miles of Norwegian lakes.[13]

The European Monitoring and Evaluation Programme has discovered that most of the sulfur-related pollutants falling in European countries are of foreign origin. Prevailing winds transport these atmospheric pollutants over great distances. Since acid rain does not respect national boundaries it is an international problem, requiring international attention. Besides the European Monitoring and Evaluation Programme, another European effort to address the acid rain problem brought about the signing by thirty-four member governments of the United Nations Economic Commission for Europe of the Convention on Long-Range Transboundary Air Pollution.[14] As an outgrowth of this convention the Protocol on the Reduction of Sulfur Emissions or Their Transboundary Fluxes by at Least 30 Percent was adopted on July 8, 1985, at Helsinki. Twenty-one countries signed the protocol immediately (including Canada but not the United States). After sixteen countries ratified the protocol it went into effect on September 2, 1987.[15]

Although solution of the acid rain problem appeared impossible at the time, experts affiliated with the UN Economic Commission for Europe argued that technological means existed for cleaning fossil fuels before they were burned and for "scrubbing" chimney gases after combustion. Some of these cleaning processes could be used economically to reduce sulfur dioxide emissions by 12 to 15 percent. More efficient methods of combustion would reduce emissions still more. In any case, given utilization of these technical-scientific controls and assuming cooperation in implementation of stringent emission controls among the nations signing the convention, scientists in the 1980s believed that sulfur dioxide emissions could be stabilized by the year 2000, and then gradually reduced thereafter.[16]

Not all technical-scientific applications lead to control of acid-related environmental problems. For example, liming acidified rivers and lakes to raise the pH of water only alleviates symptoms of acidification according to the United Nations Environment Programme. If the acidification problem is to be addressed seriously, then the only lasting solution is to reduce the emissions of the pollutants in the first place.

Various means for reducing emissions are available, from reduction of sulfur content in combustion fuels to development of alternative energy technologies. The UNEP suggests that "a more permanent solution is to use other sources of energy instead of fossil fuels and to improve energy conservation." For the short term, "the crucial issue is whether or not countries are ready to take the measures needed to cut back emissions to an acceptable level."[17]

However, getting nations to agree on binding reductions in emissions is a complex task. At meetings in the early 1990s the parties to the Convention on Long-Range Transboundary Air Pollution (LRTAP) Protocol on the Reduction of Sulphur Emissions or Their Transboundary Fluxes by at Least 30 Percent generally agreed to meet emission reduction targets by the year 2000. But in 1993 Britain announced that it was going to reduce its emission reduction from 79 percent to 70 percent and would not meet that target until 2005. That seemingly small difference amounts to the total emissions of the Netherlands, Sweden, Austria, and Norway added together.[18]

Nitrogen oxide emissions need to be reduced because they are "considered a major factor not only in acidification but also in the formation of photochemical oxidants (mainly ozone) affecting forest health."[19] Emission rates vary among countries. From 1980 to 1995 nitrogen oxide emission projections for various countries ranged from an increase of 50 percent to a decrease of 50 percent. By 1987, only eight of the countries that signed the Long Range Transboundary Air Pollution Convention had also agreed to reduce their nitrogen oxide emissions.[20]

Achieving cooperation across countries is problematic.[21] Acid rain was essentially a local problem until the advent of high smokestacks. Historically, very little was accomplished in the areas of transboundary air pollution control. This failure was due in part to uncertainties concerning the specific effects in one country of pollutants originating in another. As things are now, according to Kenneth Dahlberg, "the relationship between polluting states and those that bear the consequences of pollution is often not reciprocal. States in which transboundary pollution originates have little incentive for bearing the economic cost of controlling emissions when the benefits would be enjoyed not by them but by the downwind states."[22]

It is unclear how international conflicts raised by acid rain will be resolved. James Regens and Robert Rycroft observe that

American environmental groups, while increasingly active in the evolving debate over controls, were not very important actors in the initial process of getting acid rain on the international political agenda. Instead, the first major initiatives came from Scandinavia. This has proved to be an important factor in the evolution of interest-group strategies, because the nongovernmental environmental coalition has strong international linkages. In fact, other national governments, and particularly those of Sweden and Canada, have been important allies in articulating the seriousness of the issue.[23]

Governments like those of the United States and Great Britain have been reluctant to impose stronger emission standards on coal-burning facilities than are already in force. Still, until inequities between nations committing acid rain abuses and nations victimized by these abuses can be resolved, acid rain problems are likely to remain a source of conflict. As Caldwell has noted, "Here is a case where some nations are allowing their territory to be used for activities which result in serious environmental damage to other countries. And as long as this circumstance exists, the acid precipitation issue will remain a serious and, very likely, growing source of antagonism in international relations and environmental policy debates in industrial countries."[24] The deregulation and economic development posture adopted by the Reagan administration discouraged governmental activism in the area of acid rain control,[25] and the country, although it has since adopted stronger controls, still produces larger volumes of emissions.

The Framework Convention on Climate Change (FCCC) signed by 154 countries in June 1992 designed to reduce the emissions of all greenhouse gases (including the precursors of acid rain—sulfur dioxide and nitrogen oxides) should help considerably to address the acid rain problem. However, at the Kyoto meeting of the Conference of Parties that signed the FCCC, there were considerable disagreements as to the setting of targets for emission reductions. Furthermore, the Republican-dominated U.S. Senate under Clinton's presidency has displayed a reluctance to ratify the Kyoto commitment. The U.S. utility industry, in particular, has expressed concerns over credit for prior actions, opposes binding targets, and opposes a cap on emissions trading.[26]

Transboundary Relations:
The United States and Canada

Transboundary relations between the United States and Canada are complicated by the fact that the two nations share a common boundary running more than 5,500 miles (including many jointly shared and connecting bodies of water) from the Atlantic to the Pacific Ocean. With such a long, common boundary many opportunities occur for environmental pollution as a result of effluents from one of the nations crossing into the other. The major international body for dealing with disputes or common problems regarding the boundary waters between the United States and Canada is the International Joint Commission (IJC), which was established under the provisions of the Boundary Waters Treaty of 1909. This commission consists of six commissioners, three from Canada and three from the United States. The commissioners exercise quasi-judicial authority in water matters that affect both countries. The IJC studies and makes recommendations on issues referred to it by either government. It also has coordinating, monitoring, surveillance, and advisory responsibilities concerning actions or programs agreed on by both governments.[27]

The IJC is also responsible for alerting the two governments about emerging or potential water or air pollution problems. Although the two governments have thus far not made use of the commission's arbitration authority, the IJC is also authorized to function as an arbitrator and to make binding decisions on matters jointly referred to it for arbitration.[28] As with most international organizations, the IJC has only those powers and responsibilities that the respective national governments wish to delegate to it. Much of the commission's strength derives from water treaties that serve as bases for negotiation when other environmental difficulties arise. The IJC serves as a permanent mechanism for dealing with transboundary environmental problems, especially those having to do with the impacts of air and water pollution.[29]

Dixon Thompson of the University of Calgary identifies a number of transboundary problems and issues of mutual concern to the United States and Canada[30]—coastal and ocean fisheries, acid rain, air and water pollution in general, water quality in the Great Lakes, Columbia River developments, impacts of Skagit River reservoir, biological and hydrological impacts on the United States of Canadian coal strip-mining at Cabin Creek and elsewhere, biological and hydrological impacts on Canada of the Garrison Diversion Project.

Transboundary pollution problems between Canada and the United States have a long history. Testing of nuclear weapons in the United States during the 1960s produced measurable contamination of the Canadian environment in the form of gross beta-radiation activity in air samples, and concentrations of cesium 137 and strontium 90 in milk.[31] Furthermore, an inventory conducted by the IJC reveals that more than 1,000 chemical substances are present in the Great Lakes.[32]

Canadian officials are generally disinclined to confront their American counterparts directly with serious transboundary environmental problems of U.S. origin, tending rather to press diplomatically for solutions over a long period of time. However, this low-profile Canadian posture has changed with regard to acid rain. In 1984, noting that the United States was alone among twenty-four nations in not recognizing the need to control acid rain, Charles Caccia, Canadian minister for environment, stated:

> Canadians, and many Americans, are bitterly disappointed that the United States Administration continues to call, not for action, but for more study. There is a catch-22 here. Many of the effects of acid rain are irreversible. By the time that we get the amount of scientific knowledge that the U.S. Administration and others seem to be asking for, the time for action will be long past. . . . Many Canadians perceive the United States Administration as stalling on what has been called the greatest single irritant to good relations between our two countries. . . . We are increasing our commitment from a 25 percent to a 50 percent reduction in emissions over 1980 levels and we will attain this reduction by 1994. . . . I submit therefore that the problem has nothing to do with a lack of scientific consensus, but rather the lack of political consensus.[33]

In spite of intensified Canadian pressure, the United States was unwilling to move from a study phase to an action phase on the issue of acid rain under the Reagan administration. President Reagan agreed to "endorse a report calling for action to curb acid rain." This could be construed as an admission that atmospheric pollutants generated in the United States travel to Canada in the form of acid rain. But the admission hardly constituted an endorsement of a program of action to address this transboundary problem.[34] Still, the reality and importance of the acid rain problem gained further credence with the release of a number of scientific reports documenting the problem. Evasion and downplaying of the acid rain issue no longer remained viable options.

As the acid rain issue developed between the countries, U.S. officials noted that emission controls in the United States were in fact far more stringent than those in Canada, even though it was recognized that Canadians were attempting to tighten regulations. The Canadian commitment to strengthen controls is important, especially in view of the fact that Canadian air pollutants also drift into the United States.[35]

As a result of the acknowledgment of the movement of air pollutants across their boundaries, representatives of the United States and Canada signed a Memorandum of Intent pertaining to transboundary air pollution on August 5, 1980.[36] Implementation of this Memorandum of Intent became difficult when the Reagan administration agreed to permit thirteen states to increase their allowable emissions of sulfur dioxide, despite provisions in the Memorandum that called for strong enforcement of existing air quality standards. Furthermore, the United States was reticent to accept Canadian research reports, an ironic refusal in view of the fact that even the scientific panel established by the White House called for immediate action to control acid rain.[37]

Claiming that at least one-half the acid rain in Canada originates in the United States, Canada continued to urge the United States to deal with rather than simply talk about the acid rain problem. Canadian radio ads encouraged visitors from the United States to press for enactment of stronger U.S. legislation aimed at reducing this particular form of transboundary pollution.[38]

After thirteen years of highly acrimonious debate, the United States Congress included better acid rain control provisions in the Clean Air Act Amendments of 1990. Title IV of the Clean Air Act Amendments established a solid basis for an acid rain control program. Title IV requires the United States to reduce sulfur dioxide emissions by 10 million tons from 1980 levels by the year 2000 and to reduce nitrogen oxide emissions by 2 million tons from 1980 levels beginning in 1995.

Following passage of the Amendments, President George Bush and Prime Minister Brian Mulroney signed the Agreement between the Government of the United States of America and the Government of Canada on Air Quality (1991), ending ten years of negotiation. However, several years after the signing of the accord, significant differences still exist in how scientists of the two countries view the acid rain problem.[39]

Any approach designed to deal effectively with acid rain must have a basic orientation that recognizes that atmospheric pollutants do not respect international boundaries. High smokestacks on both sides of the border have transformed local pollution problems into regional and international ones. Arrangements between the United States and Canada, therefore, need to focus on the concept of the North American Airshed. This entails involvement of the Canadian and U.S. private sectors as well as national and subnational governments—nations, states, provinces, cities—in unified and coordinated efforts to deal with acid rain problems.

Tropical Forests: A Global Resource Problem

A global resource, tropical rain forests are closed-canopy, broad-leafed, moist forests existing in the humid tropics where temperatures are high, rainfall is abundant, and dry seasons are short.[40] Although located almost entirely in less developed countries, tropical forests are valuable for human and other forms of life everywhere. Tropical forests generate and preserve biological diversity (including genetic and species diversity); afford protection of species (including forest peoples); provide foods, medicines, and substances used in industry; aid in water conservation; affect climate; serve as living laboratories for scientific research and education; facilitate outdoor recreation; and, in general, help maintain the web of life.

Tropical forests are the richest and most diverse expressions of life that have evolved on earth. In some areas they have a continuous history of more than 50 million years. They approximate the primeval forest biomes from which they evolved originally. Tropical forests are highly complex and fragile ecosystems with a host of interlocking relationships among diverse plant species, animal species, and the nonliving environment. These unique forests contain about one-half of the world's estimated 10 to 40 million species of plants and animals.[41]

Without any possibility of recovery, many tropical forests are disappearing within our lifetimes. In 1982 the World Wildlife Fund reported that over the past one hundred years, nearly one-half of the original tropical forest areas have either been destroyed or degraded. At the current accelerating rate of destruction, only fragments of many large lowland forests will soon remain. Tropical forests yet unlogged occupy vast land areas. However, if these forests continue to be devastated at the rate of fifty acres per minute, forests the size of England, Scotland, and Wales combined will be lost each year.[42] The Food and Agriculture Organization reports that in 1990 17 million hectares of tropical forest were removed (a hectare is about 2.45 acres). In contrast, in 1981 "only" 11 million hectares of tropical forests were cut down.[43] Between 1981 and 1985 the Ivory Coast and Nigeria experienced the highest rates of deforestation in the world. The respective annual deforestation rates during this period were 6.5 percent and 5 percent. The highest rate of deforestation in tropical America was 4.7 percent in Paraguay, and in Asia, 4.3 percent in Nepal.[44]

The major causes for the destruction of tropical rain forests are (1) elimination of forested areas to make way for cultivation, (2) removal of forests to allow for cattle grazing, (3) clearance of forested areas for human settlements, and (4) extraction of timber for fuel, construction, and export for various uses to the United States, Western Europe, Japan, and South Korea.[45] There is growing recognition worldwide that destruction of tropical forests is an urgent, global problem requiring cooperative international action. This recognition stems, in part, from enhanced knowledge of the nature of tropical forests that are so complex and contain such a diversity of species that it is impossible to regenerate them, or, in most cases, even to "manage" them on a sustained yield basis.[46] Although more is generally known now about tropical forests than in the past, the public and, in particular, decisionmakers need greater awareness of tropical forest values in order to comprehend why the oldest, richest, most complex, and productive ecosystems on the planet earth should not be destroyed.

Kept intact, or as near to their natural state as possible, tropical forests contribute to the preservation and evolution of living systems including humans. Rapid deforestation of tropical areas results in human poverty and political instability. As noted earlier, soil erosion, flooding, increasing scarcity of wood for fuel, siltation of water courses, and even drought are likely consequences. These effects may be more than just temporary. Typically, in a tropical forest, many factors—soils, temperature, rainfall patterns, terrain, and species distribution—are in precarious balance. As the result of extensive disturbance, tropical forests may lose their capacity to regenerate themselves.[47]

The potential effects of tropical deforestation apply to living systems worldwide, not only to the people of less developed countries. Global carbon dioxide levels, for example, are rising. Scientists generally agree that the increase in carbon dioxide levels will change the world's climate and tropical forests are one of the most important carbon dioxide sinks in the entire world. As the amount of tropical forests decreases, the amount of carbon dioxide that they absorb will be reduced. The world's average temperature has increased about one degree in the last one hundred years. But there is growing consensus that the temperature will continue to increase by three to eight degrees within the next century, possibly within fifty years, making the earth hotter than it has been in 100,000 years. If this occurs, numerous plant and animal species will be affected, some may die out, the extent of the effect determined by their ability to disperse and how quickly climatic change occurs. Thomas Lewis notes that the dispersal rate among many North American tree species is up to twenty-five miles in a century, barring natural barriers like mountain ranges or human creations like cities that act as similar deterrents.[48]

Noting that the Amazon rain forest—one of the important mechanisms determining climate of the world—may be lost, Jose A. Lutzenberger cautions that residents in temperate or subtropical climates should perhaps be most concerned. These areas will be affected more than the tropics through small changes in the climate of the world. During the ice ages, the temperate belts moved back and forth while temperatures on the equator changed very little. Such serious disruption of the biosphere would inevitably lead to catastrophic consequences, the possibilities literally ranging from flood to famine.[49]

In 1981 the Council on Environmental Quality and The Department of State declared that "unless governments, collectively and individually, take action, much of the world's tropical forests will be scattered and highly degraded remnants by the first quarter of the 21st century."[50] They projected that if the pace of deforestation did not slacken by the year 2020 all physically accessible tropical forest in the less developed countries will have been eliminated. Vast quantities of wood will have been used just for cooking and heating.[51] In a detailed study on the Amazon tropical forests, Ans Kolk indicates that the principal causes of forest destruction are highway construction, colonization, land use for pasture, mining, wood production, hydro electricity projects, and dam construction.[52]

Decisionmakers in both public and private sectors of developed countries are in a position to shape the destiny of tropical forests. Together these persons will need to work toward preservation of these internationally significant resources in spite of resistance from representatives of the less developed nations who often express the dangerously fatalistic belief that the forests must be destroyed in order to meet growing demands for fuel, hardwoods, beef, and other forest products. A strong political determination will be necessary to overcome a climate of misunderstanding concerning the causes for rain forest destruction, the relative importance of tropical forests both locally and globally, and the positive alternatives to rain forest destruction.[53] Nicholas Guppy concludes from his observations of actual tropical forest management practices that governments and foresters are operating forests like charities that provide multibillion dollar timber and forest product industries with their raw materials at far below replacement costs. According to Guppy, the true costs of harvesting tropical timbers are unknown. The potential benefits—environmental as well as material—of tropical timbers including rain forests are also unknown. The situation has been complicated through the forces of economics. Less developed countries borrow money and this practice fosters expanded bureaucracies to plan for road, dam, and power station construction, and concentration of material wealth in the hands of wealthy elites as a result of land purchases and corruption.[54] Furthermore, both public and nongovernmental foresters are under pressure to promote short-term private interests rather than long-term public interests.

In Thailand, for example, timber companies control from 20 to 40 percent of the holdings of the Royal Forestry Department.[55] More than one-half of the trained foresters in Thailand work in the capital of Bangkok. Relatively few professional forestry personnel are in the field, where they are most needed. Additionally, there is an overall lack of skilled foresters in Asian and other developing countries. Such problems are often compounded by lack of adequate training facilities, low salaries, and poor administrative organization.[56] Guppy concludes that both public and private sectors operate according to arrangements in which "destroying rain forests is a means of avoiding tackling real problems by pursuing chimeras; a license to catastrophe." Floods, landslides, loss of life, and starvation are often caused directly by tropical forest destruction.[57]

A positive alternative to forest destruction suggested by the Office of Technology Assessment is utilization of "sustainable forestry and agriculture practices" that "generally are not being developed and applied" in the less developed countries. There is some debate concerning the feasibility of these practices in tropical areas. But even if sustained yield forestry and agriculture could be implemented, the tropical forest preservation effort would remain political and "deterioration of the forest resources seems likely to continue until combinations of improved technologies and enforced resource development policies make sustaining the forests more profitable than destroying them."[58]

International action to reverse projected trends in deforestation cannot ignore the needs of people seeking land, food, energy, and profit through the permanent removal of forest cover. Rather, these needs must be addressed for forest conservation or preservation programs to succeed. The U.S. Interagency Task Force on Tropical Forests recommended the development of an international plan of action for tropical forest management, in which responsibilities for meeting urgent needs are shared between nations and international organizations. The organization responsible for coordinating this plan would be the Forestry Division of the Food and Agricultural Organization of the UN. Both public and private contributions would be solicited for implementation of the action plan. The Interagency Task Force also recommends strengthening forest management capabilities in tropical developing countries.[59]

Recommending such a plan and implementing it is, of course, two different things. The Natural Resources Defense Council suggests that nations will have to identify their very survival with tropical forest preservation before they will be willing to act: "In the past, international efforts have succeeded on this scale when they have been recognized as vital to 'national security.' So, too, the integrity of vital natural processes must be understood by nations and individuals to be a matter of survival."[60] However justified, the argument that preservation of rain forests is necessary for survival will be a difficult one to promote, first, since the cumulative effects of rain forest devastation are not immediately visible and dramatically apparent, and second, because short-term survival for some people will be accomplished through deforestation rather than preservation.

At the United Nations Conference on Environment and Development (UNCED) held at Rio de Janeiro in 1992, many of the industrialized nations wanted to develop a treaty focusing on tropical forests. But several of the less developed countries, led by Malaysia, argued against having a treaty that only focused on tropical forests, insisting that temperate and boreal forests be included. In the end, a simple statement of seventeen nonbinding principles on forests was negotiated that merely commits the parties to take forests into consideration in their efforts at international cooperation.[61]

Preservation of tropical forests and the development of appropriate responses to other critical global resource issues may well require formation of an international research group charged with the task of providing analyses of proposals likely to have significant environmental impacts. Recognizing that research tasks pertaining to global environmental problems would be beyond the capabilities of a single research organization, Caldwell recommends the creation of "an international research institution of broad competence to which cases may be referred that exceed the capabilities of other institutions and have proved to be intractable to solution by other institutions." He believes such an institution could assist in the resolution of international conflicts while serving to mitigate environmental impacts of enterprises having global significance.[62]

Considering the extent of human influence over the fate of the earth's living systems, future international environmental policies and actions will have to incorporate recognition of the need to bring human actions into harmony with the rest of nature: "It is really no longer possible to see man as in conflict with nature—man struggling to survive against the forces of nature. The forces of nature are still very powerful but the fact is that through sheer numbers combined with technical progress humans have become the dominant influence on all life on Earth."[63] Having such profound impacts upon the environment, humans will have to accept their profound obligations as well.

The Biosphere and Environmental Administration

The basic concept of physical reality justifying international environmental policy and administration is that of the biosphere—a notion scarcely more than a century old. Conceptually, "biosphere" identifies "the thin water- and air-bathed shell covering the Earth . . . bounded on the inside by dense solids and on the outside by the near vacuum of outer space." Physically the biosphere is "a closed system of the type whose materials are cycled and reused."[64]

The Paris biosphere conference of 1968, sponsored by UNESCO and other international organizations, was the first appearance of the biosphere as an item on the agendas of official representatives of nations and international organizations. The UN Conference on the Human Environment in Stockholm in June 1972 confirmed the existence of international political awareness of the biosphere. The need for international programs to address global environmental problems on a biosphere basis was acknowledged through this conference.[65]

The idea of the biosphere was popularized as *the spaceship Earth,* a finite planetary life-support system containing an abundance of interdependent living organisms. These organisms, when conceived together with their surroundings, form a unified global ecosystem. Lynton Caldwell suggests that the spaceship concept, which originated in the late 1960s, affected the climate of opinion in most industrialized nations that took administrative or legislative action in the late 1960s and early 1970s to cope with their environmental problems. Previous to that period no nation was equipped, administratively or politically, to deal with the environment as such, let alone to deal with international environmental problems.[66] Since that time human attitudes have been evolving toward, in Caldwell's words, "an ecological view of man on earth." The spaceship model raised awareness of the human predicament early in the years of first explorations into outer space. But the model failed to describe the evolutionary self-organizing nature of the planet earth. By the mid-1980s a hypothesis was offered that provided a more plausible concept. This view, as Caldwell explains, "departs [from] the traditional perception of human dominion over nature and moves toward a more realistic appreciation of humanity's interrelationship with the biosphere. This new view has led to action in which scientific knowledge, lessons of experience, and ethical judgments have been united in public policies and international agreements."[67] This suggests that the political climate is now much better than it was a little more than two decades ago for implementing positive environmental policies at the international level.

What was and still is emerging is a new biospherical consciousness: a new, ecologically grounded, international politics that Harold and Margaret Sprout define as a "system of relationships among interdependent, earth-related communities that share with one another an increasingly crowded planet that offers finite and exhaustible quantities of basic essentials for human well-being and existence."[68] No general ecopolitical consciousness existed prior to the late 1960s. Even though the politics of ecology was already known, there was a consensus among major powers (developed, industrialized countries) to keep ecological issues from appearing on international agendas.[69] Whether or not there has been a kind of conspiracy by major powers to ignore ecopolitics, the environment as a global issue has emerged out of necessity because environmental impacts extend beyond political boundaries. The impact of environmental problems is felt directly or indirectly by large segments of the world's population, who are increasingly aware of their significance. Issues refuse to be confined by either political or geographical space.

Dahlberg et al. identify some of the characteristics of global environmental issues that prevent them from being contained. Global environmental issues (1) cannot be resolved by a single actor or body; (2) require response, thus impelling actors and institutions to compete in the design of possible solutions; (3) necessitate action both at the policymaking and implementation levels; and (4) refuse to disappear, thus forcing long-range responses.[70]

Global ecopolitics—brought into the light of day in part through pressure from the less developed countries—consists of "the use of environmental issues, control over natural resources, scarcity arguments, and related concerns of social justice to overturn the international political hierarchy and related system of rules established during the period of industrial expansion."[71] This change in international relations occurring among the world's societies can be characterized as a "third" revolution, the condition of constant environmental crisis. The first revolution was agricultural, the second industrial. The third, environmental revolution is with us because rapid population growth and increasing consumption simply cannot continue indefinitely within a closed and finite ecological system—the biosphere.[72]

We are at the outer limits of the age of material expansion. Both Dennis Pirages and Jeremy Rifkin see human society moving from an age of growth to an age of scarcity. The experience of scarcity has focused more attention than in the past on the biosphere as a finite system. International relations between developed, and less developed countries have been reexamined. In limited ways at least, and in spite of chronic avoidance of issues likely to polarize the attitudes of individuals and societies, nations have been forced to formulate international environmental policies.

Avoidance of potentially polarizing environmental issues has limited the effectiveness of international development and assistance programs. The prevailing tendency in such programs has been to stress the scientific, technical, and economic aspects of problems while avoiding consideration of environmental questions. As a result, the possible beneficial effects of these programs have often been nullified. Efforts at environmental problem solving limited exclusively to technical, scientific, or economic assistance are likely to fail precisely because beneficial impacts on the biosphere cannot take place without modification of human and institutional values. Thus, regardless of how controversial ethical explorations might be, human values must be confronted and dealt with in international forums.

Initial efforts of international governmental bodies to take the biosphere into account—exemplified by the 1949 UN Conference on Conservation and the Utilization of Resources, and the 1963 Application of Science and Technology for the Benefit of Less Developed Countries—were confined largely to technical-scientific aspects of environmental problems. Later and somewhat more innovative efforts were undertaken by the International Union for Conservation of Nature and Natural Resources (World Conservation Union), and the Scientific Committee on Problems of the Environment of the International Council of Scientific Unions. These groups found that their primary scientific missions could not be performed adequately without considering human values and related human impacts on the environment.[73]

International conferences in the 1970s on population, environment, water, and food delved into the moral aspects of ecopolitical issues. These included life-and-death questions of whether nations should be permitted to maintain high rates of population growth, whether foreign assistance should be offered to nations having no birth control policies, whether food should be sold only to those who can buy it, and whether moral obligations exist to provide food to the starving. The right of some industrialized nations to increase consumption at the expense of global environmental integrity also has been challenged.[74]

Even the language of international relations has begun to change with the introduction of such terms as "energy gap," "resource depletion," and "never-to-be-developed countries."[75] Dahlberg et al. note emerging international concern with environmental values, such as controlling pollution, preserving genetic diversity, conserving natural resources, and limiting population growth. These values are frequently correlated with other international concerns for peace and economic development.[76]

This attention given to moral, ethical, religious, and other values dramatizes a new awareness of the biosphere. Many emerging value changes are implicit in efforts to exercise a wise stewardship of the biosphere.

The 1972 Stockholm UN Conference on the Human Environment marked the beginning of a concerted effort to direct international environmental policy and administration to work "to safeguard and enhance the environment for present and future generations of man." At this conference, representatives of 113 nations drew up an action plan incorporating numerous specific recommendations. The conference also created the UNEP in order to "catalyze governments, development assistance agencies, industry, and the UN system to include an environmental component into all relevant sectors of decisionmaking." Mostafa Tolba, executive director of UNEP, observes, "On the eve of Stockholm, just nine countries possessed environmental machinery. Now [in 1982] there are 106, the majority in the Third World."[77] However, as Thomas Gladwin indicates, the mere presence of an institution does not imply anything about its ability to implement environmental policy effectively, especially where the institution is staffed by only a few people.[78]

Not long after the conclusion of the Stockholm conference, Caldwell characterized the beginning of worldwide mechanisms for addressing environmental problems:

> A structure for environmental decisionmaking is thus emerging, slowly perhaps in relation to need—but rapidly by historical precedent. This structure may in time provide a coherent system for environmental decisionmaking that links all political levels—local, national, regional, and international, and that provides regular channels for continuous communication among scientists, planners, and decisionmakers, as well as between official and nongovernmental agencies.

Although he saw structure emerging, Caldwell did not observe mechanisms by which decisions would be made. He asks, "Who makes the decisions on (world) environmental affairs? A superficial answer would be: almost everybody—or in some instances, nobody. The present disorders in our global environment reflect the inadequacy of our decision process at all jurisdictional levels."[79]

Caldwell later observed that the post-Stockholm decade (1972–82) witnessed the establishment of a widespread and complex network of international cooperation for protection against present and impending environmental threats. To a surprising degree, the internationalization of environmental policy and administration has occurred. There now exists an interconnecting system of treaties, programs, and institutions.

Despite the existence of such vehicles for policymaking and programs affecting the future of people everywhere, the implementation of cooperative international environmental programs has not yet affected the daily lives of many people, especially those living in developed countries.[80] In 1987 the architect of the 1972 Stockholm conference, Maurice Strong, was forced to lament that the quality of the global environment had deteriorated, in spite of the fact that the environment had become an issue in virtually every country of the world.[81] This would appear to substantiate the belief that development of some institutional mechanisms may not be enough. Greater emphasis on environmental values in decisionmaking is necessary to shape a global environment in which environmental quality is maintained and even improved.

International organizations experience a number of problems. First, these organizations have no resources beyond what national governments agree to delegate to them. Their power generally is limited to dealing with specific tasks. Second, these organizations are specialized, lacking adequate mechanisms for coordinating their work with the work of other organizations and entities. The result is that such organizations often work at cross purposes. Third, some international organizations often represent specific geographical or political territories, whether or not the environmental problems they address extend beyond geographical and political boundaries.[82]

Even with these limitations, there are opportunities for international organizations to become more effective. David Kay and Harold Jacobson suggest that these organizations "should seek to concentrate their efforts in areas in which they have a demonstrated advantage as a form of collaboration." Furthermore, international organizations should engage in activities that cannot be dealt with adequately by other existing mechanisms. International organizations have proven capable of operating information dissemination systems and in providing forums for the articulation of normative statements necessary for the process of obtaining consensus. As a means for making their recommendations more realizable, international organizations should develop or improve their links with economic planners and representatives of industry.[83]

Agencies of the UN, like those of national governments, may be compartmentalized and isolated from each other, addressing aspects of environmental problems without coordinating their efforts. In this, they reflect the problems of nations. "If international institutions are not well conceived or organized for the tasks of environmental management— nor endowed with an 'environmental point of view'—nor prepared for decisionmaking on the basis of alternatives or tradeoffs, it is basically because national governments suffer similar drawbacks."[84] The International Union for the Conservation of Nature and Natural Resources (IUCN) World Conservation Strategy acknowledges a need to address such structural flaws in governmental entities having environmental responsibilities: "National and international capabilities to conserve are ill-organized and fragmented. . . . [T]hey have little influence on the development process with the result that development, the principal means of tackling human problems, too often adds to them by destroying or degrading the living resource bases of human welfare."[85] Under these circumstances, it is fortunate that any vehicles exist for encouraging international cooperation.

The United Nations Environment Program is one such vehicle. Headquartered in Nairobi, Kenya, UNEP serves as a catalyst for international cooperation in environmental affairs. The program administers an environmental fund raised through voluntary contributions from various national governments. In cooperation with UNESCO and in close association with national governments, UNEP carries on environmental education and training programs. It also attempts to improve environmental awareness among governments and institutions financing various kinds of development.

Further, UNEP has promoted various important environmental treaties such as the Convention on International Trade in Endangered Species and the Protection of Wetlands of International Importance.[86] In spite of UNEP's positive environmental activities, governments have been unwilling to give the organization much real authority or power. The program, nevertheless, can perform useful functions. As Caldwell says, "To the extent that governments face common environmental problems, are disturbed by environmental threats beyond their jurisdiction, or need to harmonize environmental policies relating to international trade, UNEP may have a significant role to play."[87]

Another key nongovernmental organization is the IUCN (International Union for the Conservation of Nature and Natural Resources, or the World Conservation Union, as it now identifies itself while still employing the same acronym), which is an independent scientific organization joining together representatives from sovereign states, governmental agencies, and nongovernmental organizations to promote the wise use of natural resources throughout the world. Through its *six* commissions—for ecology; education; environmental planning; environmental policy, law, and administration; national parks and protected areas; and survival services—IUCN serves a bridging function among various governments, groups, and programs interested in international environmental issues. Through IUCN, UNEP, and UNESCO, other international organizations gain access to a body of professional expertise necessary for the pursuit of worldwide conservation efforts.[88] A major achievement of the IUCN was the completion in 1980 of the World Conservation Strategy in cooperation with UNEP and the World Wildlife Fund. This document, containing comments from more than 450 government agencies and conservation organizations, was distributed throughout the world. Subsequently the World Conservation Strategy was used by more than 50 countries in the development of national and subnational conservation strategies. A more recent document focusing on sustainable development was released by IUCN, UNEP, and WWF in 1991: *Caring for the Earth: A Strategy for Sustainable Living*. This document is intended as a guide for the use of policymakers and decision-makers who affect the course of development and the condition of the environment in individual countries around the world.[89]

International cooperation in activities associated with conservation and the maintenance of the biosphere loses some of its effectiveness without what Caldwell characterizes as the "participation and consent" of representatives of national governments. Governmental organizations such as the UN and nongovernmental agencies such as the IUCN require the economic and political support of national governments and rely heavily upon their "cooperation, organizational competence, and administrative capability." Furthermore, effective national participation in international environmental activities requires participating nations themselves to organize their domestic programs for both development and the environment to be sufficiently "well organized to promote coordination at the international level."[90]

Kay and Jacobson point to the special role in international environmental leadership played by the United States: "The inertia of the multilateral institutional system, and the general inability or unwillingness of other states to assume leadership, means that, in the absence of positive U.S. leadership, significant new international activities are unlikely to be undertaken." According to Kay and Jacobson, although the United States has unique leadership advantages in the form of its scientific and economic capabilities and active public interest groups, it has demonstrated a growing propensity to avoid taking a leadership role in international affairs. In problem areas such as soil conservation and the disposition of radioactive wastes the U.S. role has been weak. This reluctance to assume leadership in international environmental affairs is correlated with the decline of U.S. leadership in the UN system.[91]

The options that may be exercised by national governments are limited, since they operate under political and administrative constraints that are inherent in their sovereignty. Thus, national governments will adopt or support only those policies in the areas of international environmental affairs that are compatible with their national interests. Caldwell is correct in asserting that the nation-state system is ill-equipped to deal with ecological problems, since political boundaries have little or no relation to ecosystems.[92] Caldwell does not, however, support the idea that, "because the earth comprises a complex biological unity called the biosphere, it must be unified socially and politically." Rather, he believes "the world will continue to be governed by separate and often antagonistic nations." Unity among nations will not emerge from recognition of interdependence among nations. They will not "subordinate their differences to some supranational authority."[93]

The vision of the future proposed by Caldwell is one of diversity in unity or unity in diversity. Environmental protection will be achieved through "the politics of antagonistic cooperation"[94] and cooperation will be more likely to occur at technical than ethical levels. Antagonistic cooperation achieves an ominous reality when applied to the Russian nuclear disaster of April 26, 1986. At Chernobyl Reactor Four, initially at least, the ethical dimensions of public health problems associated with substantial release of dangerous radioactive materials into the environment were largely ignored. During the first few critical days the Russians provided little or no information concerning the scale of the disaster and the probable migration paths of radioactive materials. Such information had to be developed by other means.

Facts pertinent to the health and safety not only of West Europeans but also of citizens of the Soviet Union and the Soviet bloc were withheld temporarily. Some of the information released by the Russians was difficult to interpret. An editorial in the *Bulletin of Atomic Scientists* comments: "The Soviet government unconscionably delayed the release of accident information to its own citizenry and the world, and its subsequent reporting was sometimes contradictory."[95] Moreover, in those first weeks the only visibly effective pressure for release of public health information was exercised not through international bodies but rather through national entities such as the Swedish government, which reacted in alarmed response to data collected through its own radiation monitoring systems.

Although refusing some offers of technical assistance, the Russians eventually accepted limited medical aid for Chernobyl workers whose lives were in immediate jeopardy. The eventual Russian release of accurate information pertinent to public health appears to have been in reaction to political pressures coming from the United States and an array of West and East European national governments. Still, even when it became clear that the scope and danger of the Chernobyl accident could not be kept hidden from the world, the Russians were reluctant to accede to outside pressures. As viewed through Western eyes, the Russian frankness had little or nothing to do with ethical values. Rather, the Russians seemed to want to avoid international embarrassment: "The traditional, secretive elements of Soviet information policy had a powerful initial impact on the party leadership's response to the Chernobyl tragedy, and a less powerful but still considerable impact nine weeks later. Whether the nuclear accident will produce a major change in domestic and international information policy remains to be seen."[96]

Russian embarrassment is not, however, without its positive aspects, since it did engender modest but observably greater public candor later. One need only note, as Erik Hoffmann did, "the willingness of Soviet spokespersons to discuss and debate issues with foreign correspondents, to publish sharply critical letters from citizens and a few foreign officials, to provide many details about the nature and consequences of the accident, and to promise a thorough 'final report' about the causes of the disaster for international discussion."[97]

Furthermore, the Russians presented an "extremely detailed and forthright" accounting of the Chernobyl event and its aftermath at a meeting of the International Association of Atomic Energy (IAEA) in Vienna.[98] As far as the international community is concerned, most of the motion toward greater candor on the part of the Russians resulted from political pressures exercised by national governments, not international bodies. Nevertheless, communications between Russia and other nations have improved, and there now is an opportunity to facilitate dialogue on nuclear power issues at a level of depth not previously thought possible:

> The Soviet Union is potentially open, as it has never been before, to greater international involvement in its nuclear program and in the development of nonnuclear energy options. On many levels, Chernobyl is an embarrassment, and the [former] Soviet Union recognizes that it must take many extra steps domestically and internationally to put the matter to rest or to turn it to their advantage. Soviet leaders realize they must be open with Eastern and Western Europe to prevent the erosion of trust they have labored long to achieve.[99]

It remains to be seen whether or not the United States and other nations will take advantage of this Russian vulnerability, or if the Russians themselves will strive toward greater cooperation in international environmental affairs, especially as these pertain to nuclear and other forms of energy.

Guardedly optimistic, Caldwell suggests that, even if present mechanisms for achieving international environmental cooperation were to disappear, other means for addressing global issues would still be found: "The human instinct for survival and recognition of common interrelationships between the human species and its environment will find ways to enable international environmental protection to continue even in a divided world."[100] International environmental cooperation will be achieved out of necessity.

Pirages argues that dramatic changes, hastened by recognition of the finite nature of the biosphere and experience of the "new scarcity" of natural resources are already occurring in international relations. Growing interdependence among nations has fostered increasing biopolitical activity.[101] In the past, nations acted on the basis of the dominant principle of the inevitability and rightness of resource-intensive growth.[102] Faced with scarcities and diminished opportunities for growth, nations may have to stress new values as forces to shape and change international affairs. The principal goal of the new value perspective, he suggests, will be to "maximize the welfare of all human beings in both present and future generations."[103]

For this vision to become a reality, societies themselves must change. Especially in the developed countries, human behavior must be altered radically. According to Strong, "The real problems are basic societal problems stemming from the way we live, the way in which the high technology civilization orders its affairs and the way in which we act as consumers and prime actors in the technological civilization."[104]

A more optimistic image of the future has been presented by Robert Repetto in *The Global Possible* and Julian Simon and Herman Kahn in *The Resourceful Earth*. Repetto's important volume is a collection of essays presented originally at a conference of international experts that convened at Wye Island, in which the conference participants describe a sustainable world predicated upon wise use of resources. In order for the goal of sustainable development to be fulfilled, participants of the Global Possible Conference concluded that several critical transitions must occur:

1. A demographic transition to a stable world population of low birth and death rates
2. An energy transition to high efficiency in production and use and increasing reliance on renewable resources
3. A resource transition to reliance on nature's "income" without depletion of its "capital"
4. An economic transition to sustainable development and a broader sharing of its benefits
5. A political transition to a global negotiation grounded in complimentary interests between North and South, East and West.[105]

Indeed, if all of these transitions occur, we would have a far more sustainable society. Realistically speaking, these transitions will be immensely difficult to accomplish. However, the contrast that the authors describe between Ghana and Korea, which both had the same per capita income level in 1960, does provide substantial hope that a sustainable future is possible through changes in policy direction. Despite having a population density eight times greater than Ghana and fewer natural resources, by 1982, Korea's per capita income was five times greater than Ghana's, due to major changes in economic policies. The book further illustrates that different countries will require different solutions allowing for individual differences.

Simon and Kahn's *Resourceful Earth* is another collection of edited works by individuals interested in challenging the negative statistical picture portrayed by the CEQ and Department of State's *Global 2000 Report*. Simon and Kahn state that future global problems are likely to be less pressing than in the past. They believe that "environmental, resource, and population stresses are diminishing, and with the passage of time will have less influence than now upon the quality of human life on our planet." Their rosy picture of the future is based on an assumption that economic growth will continue until the year 2000 and beyond as fast as it has since World War II. Through such economic growth individual incomes will rise and prices will assure natural resource availability. In contrast, the *Global 2000* report assumes a decline in income related to a decline in worker productivity. Although Simon and Kahn and their colleagues may have successfully challenged some of the *Global 2000* statistics, it would appear that they have a relatively naive faith in the marketplace and its ability to guarantee acceptable levels of natural resources.[106]

Environmental Administration in Less Developed Countries

Environmental issues associated with less developed countries have significant impacts on international environmental politics. The principal division among the nations participating in both the Stockholm UN Conference on the Human Environment and the Rio de Janeiro UN Conference on Environment and Development was between the more developed and the less developed nations. At the Stockholm conference, industrialized nations were concerned with more traditional environmental matters of controlling pollution and reducing the negative effects of growth. But less developed countries feared that implementation of environmental control measures would impede their own economic growth and development. By linking environmental protection with development, however, conference leaders were able to reduce some of the opposition. Nevertheless, many leaders of less developed countries presumed they did not have industry-related problems and, thus, did not share environmental problems in common with developed nations.

Since the Stockholm conference, a dramatic change has occurred in

Third World attitudes toward environmental protection. This change has been paralleled by the creation of environmental management agencies in many less developed countries. The existence of such agencies has facilitated interaction between the less developed countries and international institutions concerned with the Third World. The United Nations Environment Program serves as a catalyst for environmental action through its funding and ideas, while UNESCO, the World Health Organization, and the Food and Agriculture Organization, along with various national and nongovernmental organizations, address environmental needs in the less developed countries.

The Role of Funding in International Environmental Administration

Many negative environmental impacts in less developed countries may be correlated directly with developmental activities that have taken place without due regard for the environment. As a consequence, a number of these projects have failed. Underscoring this point, the Center for International Environmental Information highlights that

> modern technologies and machinery are being thrust into many underdeveloped areas of the world. Industrial and infrastructural development that occurred over the course of many decades in advanced countries is being telescoped into a few short years in many Third World countries. The potential for development-related environmental destruction is thus extremely high in the developing world. The overwhelming nature of these problems has forced environmental awareness upon developing countries.[107]

Nine major sources of development funding, including the World Bank, have signed a declaration of intent to make environmental considerations an integral part of their lending programs.[108] Despite such agreements, however, international financing agencies remain a major source of environmental destruction in less developed countries. The United States, particularly, is the principal source of funding for many unnecessary and counterproductive development projects that often serve the needs of multinational corporations more than local citizens. The Volta Dam in Ghana, for example, displaced 80,000 people and spread serious diseases like schistosomiasis and onchocerciasis in order to provide cheap electricity for Kaiser Aluminum. Other examples include

1. The first major reservoir in a tropical forest, Brokopondo Dam in Surinam, which drowned tropical vegetation and produced massive quanti-

ties of hydrogen sulfide that required dam operators to wear gas masks
2. Brazil's Tucuri Dam, which drowned 2,000 square kilometers of tropical forests before creating similar problems closed down in 1986
3. A World Bank irrigation project that displaced Ethiopians in Awash Valley in order to produce cash crops like carnations and strawberries for export to Europe while other Ethiopians experienced famine
4. The World Bank's provision of more than one-half billion dollars for the Polonoroeste Project in Brazil to build a road deep into remote parts of the Amazon basin, where widespread slash and burn agriculture projects have been initiated on soils unsuited for long-term agriculture and are doomed to failure
5. The U.S. Bureau of Reclamation's assistance in the design of the world's largest dam, the Three Gorges Dam on the Yangtze River, which is to back up water for 500 kilometers, fill in China's equivalent to the U.S. Grand Canyon, and displace over 1 million people—a project that will produce tremendous wildlife and fishery impacts as the entire hydrologic balance of the Yangtze estuary is changed.[109]

In the context of developmental possibilities, critical environmental resources in the less developed countries are subject to stresses of unprecedented magnitude. The survival and well-being of the economically disadvantaged majority in less developed countries depend on the capability of government to manage environmental resources effectively over the long term.[110]

Although some spokespersons for less developed countries argue for redistribution of wealth and technology and rectification of social and economic inequalities prior to implementation of measures to protect the environment, there appears to be more support in the less developed countries for the promotion of technical-scientific assistance from the developed countries than there is for political reforms within the less developed countries. The developed countries thus have some responsibility for assisting the less developed countries to take measures to protect their environmental life-support systems.[111]

The Global Environmental Facility (GEF) was established in May 1991 to provide help from industrialized countries (the "North") to less developed countries (the "South") to finance environmental projects. The GEF was an initiative of France, Germany, and the World Bank. Initially it was a three-year trial project, but in 1994 the GEF was restructured and funded on a permanent basis through funding from the industrialized countries. The GEF was originally designed to address four major environmental problem areas: reduction of global warming gas emissions, protection of the biosphere, protection of international waters, and protection of the ozone layer.[112] The GEF started with $1.3 billion, which was controlled by the World Bank. However, at the United Nations Conference on Environment and Development the less developed countries demanded that the World Bank's system for allocating funds be changed. Basically, they distrusted the World Bank. They agreed to support the continued existence of GEF if the industrialized nations would agree to a restructuring of the GEF and the use of a new voting system for allocating aid funds.[113]

The United Nations Environment Program stresses the need for continuous communication between developed and less developed countries: "It cannot be too much or too often emphasized how crucial the dialogue between and amongst developed and developing countries is for the appropriate solutions of environmental problems." The program sees dialogue among nations concerning environmental problems serving as a vehicle for achieving international cooperation. Working together to solve environmental problems will engender greater confidence among different peoples and nations. Furthermore, cooperation in noncontroversial areas might "help to create a propitious background and favorable climate for attacking more difficult and more controversial problems."[114]

One of the greatest dialogues on international environmental policy issues occurred at the 1992 United Nations Conference on Environment and Development (UNCED) at Rio de Janeiro. This conference brought together the greatest number of governmental and nongovernmental agency representatives ever to meet on the subject of the environment. The Earth Summit was heralded as a meeting of the North and the South. There were great expectations for this meeting, perhaps too great, as indicated by some of the negative fall out after the meeting.[115] Five major documents were signed by many of the nations at UNCED: (1) the Rio Declaration, which calls for the creation of individual national plans of action for sustainable development; (2) the Framework Convention on Climate Change, which calls for the reduction of greenhouse gases by both industrialized and less developed countries; (3) the Biodiversity Convention, which calls for a pledge to reduce the rate of destruction of plant and animal species of the world; (4) the Statement of Forest Protection Principles, which, as noted earlier, is a nonbinding agreement of many nations of the world to embark upon sustainable forest management practices for both temperate and tropical forests; and (5) Agenda 21, which is a highly comprehensive eight-hundred-page blueprint to address practically every conceivable environmental problem with an emphasis on sustainability.[116] The United Nations General Assembly convened a special session in September 1997 to assess progress toward the implementation of Agenda 21 and to develop strategies to accelerate progress toward sustainable development. The overall conclusion was that a stronger political will needs to be mobilized, along with an invigoration of the new global partnership, with special consideration for the developing countries. The Assembly confirmed that there needs to be a greater integration of economic, social, and environmental objectives, that trade and environment be made mutually supportive, and that the implementation and application of the principles of the Rio Declaration on Environment and Development should be the subject of regular assessment.[117]

Humans and Sustainability

Problems in the Third World cannot be analyzed only from an environmental point of view: international environmental policy and administration must incorporate human as well as environmental aspects. Particular attention must be paid to preservation of those living resources serving as bases for the survival and well-being of Third World people. This does not, however, minimize the importance of environmental perspectives.

All people, but particularly those residing in the less developed countries, face severe survival problems in the coming decades. As the CEQ and Department of State predict, "Barring revolutionary advances in technology, life for most people on earth will be more precarious . . . unless the nations of the world act to alter current trends." And altering current trends is not a short-term proposition. "Long lead times are required for effective action. If decisions are delayed until the problems become worse, options for effective action will be severely reduced." The United States, which has the world's largest economy, needs to reassess its foreign and domestic policies relating to population, resources, and environment. The United States should also cooperate with less developed countries in efforts to relieve poverty and hunger, stabilize populations, enhance economic productivity, and encourage environmental protection.[118] Many of these ideas have been incorporated into Agenda 21 at UNCED, but progress toward their attainment needs to be accelerated.

Massive redistribution of total world output would not be enough to provide a decent material existence for the world's human population. The current rate of growth in resource consumption worldwide cannot be sustained indefinitely without causing irreversible damage to the biosphere: "We face the specter of a world unable to feed its millions, yet unable to tolerate the environmental impact of its inadequate efforts to do so . . . as a serious threat to our existence."[119] Lester Brown points out that there has been more population growth since 1950 than in the previous four million years. The additional population will demand more resources from an already heavily diminished base. With regard to food alone, Brown reports that all seventeen oceanic fisheries are being fished at or beyond capacity. As fishery production has declined, so too has agricultural production. From 1950 to 1990 the world's grain harvest grew at 3 percent per year, but in the six succeeding years the grain harvest only grew at an average of 0.5 percent per year.[120]

Water mismanagement is reducing water tables, shrinking lakes, and destroying wetlands worldwide. And yet, because of improved living standards, the demand for water has increased rapidly. Per capita water supplies are one-third lower now than in 1970 and will continue to decline as population grows.[121] In fact, Peter Gleick correctly points out that "the world faces a wide range of ecological and human health crises, related to inadequate access to, or inappropriate management of clean fresh water. As human populations continue to grow, regional conflicts over water, ecological degradation, and human illness and death are becoming more frequent and serious. . . . [N]ew approaches to long-term water planning and management that incorporate principles of sustainability and equity are required."[122]

The less developed countries face the additional threat of technological and resource exploitation by the developed countries themselves, who are experiencing resource scarcity as they attempt to maintain high consumption levels.[123] Donella and Dennis Meadows believe that the world as a whole actually has more than enough basic resources to meet all human needs for the present and immediate future. They also indicate that technologies are available that would permit resource use on a sustainable and economical basis. However, these standards cannot be met so long as the nations of the world continue to use resources improperly: "For nearly every nation in the world, the resource base is now being used wastefully, not efficiently, and the whole population is not able to satisfy its basic material needs."[124]

Donella Meadows indicates that various studies, regardless of the methodology employed, agree with conclusions drawn in the book *The Limits to Growth,* of which she was a coauthor. While there is enough food to feed the population of the world for some time into the future, this will happen only if there are dramatic changes in political, social, and economic systems affecting food production and distribution.[125]

The United Nation Environment Program identifies a variety of economic instruments for environmental protection and natural resource management with an emphasis on sustainability. Used properly, such instruments should facilitate sustainable development while protecting the environment. The program divides these instruments into seven interrelated categories: (1) property rights (ownership rights, use rights, development rights), (2) fiscal instruments (emission and effluent taxes, product taxes, royalties and resource taxes, subsidies), (3) liability systems (noncompliance charges, joint and several liability, enforcement incentives), (4) charge systems (pollution charges, impact fees, user charges), (5) market creation (tradeable emission permits, tradeable catch quotas, tradeable development quotas, tradeable water shares, tradeable resource shares), (6) financial instruments (grants, subsidies, revolving funds, relocation incentives, environmental funds), and (7) bonds and deposit refund systems (environmental performance bonds, land reclamation bonds, waste delivery bonds, environmental accident bonds, deposit refund systems).[126] Individual national governments must be willing to use such instruments if progress toward environmental and economic sustainability is to be made. Furthermore, no supranational organization can effectively force any sovereign nation to employ any of the above-mentioned techniques, though they can provide technical assistance to those nations willing to utilize them.

International environmental policy and administration in the future will have to confront problems of resource waste and mismanagement while providing alternatives to present practices that might smooth a transition into a steady-state society based on maintaining stable human populations and sustainable levels of resource development. Environmental values and considerations will have to be honored by developed and less developed countries alike. Governmental institutions and mechanisms will have to be strong enough to ensure sustainable resource development while protecting ecosystems and promoting public health, education, and welfare.

Some of these objectives are reflected in changes in international environmental policy and administration. The United Nations Environment Program points to (1) decisionmaking by UN bodies revealing governmental acceptance of new environmental principles; (2) wide public acceptance of scientific findings on environmental problems; and (3) a growing body of practical experience in dealing with worldwide environmental issues. Caldwell expresses worry that time delays may precipitate unnecessary environmental damage: "The social and political divisions of the world burden and delay the implementation of environmental policies to which nations have in principle agreed. . . . The question for many environmental issues is whether the action will be timely enough to avoid the consequences of delay."[127]

Both the World Conservation Strategy and the Caring for the Earth strategy developed by the IUCN, UNEP, and WWF provide intellectual frameworks and practical guides for implementing international and national policies aimed at integrating conservation with development. These frameworks serve as blueprints for changing governments and private organizations by bringing environmental values into developmental agendas so that development is sustainable while also meeting social and economic objectives. The three main goals of the World Conservation Strategy are maintenance of essential ecological processes and life-support systems; preservation of genetic diversity; and utilization of species and ecosystems at sustainable levels.[128]

Perhaps no concept is more compelling than sustainability. But the finite nature of the world's resources limits the capacity of humans to continue on a path of infinite growth. This means that humans must impose constraints on their behavior and follow a path of enlightened management, or economic as well as environmental catastrophe will occur.[129] Furthermore, as Becky Brown et al. conclude after attempting to define the meaning of global sustainability, "In stating the goal of sustainability to be survival of virtually all humans to adulthood, the caveat 'once [they have been] born' recognizes that, without controlling birth rates, sustainability is unlikely."[130] Overpopulation is widely recognized as one of the leading causes of environmental degradation. Until human population growth is brought under control, the environmental crisis will not be managed.

Notes

1 Forces Shaping Environmental and Natural Resource Administration

1 Interagency Ecosystem Management Task Force, *The Ecosystem Approach: Healthy Ecosystems and Sustainable Economies,* vols. 1, 2, and 3 (Washington, D.C.: U.S. Government Printing Office [GPO], June 1995).

2 Timothy N. Cason, "An Experimental Investigation of the Seller Incentives in the EPA's Emission Trading Auction," *American Economic Review* 85 (September 1995): 905–22; and Gary C. Bryner, *Blue Skies Green Politics: The Clean Air Act of 1990 and Its Implementation* (Washington, D.C.: Congressional Quarterly [CQ] Press, 1995).

3 Keith Willett and Ramesh Sharda, "Alternative Control Policies for Water Quality Management: An Experimental Economics Approach," *Journal of Environmental Planning and Management* 40 (July 1997): 507–26.

4 United States Environmental Protection Agency (EPA), *Managing for Better Environmental Results,* EPA 100-R-97-004 (Washington, D.C.: Office of the Administrator, March 1997).

5 See Council of State Governments, *Synthesizing Devolution: Ecosystem Protection and State Environmental Management Programs* (Lexington, Ky.: Council of State Governments, 1997).

6 Barry Rabe, "Power to the States: The Promise and Pitfalls of Decentralization," in Norman J. Vig and Michael E. Kraft, eds., *Environmental Policy in the 1990s,* 3rd ed. (Washington, D.C.: CQ Press, 1997), 31–52.

7 U.S. Bureau of Land Management (BLM), *Public Land Statistics 1996* (Washington, D.C.: GPO, 1996).

8 Peter Savage, "Optimism and Pessimism in Comparative Administration," *Public Administration Review* 36 (1976): 418.

9 William R. Mangun and Jean C. Mangun, "An Ecological Approach to Decision-Making in Renewable Resource Management," *Policy Studies Review* 12 (autumn/winter 1993): 197–210; Ecological Society of America ad hoc Committee on Ecosystem Management, "The Report of the Ecological Society of America Committee on the Scientific Basis for Ecosystem Management," available: http://www.sdsc.edu/~ESA/ecmtext.htm (16 January 1997).

10 "Seeing Both Forests and Trees," *Resources for the Future* (spring 1981): 1.

11 See Lance H. Gunderson, C. S. Holling, and Stephen S. Light, eds., *Barriers and Bridges to the Renewal of Ecosystems and Institutions* (New York: Columbia University Press, 1995).

12 See Kai N. Lee, ed., *Compass and Gyroscope: Integrating Science and Politics for the Environment* (Washington, D.C.: Island Press, 1993).

13 See Andrew Goudie, *The Human Impact on the Natural Environment,* 4th ed. (Cambridge, Mass.: MIT Press, 1994).

14 David Ehrenfeld, *The Arrogance of Humanism* (New York: Oxford University Press, 1981).

15 Daniel H. Henning, "Comments on an Interdisciplinary Social Science Approach for Conservation Administration," *Bioscience* 1 (January 1970): 11–12.

16 Paul R. Ehrlich and Anne H. Ehrlich, *Betrayal of Science and Reason: How Anti-Environmental Rhetoric Threatens Our Future* (Washington, D.C.: Island Press, 1996).

17 Wayne A. Morrissey, *Ecosystem Management: Federal Agency Activities,* Report 94-339 ENR (Washington, D.C.: Congressional Research Service, April 19, 1994).

18 Eduardo La Chica, "Saving the Earth: U.S. Asks World Bank to Make Safeguarding a Priority," *Wall Street Journal,* July 3, 1987.

19 John C. Hendee, Richard P. Gale, and Joseph Harry, "Conservation, Politics, and Democracy," *Journal of Soil and Water Conservation* 24 (November/December 1969): 212–15.

20 United States Department of the Interior, *1996 National Survey of Fishing, Hunting, and Wildlife-Associated Recreation* (Washington, D.C.: U.S. Fish and Wildlife Service, 1997), 5.

21 International Union for the Conservation of Nature and Natural Resources (IUCN; also known as World Conservation Union), United Nations Environment Programme (UNEP), and World Wide Fund for Nature (WWF), *Caring for the Earth: A Strategy for Sustainable Living* (Gland, Switzerland: IUCN, 1991).

22 Charles E. Lindblom, "The Science of Muddling Through," *Public Administration Review* (spring 1959): 84.

23 Garrett Hardin, "The Tragedy of the Commons," *Science* 162, no. 3859 (1968): 1243.

24 Lynton K. Caldwell, *Science and the National Environmental Policy Act: Redirecting Policy through Procedural Reform* (Tuscaloosa: University of Alabama Press, 1982).

25 "World Environment Day," *Environmental News* 4 (Washington, D.C.: EPA, April 19, 1978).

26 Gabriel A. Almond, "The General Value Orientations of the American People," in L. Earl Shaw and John C. Pierce, eds., *Readings in the American Political System* (Lexington, Mass.: Heath, 1970), 24–27.

27 Ibid., 25–27.

28 Hendee et al., "Conservation, Politics, and Democracy," 212–15.

29 U.S. Department of Commerce, Bureau of the Census, *Statistical Abstract of the United States—1997* (Washington, D.C.: Bureau of the Census, 1997).

30 Hendee et al., "Conservation, Politics, and Democracy," 214.

31 William R. Mangun, "The Role of Scientists, Lobbyists, and the Courts: Why

Acid Rain Bills Don't Become Law," *Research in Public Policy Analysis and Management,* vol. 6 (Greenwich, Conn.: JAI Press, 1995): 25–90.

32 See Anthony M. H. Clayton and Nicholas J. Radcliffe, *Sustainability: A Systems Approach* (Boulder, Colo.: Westview Press, 1996; Leonard W. Doob, *Sustainers and Sustainability: Attitudes, Attributes and Actions for Survival* (Westport, Conn.: Praeger, 1995).

33 Norman J. Vig and Michael E. Kraft, eds., *Environmental Policy in the 1980s: Reagan's New Agenda* (Washington, D.C.: CQ Press, 1984).

34 U.S. Council on Environmental Quality (CEQ) *Environmental Quality, 1981: Twelfth Annual Report of the CEQ* (Washington, D.C.: GPO, 1982), 16–17.

35 Marshall McLuhan, *The Medium Is the Message* (New York: Bantam, 1970), 1–25.

36 Riley E. Dunlap, "Polls, Pollution, and Politics Revisited: Public Opinion on the Environment in the Reagan Era," *Environment* 29 (July/August 1987): 32–33.

37 Ehrlich and Ehrlich, *Betrayal of Science and Reason.*

38 See Emery Roe, "Why Ecosystem Management Can't Work without Social Science: An Example from the California Northern Spotted Owl Controversy," *Environmental Management* 20 (1996): 667–74; Dennis L. Soden, Berton Lee Lamb, and John R. Tennert, eds., *Ecosystems Management: A Social Science Perspective* (Dubuque, Iowa: Kendall/Hunt, 1998).

39 Stanley Milgram, "The Experience of Living in Cities," *Science* 167, no. 13 (March 1970): 1461.

40 Robert Hamrin, *Environmental Quality and Economic Growth* (Washington, D.C.: Council of State Planning Agencies, 1981), 17.

41 Alexis de Tocqueville, *Democracy in America* (New York: New American Library, 1965), 196–200.

42 Lynn White Jr., "The Historical Roots of Our Ecological Crisis," *Science* 155, no. 3767 (March 10, 1967): 1203–7.

43 "A Theology of Ecology," *Time* (June 8, 1970): 49.

44 Joseph Sittler, "Theology," in Albert E. Utton and Daniel H. Henning, eds., *Interdisciplinary Environmental Approaches* (Costa Mesa, Calif.: Educational Media Press, 1974), 63–77.

45 Chuck D. Barlow, "Why the Christian Right Must Protect the Environment: Theocentricity in the Political Workplace," *Boston College Environmental Affairs Law Review* 23 (1996): 781–828.

46 Thomas C. Donnally, *Rocky Mountain Politics* (Albuquerque: University of New Mexico Press, 1940), 1–30; Frank Jonas, *Western Politics* (Salt Lake City: University of Utah Press, 1961), 357–73.

47 See Robert H. Nelson, *Public Lands and Private Rights: The Failure of Scientific Management* (Lanham, Md.: Rowman and Littlefield, 1995.

48 Frederick Jackson Turner, *The Frontier in American History* (New York: Holt, Rinehart, and Winston, 1947).

49 Curt Tarnoff, *Population and Development: The 1994 Cairo Conference* Congressional Research Service (CRS) Report 94-533 (Washington, D.C.: CRS, August 25, 1994).

50 "Choices and Responsibilities: Finding the Balance," *UN Chronicle* 31 (September 1994): 40–44.

51 "Population Growth Threatens Natural Resources Renewal," *Public Health Reports* 107 (September/October 1992): 608.

52 Paul R. Ehrlich, *The Population Bomb* (New York: Ballantine Books, 1968).

53 Jennifer D. Williams, *The U.S. Population: A Factsheet,* Report 95-705 GOV (Washington, D.C.: CRS, June 12, 1995), 1.

54 David Pimentel, "Natural Resources and an Optimum Human Population," *Earth Island Journal* 9 (Summer 1994): 26–28.

55 CEQ and Department of State, *Global Future: Time to Act* (Washington, D.C.: GPO, 1981), xiii–xviii.

56 Lynton K. Caldwell, *U.S. Interests and the Global Environment* (Muscatine, Iowa: Stanley Foundation, 1985), 9–17.

2 Environmental and Natural Resource Policy

1 Norman Wengert, "The Ideological Basis for Conservation and Natural Resources Policies and Programs," *Annals of the American Academy of Political and Social Sciences* 344 (November 1962): 65.

2 See Robert Paehlke, *The Center That Holds* (New Haven, Conn.: Yale University Press, 1988).

3 Wengert, "The Ideological Basis for Conservation," 69.

4 CEQ, *First Annual Report* (Washington, D.C.: GPO, 1970), 243–53.

5 Walter A. Rosenbaum, *Environmental Politics and Policy* (Washington, D.C.: Congressional Quarterly Press, 1985), 6.

6 Gaylord Nelson, "Protect Environment from GOP Radicals," *Wisconsin State Journal,* January 17, 1996.

7 V. Kerry Smith, ed., *Environmental Policy under Reagan's Executive Order* (Chapel Hill: University of North Carolina Press, 1984), 6. See also Gregory A. Daneke, "Reassessing Attempts to Reform Environmental Regulation," in Gregory A. Daneke and David J. Lemak, eds., *Regulatory Reform Reconsidered* (Boulder, Colo.: Westview, 1985), 83–95.

8 Paul R. Portney, "Natural Resources and the Environment," in John L. Palmer and Isabel W. Sawhill, eds., *The Reagan Record* (Washington, D.C.: Urban Institute, 1984), 141–75.

9 Rogelio Garcia, *Federal Regulatory Reform: An Overview,* Issue Brief GD 95-035 (Washington, D.C.: CRS, September 22, 1997).

10 Joseph Gusfield, "The Study of Social Movements," in David Sills, ed., *International Encyclopedia of the Social Sciences* (New York: Macmillan, 1968), 1.

11 See Lettie McSpadden Wenner, *U.S. Energy and Environmental Interest Groups: Institutional Profiles* (New York: Greenwood Press, 1990).

12 Guy-Harold Smith, *Conservation of Natural Resources* (New York: John Wiley and Sons, 1971), 6–7.

13 John C. Hendee, Richard P. Gale, and Joseph Harry, "Conservation, Politics, and Democracy," *Journal of Soil and Water Conservation* 24 (November/December 1969): 3.

14 Robert Hamrin, *Environmental Quality and Economic Growth* (Washington, D.C. Council of State Planning Agencies, 1981), 15–16.

15 Riley E. Dunlap, "Public Opinion and Environmental Policy," in James P.

Lester, ed., *Environmental Politics and Policy,* 2nd ed. (Durham, N.C.: Duke University Press, 1995).

16 Ibid., 32.

17 Ibid., 10–11.

18 Riley E. Dunlap, George H. Gallup Jr., and Alec M. Gallup, "Of Global Concern: Results of the Health of the Planet Survey," *Environment* 35 (9): 7–15, 33–39.

19 Charles Prysby and Carmine Scavo, *Voting Behavior: The 1996 National Election* (Washington, D.C.: American Political Science Association and the Inter-University Consortium for Political and Social Research, 1997), 41.

20 Cited in Nancy K. Kubasek and Gary S. Silverman, *Environmental Law,* 2nd ed. (Upper Saddle River, N.J.: Prentice Hall, 1997), 108.

21 National Wildlife Federation (NWF), *Conservation Directory* (Washington, D.C.: NWF, 1998).

22 Arne Naess, "The Shallow and the Deep, Long-Range Ecology Movements," *Inquiry* 1, no. 16 (1973): 95–100.

23 George Sessions, *Deep Ecology for the 21st Century: Readings on the Philosophy and Practice of the New Environmentalism* (Boston, Mass.: Shambhala Publications, 1995).

24 *Time* (August 3, 1970): 42.

25 William L. Bryan Jr., "Toward a Viable Environmental Movement," *Journal of Applied Behavioral Science* 10, no. 3 (1974): 387.

26 Edward J. Burger, *Health Risks: the Challenge of Informing the Public* (Washington, D.C.: Media Institute, 1984).

27 Ibid.

28 "Popular 'Superfund' Gets Reagan's Signal," Associated Press (October 18, 1986).

29 Vladimir Bencko, T. Geist, D. Arbetova, D. M. Dharmadikari, and E. Svandova, "Biological Monitoring of Environmental Pollution and Human Exposure to Some Trace Elements," *Journal of Hygiene, Epidemiology, Microbiology, and Immunology* 30, no. 1 (1986): 1.

30 EPA, *Risk Assessment and Risk Management: Framework for Decisionmaking* (Washington, D.C.: GPO, 1984).

31 Stephen Breyer, *Breaking the Vicious Circle: Toward Effective Risk Regulation* (Cambridge, Mass.: Harvard University Press, 1993).

32 Stephen P. Duggan, "Whither the Environmentalist," *Amicus* (fall 1980): 20–21.

33 Gregg Easterbrook, *A Moment on the Earth: The Coming Age of Environmental Optimism* (New York: Viking Press, 1995).

34 Dixy Lee Ray, with Louis R. Guzzo, *Trashing the Planet: How Science Can Help Us Deal with Acid Rain, Depletion of the Ozone, and Nuclear Waste (Among Other Things)* (New York: Harper Collins, 1992); Dixy Lee Ray, with Louis R. Guzzo, *Environmental Overkill: Whatever Happened to Common Sense* (Washington, D.C.: Regnery Gateway, 1993).

35 Ronald Bailey, *Eco-Scam: The False Prophets of Ecological Apocalypse* (New York: St. Martin's Press, 1993); Ronald Bailey, ed., *The True State of the Planet* (New York: Free Press, 1995); J. Bast, P. Hill, and R. Rue, *Ecosanity: A Common-Sense Guide to Environmentalism* (Lanham, Md.: Madison Books, 1994).

36 Paul R. Ehrlich and Anne H. Ehrlich, *Betrayal of Science and Reason: How Anti-*

Environmental Rhetoric Threatens Our Future (Washington, D.C.: Island Press, 1996).

37 George Cameron Coggins and Doris K. Nagel, " 'Nothing Beside Remains': The Legal Legacy of James G. Watt's Tenure as Secretary of the Interior on Federal Land Law and Policy," *Boston College Environmental Affairs Law Review* 17 (spring 1990): 473–561.

38 NWF, *The Arctic National Wildlife Refuge Coastal Plain: A Perspective for the Future* (Washington, D.C.: NWF, 1987).

39 Donald L. Rheem, "Environmental Action: The Nature Lobby Moves Off the Mall and onto Capitol Hill," *Christian Science Monitor* 13 (January 1987): 17–18.

40 Wilderness Society, "Public Reaction," *The Watt Book* (Washington, D.C.: Wilderness Society, 1981) (*The Watt Book Update,* 12/81).

41 See Riley E. Dunlap and Angela G. Mertig, eds., *American Environmentalism: The U.S. Environmental Movement, 1970–1990* (Philadelphia: Taylor and Francis, 1993).

42 Wengert, "The Ideological Basis for Conservation," 8.

43 Ibid., 11.

44 Christine Maurer and Tara E. Sheets, eds., *Encyclopedia of Associations: An Associations Unlimited Reference,* 34th ed., vol. 1, pt. 1 (Detroit, Mich.: Gale Research, 1999).

45 Ibid.

46 Paul K. Wapner, *Environmental Activism and World Civic Politics* (Albany: State University Press of New York, 1996).

47 William Ophuls, *Ecology and the Politics of Scarcity* (San Francisco: W. H. Freeman, 1977), 190.

48 Daniel H. Henning, "Natural Resources Administration and the Public Interest," *Public Administration Review* 30, no. 2 (March/April 1970): 139.

49 Henry W. Ehrmann, "French Bureaucracy and Organized Interests," *Administrative Science Quarterly* 5 (March 1961): 539.

50 "Always Right and Ready to Fight," *Time* (April 23, 1982): 27.

51 Duggan, "Whither the Environmentalist," 23.

52 Roger W. Findley and Daniel A. Farber, *Environmental Law* (St. Paul, Minn.: West Publishing, 1983), 9.

53 Anthony Tarlock, "Environmental Law: What It Is, What It Should Be," *Environmental Science and Technology* 13, no. 11 (1979): 1347–50.

54 Walter A. Rosenbaum, *The Politics of Environmental Concern* (New York: Praeger, 1977), 72–73.

55 Christopher D. Stone, *Should Trees Have Standing? Toward Legal Rights for Natural Objects* (Los Altos, Calif.: William Kaufmann, 1974), 1–42.

56 Tarlock, "Environmental Law: What It Is, What It Should Be," 1347–48.

57 Robert V. Bartlett, "Rationality and the Logic of the National Environmental Policy Act," *Environmental Professional* 8 (1986): 105–11.

58 Richard L. Revesz, "Environmental Regulation, Ideology, and the D.C. Circuit," *Virginia Law Review* 83 (1997): 1717–72.

59 Sheldon M. Novick, "The 20-Year Evolution of Pollution Law: A Look Back," *Environmental Forum* (January 1986): 12–18.

60   Gregory Wetstone and Sarah A. Forster, "Acid Precipitation. What Is It Doing to Our Forests?" *Environment* 25, no. 4 (1983): 10–12.

61   *International Environmental Law Reporter* 7 (1984): 8. On July 10, 1976, a major chemical accident occurred in Seveso, Italy. A Roche Group chemical company released a mixture of chemicals containing dioxin into the Seveso area. In 1983, forty-one barrels of dioxin-contaminated waste from the accident mysteriously disappeared. The barrels, it turned out, were illegally transported around Europe and ended up in a northern French town in 1985, where they were eventually located. These events raised numerous concerns about the transportation of hazardous wastes, and the European Union subsequently established an environmental directive on the transboundary shipment of hazardous materials.

62   Hao-Nhien Q. Vu, "The Law of Treaties and the Export of Hazardous Waste," *UCLA Journal of Environmental Law and Policy* 12 (1994): 389.

63   Wolfgang Burhenne, Alexandre Kiss, and Malcolm Forster, "Environmental Policy and Law," in Francis R. Thibodeau and Hermann H. Field, eds., *Sustaining Tomorrow: A Strategy for World Conservation and Development* (Hanover, N.H.: University Press of New England, 1984), 111–18.

64   David Hunter, James Salzman, and Durwood Zaelke, *International Environmental Law and Policy* (New York: Foundation Press, 1998), 198–247.

3   Environmental Administration in a Decisionmaking Context

1   Lynton K. Caldwell, "The Public Administration of Environmental Policy," in Stuart W. Nagel, ed., *Environmental Politics* (New York: Praeger, 1974), 297.

2   Talcott Parsons, *Structure and Process of Modern Organization* (New York: Free Press, 1969), 1–15.

3   Carl H. Reidel, "Wasteful Organizational Practices," in R. Rodney Foil, ed., *Organizational Management in Forestry* (Baton Rouge: Louisiana State University Press, 1969), 71.

4   Jack H. Knott and Greg J. Miller, *Reforming Bureaucracy: The Politics of Institutional Choice* (Englewood Cliffs, N.J.: Prentice-Hall, 1987), 1.

5   Robert V. Bartlett, "The Budgetary Process and Environmental Policy," in Norman J. Vig and Michael E. Kraft, eds., *Environmental Policy in the 1980s: Reagan's New Agenda* (Washington, D.C.: Congressional Quarterly Press, 1984), 121–42. See also NWF, *The Full Story Behind the EPA Budget Cuts* (Washington, D.C., NWF 1982).

6   Douglas Price, "Organizations: A Social Scientist's View," *Public Administration Review* (spring 1959): 127.

7   EPA, *Getting Back to Basics: Reinventing Environmental Regulations*, EPA 100-R-97-005 (Washington, D.C.: Office of the Administrator, May 1997).

8   Joel Fleishman and Bruce L. Payne, *Ethical Dilemmas and the Education of Policymakers* (Hastings-on-Hudson, N.Y.: Hastings Center, 1980), 45.

9   Gordon Donaldson and Jay W. Lorch, *Decisionmaking at the Top* (New York: Basic Books, 1983), 111.

10   Fleishman and Payne, *Ethical Dilemmas,* 45–46.

11 See Geoffrey Vickers, *The Art of Judgment: A Study of Policy Making* (New York: Harper and Row, 1985).

12 Phillip H. Abelson, "Science and Immediate Social Goals," *Science* 3947 (August 21, 1970): 722.

13 Felix A. Nigro and Lloyd G. Nigro, *Modern Public Administration* (New York: Harper and Row, 1980), 72.

14 Raymond E. Wolfinger, Martin Shapiro, and Fred I. Greenstein, *The Dynamics of American Politics* (Englewood Cliffs, N.J.: Prentice-Hall, 1980), 518–19.

15 Herbert Simon, *Administrative Behavior* (New York: Free Press, 1947).

16 Ibid., 462–63.

17 Robert D. Behn and James W. Vaupel, *Quick Analysis for Busy Decision Makers* (New York: Basic Books, 1982), 16.

18 Carl V. Patton and David S. Sawicki, *Basic Methods of Policy Analysis and Planning* (Englewood Cliffs, N.J.: Prentice-Hall, 1986).

19 Henry Mintzberg, *Designing Effective Organizations* (Englewood Cliffs, N.J.: Prentice-Hall, 1983), 101.

20 Price, "Organizations: A Social Scientist's View," 127–28.

21 Herbert Kaufman, *The Forest Ranger: A Study of Administrative Behavior* (Baltimore, Md.: Johns Hopkins University Press, 1960), 91.

22 Ibid., 96.

23 Knott and Miller, *Reforming Bureaucracy,* 108.

24 Chris Argyris, "Understanding and Increasing Organizational Effectiveness," *Commercial Letter* (October 1968): 1–4.

25 Daniel H. Henning, "National Park Wildlife Management Policy: A Field Administration and Political Study at Rocky Mountain National Park," Ph.D. dissertation, Syracuse University, 1965.

26 Ben Heirs and Gordon Pehrson, *The Mind of the Organization* (New York: Harper and Row, 1982), 55.

27 Nigro and Nigro, *Modern Public Administration,* 216–21.

28 Daniel H. Henning, "Natural Resources Administration and the Public Interest," *Public Administration Review* 30, no. 2 (March/April 1970): 134–40.

29 Richard W. Behan, "The Myth of the Omnipotent Forester," *Journal of Forestry* (June 1966): 400.

30 Rosenbaum, *The Politics of Environmental Concern,* 96.

31 EPA, *Getting Back to Basics: Reinventing Environmental Regulations,* 1–2.

32 EPA, *Performance Partnerships: Building a Stronger Relationship between EPA and the States,* EPA 100-R-97-001 (Washington, D.C.: Office of the Administrator, January 1997), 1.

33 Interagency Ecosystem Management Task Force, *The Ecosystem Approach: Healthy Ecosystems and Sustainable Economies,* vol. 2: *Implementation Issues* (Washington, D.C.: GPO, November, 1995), 37–46.

34 Emmette S. Redford, *Administration of National Economic Control* (New York: Macmillan, 1952), 262.

35 EPA, *Harnessing the Power of the Internet: EPA Responds to the Rising Public Demand for Environmental Information,* EPA 100-R-98-04 (Washington, D.C.: Office of Reinvention, August 1998).

36 Peter Navarro, *The Policy Game: How Special Interests and Ideologues Are Stealing America* (Lexington, Mass.: Lexington Books, 1984), 6.

37 *Federal Register* 45, no. 85 (April 30, 1980): 28916.

38 UNESCO Regional Office for Education in Asia and the Pacific, *Environmental Education in Asia and the Pacific: Report of a Regional Workshop* (Bangkok: UNESCO, 1980), 1–20.

39 Quoted in Edwin N. Winge, "Involving the Public in Park Planning: USA," *Parks* 3, no. 1 (April/May 1978): 2.

40 David Clary, *Timber and the Forest Service* (Lawrence: University of Kansas Press, 1986).

41 Barry Commoner, "Informing the Public: The Duty of the Scientist in the Environmental Crisis," paper delivered to the Secrecy, Privacy, and Public Information Symposium, Annual Meeting of the American Association for the Advancement of Science, New York, December 29, 1967, p. 13.

42 David M. Lenny, "The Case for Funding Citizen Participation in the Administrative Process," *Administrative Law Review* (Summer 1976): 490–93.

43 Daniel Heinz, personal interview, October 23, 1986.

44 Harold Eidsvik, "Involving the Public in Park Planning: Canada," *Parks* 3, no. 1 (April/May 1978): 5.

45 W. R. Derrick Sewell and Susan D. Philips, "Models for the Evaluation of Public Participation Programmes," *Natural Resources Journal* 19 (1979): 337.

46 Paul Mohai, "Public Participation and National Decisionmaking: The Case of the RARE II Decisions," *Natural Resources Journal* 27 (1987): 123–55.

47 Ben W. Twight, *Organizational Values and Political Power: The Forest Service Versus the Olympic National Park* (University Park: Pennsylvania State University Press, 1983).

48 Paul J. Culhane, *Public Lands Politics: Interest Group Influence on the Forest Service and the Bureau of Land Management* (Baltimore, Md.: Johns Hopkins University Press, 1981).

49 Mohai, "Public Participation and National Decisionmaking," 155.

50 U.S. Congress, House Subcommittee of the Committee on Government Operations, 91st Cong., 1st sess., *Transferring Environmental Evaluation Functions to the Environmental Quality Council* (Washington, D.C.: GPO, July 9, 1969).

51 Lynton K. Caldwell, *Science and the National Environmental Policy Act: Redirecting Policy through Procedural Reform* (Tuscaloosa: University of Alabama Press, 1982).

52 Public Law 91-190, 91st Cong. (January 1, 1970), 83 Stat. 853, p. 2.

53 Caldwell, "The Public Administration of Environmental Policy," 285.

54 Daniel A. Farber, "Disdain for 17-Year-Old Statute Evident in High Court Filings," *National Law Journal* 9 (May 4, 1987): 20.

55 Serge Taylor, *Making Bureaucracy Think: The Environmental Impact Statement Strategy of Administrative Reform* (Stanford, Calif.: Stanford University Press, 1984), 136–56.

56 Caldwell, *Science and the National Environmental Policy Act,* 1–12.

57 Lynton K. Caldwell, Robert V. Bartlett, and David L. Keys, *A Study of Ways to Improve the Scientific Content and Methodology of Environmental Impact Analysis*

(Bloomington: Indiana University, Advanced Studies in Science, Technology, and Public Policy, School of Public and Environmental Affairs, 1982), 412.

58 Taylor, *Making Bureaucracy Think*, 160.

59 See William R. Mangun, ed., *American Fish and Wildlife Policy: The Human Dimension* (Carbondale: Southern Illinois University Press, 1991), 1–30.

60 Theodore J. Lowi, *Incomplete Conquest: Governing America* (New York: Holt, Rinehart, and Winston, 1981), 428.

61 Philip J. Cooper, *Public Law and Public Administration,* 2nd ed. (Englewood Cliffs, N.J.: Prentice-Hall, 1988), 244.

62 Walter Gellhorn, Clark Byse, Peter L. Strauss, Todd Rakoff, and Roy A. Schotland, *Administrative Law* (Mineola, N.Y.: Foundation, 1987), 161.

63 Walter A. Rosenbaum, *The Politics of Environmental Concern* (New York: Praeger, 1977), 136–39.

64 Ira Sharkansky, *Public Administration: Agencies, Policies, and Politics* (San Francisco: W. H. Freeman, 1982), 333.

65 Ibid., 343.

66 Alan Stone, *Regulations and Environmental Quality* (Washington, D.C.: CQ Press, 1982), 10–11.

67 William Drayton, "Economic Law Enforcement," *Harvard Environmental Law Review* 4, no. 1 (1980): 1.

68 Thomas P. Murphy, "Regulation: Have We Gone Overboard on It?" *Think* (September/October 1980): 23.

69 Marver Bernstein, *Regulating Business by Independent Regulatory Commission* (Princeton, N.J.: Princeton University Press, 1955).

70 Murphy, "Regulation," 21–25.

71 Kent A. Price, "Environmental Regulation—Direct and Indirect Costs," *Resources* (March 1982): 19, 23.

72 Michael Hazilla and Raymond J. Kopp, "Social Cost of Environmental Quality Regulations: A General Equilibrium Analysis," *Journal of Political Economy* 98 (1990): 854.

73 Ibid., 19.

74 Paul H. Weaver, "Regulation, Social Policy, and Class Conflict," in Chris Argyris et al., *Regulating Business* (San Francisco: Institute for Contemporary Studies, 1978), 207–13.

75 J. Daniel Nyhart and Milton M. Carnow, eds., *Law and Science in Collaboration* (Lexington, Mass.: Lexington Books, 1983).

76 Friends of the Earth et al., "National Conservation Groups Indict Reagan for 'Systematic Destruction' of Environmental Programs" (news release, March 30, 1982).

77 Craig E. Reese, *Deregulation and Environmental Quality* (Westport, Conn.: Quorum Books, 1983), 308.

78 Vig and Kraft, eds., *Environmental Policy in the 1980s: Reagan's New Agenda,* 4.

79 Gregory A. Daneke, "The Future of Environmental Protection: Reflections on the Difference between Planning and Regulating," *Public Administration Review* (May/June 1982): 227–28.

80 Ibid., 227–29.

81 Richard N. Andrews, "Deregulation: The Failure of EPA," in Vig and Kraft, eds., *Environmental Policy in the 1980s: Reagan's New Agenda,* 178.

82 Daneke, "The Future of Environmental Protection," 229–32.

83 Gail Bingham, "Does Negotiation Hold a Promise for Regulatory Reform?" *Resolve* (Conservation Foundation Newsletter, fall 1981): 1.

84 U.S. Office of Planning and Management, *Smarter Regulations* (Washington, D.C.: EPA, 1981), 1–2.

85 BLM, "BLM and the Environment," *Public Administration Review* (November/December 1982), 584–86.

86 Sabine Kremp, "A Perspective on BLM Grazing Policy," in J. Baden and R. L. Stroup, eds., *Bureaucracy in Environment* (Ann Arbor: University of Michigan Press, 1981), 142.

87 Howard E. McCurdy, "Public Administration in the Wilderness: The New Environmental Management," *Public Administration Review* (November/December 1982): 584–86.

88 Guy Benveniste, *Regulations and Planning: The Case of Environmental Politics* (San Francisco: Boyd and Fraser, 1981), 104–06.

89 Public Law 91-190, Stat. 853, p. 2.

90 Norman Wengert, "Perennial Problems of Federal Coordination," in Lynton K. Caldwell, ed., *Political Dynamics of Environmental Control* (Bloomington: Indiana University, Institute of Public Administration, 1967), 42.

91 Henning, "National Park Wildlife Management Policy."

92 See R. Edward Grumbine, "What Is Ecosystem Management?" *Conservation Biology* 8 (March 1994): 27–38; and M. Lynne Corn, *Ecosystems, Biomes, and Watersheds: Definitions and Use,* Report 93-655 ENR (Washington, D.C.: CRS, July 14, 1993).

93 Wolfinger et al., *The Dynamics of American Politics,* 466.

94 IUCN, *World Conservation Strategy: Living Resource Conservation for Sustainable Development* (Gland, Switzerland: IUCN, 1980), sec. 11.

95 Ibid.

96 Climos A. Davos, "Environmental Management: Can We Afford Established Trends?" *Environmental Professional* 8 (1986): 205–310.

4 Energy Considerations in Environmental Policy

1 U.S. Department of Energy (DOE), *A Strategic Plan for Solar Thermal Energy: A Bright Path to the Future* (Washington, D.C.: Solar Thermal Electric Program, December 1996).

2 DOE, *Sustainable Energy Strategy: Clean and Secure Energy for a Competitive Economy* (Washington, D.C.: GPO, July, 1995).

3 Congressional Quarterly, *Energy and Environment: The Unfinished Business* (Washington, D.C.: CQ Press, 1985), 20.

4 Walter A. Rosenbaum, *Energy, Politics, and Public Policy* (Washington, D.C.: CQ Press, 1981), 8.

5 Don E. Kash and Robert W. Rycroft, *U.S. Energy Policy* (Norman: University of Oklahoma Press, 1984), 21.

6   James E. Katz, *Congress and National Energy Policy* (New Brunswick, N.J.: Transaction Books, 1984), vii.

7   Walter A. Rosenbaum, *Energy, Politics, and Public Policy* (Washington, D.C.: CQ Press, 1981), 5.

8   Robert Stobaugh and Daniel Yergin, eds., *Energy Future* (New York: Ballantine Books, 1980), 271–94.

9   Michael D. Reagan, "Energy: Government Policy or Market Result?" *Policy Studies Journal* 11, no. 3 (1982): 383.

10 Hans Landsberg, "U.S. Energy Priorities," *Resources* (spring 1981): 2.

11 Katz, *Congress and National Energy Policy*, 7–8.

12 Rosenbaum, *Energy, Politics, and Public Policy* (1981), 130–31.

13 Milton Russell, "Energy in America's Future: The Choices Before Us," *Resources* (September–December 1979): 1–4.

14 Ibid., 2–4.

15 Amory B. Lovins, *Soft Energy Paths: Toward a Durable Peace* (Cambridge, Mass.: Ballinger, 1977), 4.

16 Albert J. Bartlett, "Forgotten Fundamentals of the Energy Crisis," *American Journal of the Energy Crisis* (September 1978): 876–87.

17 *Interim Report of the National Research Council Committee on Nuclear and Alternative Energy Systems* (Washington, D.C.: National Academy of Sciences, 1977).

18 Jeremy Rifkin, *Entropy: A New World View* (New York: Viking, 1980), 6.

19 Pietro S. Nivola, *The Politics of Energy Conservation* (Washington, D.C.: Brookings Institution, 1986), 1.

20 John H. Adams et al., *An Environmental Agenda for the Future* (Washington, D.C.: Island Press, 1985), 43–44.

21 Lovins, *Soft Energy Paths*, 1–24.

22 Rosenbaum, *Energy, Politics, and Public Policy* (1981), 194–95.

23 Joel Darmstadter, Hans H. Landsberg, Herbert C. Morton, "Research and Development: Widening the Energy Horizon," *Environment* 26, no. 3 (April 1984): 36.

24 Stobaugh and Yergin, *Energy Future*, 216.

25 Daniel Yergin, "Conservation: The Key Energy Source," in Stobaugh and Yergin, eds., *Energy Future*, 136–37.

26 Lovins, *Soft Energy Paths*, 1–41.

27 Stobaugh and Yergin, *Energy Future*, 137–41.

28 DOE, *International Energy Outlook, 1995* (Washington, D.C.: Energy Information Administration, 1995).

29 DOE, *Sustainable Energy Strategy*, 19.

30 Jeremy Rifkin, "The End of the Modern Age," *Sojourners* (September 1979): 1–5.

31 Rosenbaum, *Energy, Politics, and Public Policy* (1981), 89.

32 Mark Crawford, "The Electricity Industry's Dilemma," *Science* 229 (July 19, 1984): 248–50.

33 David H. Davis, *Energy Politics*, 3rd ed. (New York: St. Martin's Press, 1982), 293.

34 James E. Katz, *Congress and National Energy Policy*, 201.

35 Natural Resources Defense Council (NRDC), *Indictment: The Case against the Reagan Environmental Record* (Washington, D.C.: NRDC, March 1982), 1.

36 Ibid., 1–35.

37 "Fort Union Coal Sale Scrutinized," Associated Press (November 7, 1981); see U.S. House of Representatives, *Powder River Basin Regional Coal Lease Sales: Was Fair Market Value Received? Hearing before Subcommittee on Oversight and Investigations of the Committee on Interior and Insular Affairs, May 16, 1983* (Washington, D.C.: GPO, 1984).

38 Jeffrey St. Clair and Alexander Cockburn, "Teapot Dome, Part II: The Rush for Alaskan Oil," *The Nation* 264 (April 7, 1997): 20–24.

39 Rosenbaum, *Energy, Politics, and Public Policy* (1981), 184–186.

40 "Will Drilling Affect Geysers?" Associated Press (July 22, 1982).

41 Rosenbaum, *Energy, Politics, and Public Policy* (1981), 112–23.

42 Jim Harding, "The Myths of Chernobyl," *Earth Island Journal* (fall 1986): 23.

43 See Boris N. Porfieriev, "Environmental Aftermath of the Radiation Accident at Tomsk-7," *Environmental Management* 20 (1996): 25–33.

44 Congressional Quarterly, *The Nuclear Age: Power, Proliferation, and the Arms Race* (Washington, D.C.: CQ Press, 1981), 89.

45 "141 N-Plant Mishaps Singled Out," Associated Press (July 6, 1982).

46 Rosenbaum, *Energy, Politics, and Public Policy* (1981), 111–16.

47 Darmstadter et al., "Research and Development," 32.

48 Mark Holt and Zachary Davis, *Nuclear Energy Policy,* CRS Report IB88090 (Washington, D.C.: CRS, January 30, 1996), 1–2.

49 U.S. Department of Energy, *International Energy Outlook, 1995.*

50 See James N. Galloway, "Anthropogenic Mobilization of Sulphur and Nitrogen: Immediate and Delayed Consequences," *Annual Review of Energy and Environment* 21 (1996): 261–91.

51 See U.S. House of Representatives, *Acid Rain in the West, Hearing before the Subcommittee on Health and the Environment of the Committee on Energy and Commerce, June 28, 1985* (Washington, D.C.: GPO, 1986).

52 See U.S. House of Representatives, *Carbon Dioxide and Climate: The Greenhouse Effect, Hearing before Subcommittee on Natural Resources, Agricultural Research, and Environment, July 31, 1981* (Washington, D.C.: GPO); also see U.S. Senate, *Ozone Depletion, The Greenhouse Effect, and Climate Change; Hearing before Subcommittee on Environmental Protection and Hazardous Waste and Toxic Substances, January 28, 1987* (Washington, D.C.: GPO); also see Office of Technology Assessment, *Preparing for an Uncertain Climate Report to Congress,* OTA-O-567 (Washington, D.C.: GPO, 1993).

53 Dennis Pirages, *Global Ecopolitics: The New Context for International Relations* (North Scituate, Mass.: Duxbury, 1978), 111. Also see U.S. Senate, *World Petroleum Outlook—1984, Hearing before Committee on Energy and Natural Resources, January 30, 1984* (Washington, D.C.: GPO, 1984).

54 CEQ and Department of State, *Global Future: Time to Act* (Washington, D.C.: GPO, 1981), 59.

55 EIA, *International Energy Outlook* 1998, DOE/EIA-0484 (98) (Washington, D.C.: Energy Information Administration, 1998).

56 Joseph P. Riva Jr., *World Oil Production after Year 2000: Business As Usual or Crises?* CRS Report 35-925 ENR (Washington, D.C.: CRS, August 18, 1995), 2–3.

57 DOE, *International Energy Outlook, 1995;* and U.S. Energy Information Administration (EIA), "United States" (country profile) (Washington, D.C.: EIA, April, 1998), 7–9.

58 World Resources Institute and International Institute for Environment and Development, *World Resources 1986: An Assessment of the Resource Base that Supports the Global Economy* (New York: Basic Books, 1986), 106–9.

59 U.S. Senate, *World Petroleum Outlook—1984,* 2.

60 Robert W. Fri, "New Directions for Oil Policy," *Environment* 29, no. 5 (1987): 19–20.

61 UNEP, *The State of the World Environment, 1972–1982* (Nairobi, Kenya: UNEP, 1982), 43.

62 CEQ and Department of State, *Global Future,* 32–37.

63 World Resources Institute and International Institute for Environment and Development, *World Resources 1986.*

64 Tropical Conservation Newsbureau, "Honduras: Fuelwood Shortage Sparks Strategies." Available at http://www.txinfinet.com/mader/planeta/1195/1195 hon.html.

65 Food and Agricultural Organization (FAO), *Fuelwood Suppliers in Developing Countries* (Rome: FAO, 1983), 117–18.

66 FAO, Committee on Forest Development in the Tropics, *Tropical Forestry Action Plan* (Rome: FAO, 1985); also see U.S. Congress, Office of Technology Assessment, *Technologies to Sustain Tropical Forest Resources* (Washington, D.C.: GPO, 1984).

67 Robert C. Stowe, "United States Foreign Policy and the Conservation of Natural Resources: The Case of Tropical Deforestation," *Natural Resources Journal* 27 (winter 1987): 55–101.

68 World Resources Institute and International Institute for Environment and Development, *World Resources 1986,* 70.

69 Stowe, "U.S. Foreign Policy and the Conservation of Natural Resources," 58.

70 Ibid., 59.

71 Ibid., 60.

72 U.S. Department of State, *The World's Tropical Forests: A Policy, Strategy, and Program for the United States* (Washington, D.C.: GPO, 1980), 4.

73 FAO, *Tropical Forest Resources,* Forestry Paper no. 30 (Rome: FAO, 1982), 50.

74 FAO, *State of the World's Forests—1997* (Rome: FAO, 1997).

75 CEQ and Department of State, *Global Future,* 42–44.

76 "Tree Conservation Profitable, Experts Tell Third World Nations," Associated Press (October 23, 1985).

77 CEQ and Department of State, *Global Future,* 42–65.

78 Lynton K. Caldwell, *International Environmental Policy,* 3rd ed. (Durham, N.C.: Duke University Press, 1996), 268–69.

5 Renewable Resource Management

1 Douglas S. Powell, Joanne L. Faulkner, David R. Darr, Zhiliang Zhu, Douglas W. MacCleery, *Forest Resources of the United States, 1992,* USDA Forest Service Gen-

eral Technical Report RM-234, rev. (Fort Collins, Colo.: Rocky Mountain Forest and Range Experiment Station, 1993), 4.

2   U.S. Forest Service, *An Assessment of the Forest and Range Land Situation in the U.S.* (Washington, D.C.: GPO, 1980), 25–33.

3   CEQ, *Environmental Trends* (Washington, D.C.: GPO, 1981), 126–140.

4   Roberta A. Moltzen, "New Demands for Public Lands," fourth lecture in the Distinguished Lecturer Series presented at the School of Forest Resources and Conservation, University of Florida, Gainesville, Florida, April 14, 1998, 8.

5   Michael Williams, *Americans and Their Forests* (Cambridge: Cambridge University Press, 1989), 4.

6   Moltzen, "New Demands for Public Lands," 8.

7   Powell et al., *Forest Resources of the United States, 1992*, 3.

8   Ibid., 3.

9   Greg Brown and Charles C. Harris Jr., "National Forest Management and the 'Tragedy of the Commons': A Multidisciplinary Perspective," *Society and Natural Resources* 5 (January 1992): 75.

10   Conservation Foundation, *State of the Environment, 1982* (Washington, D.C.: Conservation Foundation, 1982), 266.

11   Multiple Use Sustained Yield Act of 1960, Public Law 86-517, 86th Cong. (June 12, 1960) 16 USCA 528-31.

12   Moltzen, "New Demands for Public Lands," 11.

13   CEQ, *Environmental Quality, 1980: Eleventh Annual Report of the CEQ* (Washington, D.C.: GPO, 1980), 297–309.

14   Ibid., 150.

15   Katherine Barton and Whit Fosburgh, "The U.S. Forest Service," in A. S. Eno, R. L. DiSilvestro, and W. J. Chandler, eds., *Audubon Wildlife Report 1986* (New York: National Audubon Society, 1986), 19–29.

16   John C. Gordon, "The New Face of Forestry: Exploring a Discontinuity and the Need for a Vision," Pinchot Lecture Series, Pinchot Institute for Conservation (Milford, Pa.: Grey Towers Press, 1996), 7.

17   CEQ, *Environmental Quality, 1982* (Washington, D.C.: GPO, 1982), 144–45.

18   Barton and Fosburgh, "The U.S. Forest Service," 19–29.

19   Donald W. Floyd, "The Uncertainty of Production, the Certainty of Conflict," *Journal of Forestry* 96 (September 1998): 5.

20   Brown and Harris Jr., "National Forest Management and the 'Tragedy of the Commons,'" 75.

21   Patrick Moore, "Green Bans Won't Save Forests," reprinted from the *Canberra Times* (Australia): June 14, 1997, *Wood Products Bulletin* 14, no. 6 (1998): 5–8.

22   Mike Dombeck, "Sustaining the Health of the Land through Collaborative Stewardship," address to USDA Forest Service employees, Washington, D.C., January 6, 1997.

23   Powell et al., *Forest Resources of the United States, 1992*, 8.

24   Thomas Barlow, "The Giveaway in the National Forests," mimeograph, Natural Resources Defense Council (1980), 1–4.

25   "$100,000 Road to $50,000 Wood Upsets FS Critics," Associated Press (September 19, 1982).

26 Joseph A. Davis, "Congress Starting to Review Federal Policy on Timber Sales," *Congressional Quarterly Almanac* (February 1985): 337–53.

27 "Forest Service Ordered to Revise Timber Policy," *Washington Post* (August 13, 1985).

28 Thomas Arrandale, *The Battle for Natural Resources* (Washington, D.C.: CQ Press, 1983), 138.

29 Randal O'Toole, "Memo to President Clinton: The Forest Service Has Already Been Reinvented," *Different Drummer Magazine* (spring 1995), available at the Thoreau Institute website <http://www.teleport.com/~rot/new.html> (January 29, 1998).

30 Ibid.

31 Floyd, "The Uncertainty of Production, the Certainty of Conflict," 5.

32 The Thoreau Institute, "1997 National Forest Timber Sale Receipts and Costs," *Electronic Drummer,* available at the Thoreau Institute website: <http://www.teleport.com/~rot/new.html> (January 29, 1998).

33 National Association of State Foresters, "USDA Forest Service Works on Improving Accountability," *Washington Update* 14, no. 3 (1998): 1.

34 Timothy J. Farnham, Cameron P. Taylor, and Will Callaway, "A Shift in Values: Non-Commodity Resource Management and the Forest Service," *Policy Studies Journal* 23, no. 2 (1995): 292.

35 Powell et al., *Forest Resources in the United States,* 9.

36 Thomas W. Birch, "Private Forest Landowners of the United States, 1994," in *Proceedings: Symposium on Nonindustrial Private Forests: Learning from the Past, Prospects for the Future* (St. Paul: University of Minnesota Extension Service, 1996), 10–18.

37 Jean C. Mangun and William R. Mangun, "Reciprocal Relations Between Humans and Ecosystems: Toward a Social Science Exchange Perspective," in Dennis L. Soden, Berton L. Lamb, and John R. Tennert, eds., *Ecosystems Management: A Social Science Perspective* (Dubuque, Iowa: Kendall/Hunt, 1998), 66.

38 Thomas J. Barlow, "Clearcutting the National Forests," mimeograph, Natural Resources Defense Council (1980), 1–4.

39 Society of American Foresters (SAF), "Clearcutting: The Position of the Society of American Foresters," available at the SAF website: <http://www.safnet.org/news/clearcut.html> (September 28, 1998).

40 Malcolm L. Hunter Jr., *Wildlife, Forests, and Forestry* (Englewood Cliffs, N.J.: Prentice-Hall, 1990), 84.

41 Ibid., 96.

42 Jeff Romm, "Sustainable Forests and Sustainable Forestry," *Journal of Forestry* 92 (July 1994): 35.

43 Gordon Robinson, "The Sierra Club Position on Clearcutting and Forest Management," Sierra Club Policy Paper no. 2 (San Francisco: Sierra Club, n.d.), 1–4.

44 SAF, "General Policy Information," available at the SAF website: <http://www.safnet.org/policy/goals.html> (September 28, 1998).

45 Romm, "Sustainable Forests and Sustainable Forestry," 38.

46 NWF, *1985 Environmental Quality Index,* reprint (1985), 5.

47 American Forestry Association, "The Forest Effects of Air Pollution," *American Forests* 93, nos. 11–12 (1987): 39–41.

48 Jim Conrad, "An Acid Rain Trilogy," *American Forests* 93, nos. 11–12 (1987): 78–79.

49 As quoted in ibid., 78.

50 Arthur H. Johnson and T. G. Siccama, "Acid Deposition and Forest Decline," *Environmental Science and Technology* 17 (1983): 294A.

51 William M. Ciesla and Jorge Samano, "Desierto de los Leones, A Forest in Crisis," *American Forests* 93, nos. 11–12 (1987): 30–32.

52 Johnson and Siccama, "Acid Deposition and Forest Decline," 294A.

53 William R. Pierson and T. Y. Chang, "Acid Rain in Western Europe and Northeastern United States—A Technical Appraisal," *Critical Reviews in Environmental Control—1986* 16, no. 2 (1986): 174.

54 Ibid., 188.

55 American Forestry Association, "The Forest Effects of Air Pollution," 41.

56 T. E. Kolb, M. R. Wagner, and W. W. Covington, "Concepts of Forest Health," *Journal of Forestry* 92 (July 1994): 10.

57 Alan A. Lucier, "Criteria for Success in Managing Forested Landscapes," *Journal of Forestry* 92 (July 1994): 22.

58 Dennis Meadows and Donella H. Meadows, lecture presented at the Workshop on Systems Dynamics of Sustainable Resource Attainment, Turrialba, Costa Rica, July 1–5, 1985, pp. 9–10.

59 Sandra Postel, "Protecting Forests," in Lester R. Brown, Edward C. Wolf, and Linda Starke, eds., *State of the World 1984* (New York: Norton, 1984), 94.

60 CEQ, *Environmental Quality, 1994–1995* (Washington, D.C.: GPO, 1995), 284.

61 Ibid., 285.

62 Ibid., 284.

63 Ibid.

64 BLM, *Public Land Statistics* (Washington, D.C.: 1994).

65 CEQ, *Environmental Quality, 1984* (Washington, D.C.: GPO, 1984), 259–60.

66 Duane Whitmere, interview, March 22, 1983, with Daniel Henning, BLM Regional Office, Billings, Mont.

67 Phillip O. Foss, *Politics and Grass: The Administration of Grazing on the Public Domain* (Seattle: University of Washington Press, 1960).

68 Betsy A. Cody, *Grazing Fees: An Overview,* CRS Report 96-450 ENR (Washington, D.C.: CRS, May 21, 1996).

69 Katherine Barton, "Wildlife and the Bureau of Land Management," in Eno et al., eds., *Audubon Wildlife Report 1986,* 513.

70 "New Reagan Land Rules Will Give Ranchers Control," Associated Press (February 15, 1983).

71 "BLM 'Cooperation' Ends," Associated Press (November 20, 1985).

72 Public Land Law Review Commission (PLLRC), *One Third of the Nation's Land* (Washington, D.C.: GPO, 1970), 289.

73 Natural Resources Defense Council v. Morton, 388 F. Supp. 829, 7ERC1298 (D.D.C. 1974).

74 CEQ, *Environmental Quality, 1980,* 310–11.

75   CEQ, *Environmental Quality, 1981* (Washington, D.C.: GPO, 1981), 147.

76   CEQ, *Environmental Quality, 1980,* 311.

77   Raymond F. Dasmann, *Environmental Conservation* (New York: John Wiley and Sons, 1976), 206–207.

78   J. A. Mabbutt, "A New Global Assessment of the Status and Trends of Desertification," *Environmental Conservation* 11, no. 2 (1984): 106.

79   World Resources Institute and International Institute for Environment and Development, *World Resources 1986,* 45.

80   UNEP, *The State of the Environment, 1984* (Nairobi, Kenya: UNEP, 1984), 29.

81   U.S. Forest Service, "Custer National Forest Plan," mimeograph, Custer National Forest, Billings, Mont. (May 20, 1982), 1.

82   Harvey Doerksen, "Water, Politics, and Ideology: An Overview of Water Resources Management," *Public Administration Review* (September/October 1977): 444.

83   Dean E. Mann, *The Politics of Water in Arizona* (Tucson: University of Arizona Press, 1963); and Helen M. Ingram, *Patterns of Politics in Water Resource Development: A Case Study of New Mexico's Role in the Colorado River Basin Bill* (Albuquerque: University of New Mexico Press, 1969).

84   Christine Olsensius, "Tomorrow's Water Manager," *Journal of Soil and Water Conservation* 42, no. 5 (1987): 312.

85   Steve H. Hanke and Robert K. Davis, "Demand Management through Responsive Pricing," *Journal of the American Water Works Association* (September 1971): 555–56.

86   Emery Castle, "Water Availability—The Crisis of the Eighties?" *Resources* (June 1983): 8–10.

87   "The Big Thirst," *Commonweal* 121, no. 6 (1994): 4.

88   Glenn Schaible, "Water Conservation Policy Analysis: An Interregional, Multi-Output, Primal-Dual Optimization Approach," *American Journal of Agricultural Economics* 79 (February 1997): 163.

89   David L. Shapiro, "Water Rights and Wrongs," *Annals of Regional Science* (December 1969): 139–43; also see George Cameron Coggins, Charles F. Wilkinson, and John D. Leshy, *Federal Public Land and Resources Law,* 3rd ed. (New York: Foundation Press, 1993), 365.

90   Jack Hirschleifer, James C. DeHaven, and Jerome W. Milliman, *Water Supply: Economics, Technology, and Policy* (Chicago: University of Chicago Press, 1960), 362.

91   "UN Dam Researcher Advises World Bank," Associated Press (November 20, 1985).

92   Ibid.

93   UNEP, *Annual Report of the Executive Director, 1984* (Nairobi, Kenya: UNEP, 1984), 59.

94   Robert V. Bartlett, "Adapt or Get Out: The Garrison Diversion Project and Controversy," paper presented at "Forests, Habitats, and Resources: A Conference in World Environmental History," Duke University, Durham, N.C., 1987.

95   Garrison Diversion Conservancy District, "Garrison Diversion Conservancy District Abbreviated History" (1997), available at http://www.GARRISONDIV. ORG/HISTORY.HTM.

96 NWF, "Another Low Discount Rate," *National Wildlife* 19, no. 2 (February/ March 1981): 24A.

97 Conservation Foundation, *State of the Environment: An Assessment,* 381–82.

98 NWF, *1985 Environmental Quality Index,* 4.

99 Friends of the Earth, "Jurassic Pork Savings = $450 million: Animas–La Plata Water Project (Colorado), 1997, available at http://www.foe.org/eco/scissor97/ alp.html.

100 John M. Bruner and Martin T. Farris, "The Conventional Wisdom in Water Philosophy," paper presented at the eighth annual meeting of the Western Regional Science Association, Newport Beach, California, February 1969, 10–11.

101 National Environmental Policy Act of 1969 (NEPA), Public Law 91-190, 91st Congress (January 1, 1970), 83 Stat. 853.

102 U.S. Geological Survey, *Water for Recreation—Values and Opportunities, Outdoor Recreation Review Commission Study Report 10* (Washington, D.C.: GPO, 1962), 11.

103 Jeanne Nienaber Clarke and Daniel McCool, *Staking Out the Terrain: Power Differentials among Natural Resource Management Agencies* (Albany: State University of New York Press, 1985).

104 CEQ, *Environmental Quality, 1978* (Washington, D.C.: GPO, 1978), 285.

105 Doerksen, "Water, Politics, and Ideology," 447.

106 CEQ, *Environmental Trends,* 217.

107 A. H. Chan, "The Structure of Federal Water Resources Policy Making," *American Journal of Economics and Sociology* (April 1981): 115.

108 Daniel A. Mazmanian and Jeanne Nienaber, *Can Organizations Change?* (Washington, D.C.: Brookings Institution, 1974), 193–94.

109 EO 12113 of January 4, 1979, "Independent Water Review," *Federal Register* 44, no. 6 (January 9, 1979): 1955.

110 Interagency Floodplain Management Review Committee, *Sharing the Challenge: Floodplain Management into the 21st Century. Report of the Administration Floodplain Management Task Force* (Washington, D.C.: GPO, 1994).

111 Richard A. Haeuber and William K. Michener, "Policy Implications of Recent Natural and Managed Floods," *Bioscience* 48 (September 1998): 4–5.

112 Doug Snyder and Stephanie Polsley Bruner, "The Galloway Report: New Floodplain Management or Business as Usual?" *Journal of Soil and Water Conservation* 49 (November–December 1994): 531.

113 W. B. Solley, R. R. Pierce, and H. A. Perlman, *Estimated Water Use in the United States in 1990,* Circular 1081 (Reston, Va.: U.S. Department of the Interior, Geological Survey, 1993).

114 CEQ, *Environmental Quality, 1981,* 154.

115 Solley et al., *Estimated Water Use in the United States in 1990.*

116 Sandra Postel, "Dividing the Waters," *Technology Review* (April 1997): 54; also see Sandra Postel, *Dividing the Waters: Food Security, Ecosystem Health, and the New Politics of Scarcity* (Washington, D.C.: Worldwatch Institute, 1996).

117 NWF, *The Full Story behind the EPA Budget Cuts* (Washington, D.C.: NWF, 1982), 5.

118 CEQ, *Environmental Quality, 1981,* 154.

119 U.S. Bureau of the Census. International Electronic Database (Suitland, Md.: U.S. Bureau of the Census, 1996).

120 Andrew C. Gross, "Water Quality Management Worldwide," *Environmental Management* 10, no. 1 (1986): 27; also see Sandra Postel, Gretchen C. Daily, and Paul R. Ehrlich, "Human Appropriation of Renewable Freshwater," *Science* 271: 785–88.

121 Meadows and Meadows, lecture presented at the Workshop on Systems Dynamics of Sustainable Resource Attainment, 7; also see Sandra Postel, *Last Oasis: Facing Water Scarcity* (New York: W. W. Norton, 1992).

122 Gross, "Water Quality Management Worldwide," 26.

123 CEQ and Department of State, *Global Future: Time to Act* (Washington, D.C.: GPO, 1981), 118–21.

124 Olsensius, "Tomorrow's Water Manager," 315.

125 Carl F. Myers, "A New Nonpoint-Source Water Pollution Control Challenge," *Journal of Soil and Water Conservation* 42, no. 4 (1987): 222.

126 "National Water Policy Unveiled," *Journal of Environmental Health* 58 (June 1996): 30.

127 U.S. Fish and Wildlife Service (FWS), "History and Activities," available at the FWS website: <http://www.fws.gov/who/usfws.html> (August 8, 1998).

128 Wildlife Management Institute, *Wildlife: The Environmental Barometer* (Austin: Texas Parks and Wildlife Department, 1978); *World Conservation Strategy*, sec. 4.

129 Carl Reidel, "Environment: New Imperatives for Forest Policy," paper presented to Divisions of Economic and Policy Division Meeting, National Convention of the Society of American Foresters, October 13, 1970, 5.

130 S. M. Campbell and D. B. Kittredge Jr., "Woodscape Crew for Small Woodlot Management in Southeastern Massachusetts," *Northern Journal of Applied Forestry* 3, no. 9 (1992): 116.

131 U.S. Department of the Interior, *1996 National Survey of Fishing, Hunting, and Wildlife-Associated Recreation* (Washington, D.C.: FWS, 1997), 4–8.

132 William R. Mangun, "Fiscal Constraints to Nongame Management Programs," in J. B. Hale, L. B. Best, and R. L. Clawson, eds., *Management of Nongame Wildlife in the Midwest* (North Central Section of the Wildlife Society, 1986), 24.

133 See William W. Shaw and William R. Mangun, *Nonconsumptive Use of Wildlife in the United States,* Resource Publication 154 (Washington, D.C.: FWS, 1984).

134 William R. Mangun and William W. Shaw, "Alternative Mechanisms for Funding Wildlife Conservation," *Public Administration Review* 44, no. 5 (1984): 407–13.

135 Reuben E. Trippensee, *Wildlife Management,* vol. 1: *Upland Game and General Principles* (New York: McGraw-Hill, 1948), v.

136 Reuben E. Trippensee, *Wildlife Management,* vol. 2: *Fur Bearers, Waterfowl, and Fish* (New York: McGraw-Hill, 1953), vii.

137 Interagency Ecosystem Management Task Force, *The Ecosystem Approach: Healthy Ecosystems and Sustainable Economies* (Washington, D.C.: GPO, September 1996), 1; also see Wayne A. Morissey, Jeffrey A. Zinn, and M. Lynne Corn, *Ecosystem Management: Federal Agency Activities,* CRS Report 94-339 ENR (Washington, D.C.: CRS, April 19, 1994).

138 FWS, "America's National Wildlife Refuges . . . Where Wildlife Comes Naturally," available at the FWS National Wildlife Refuge System website: <http://

bluegoose.arw.r9.fws.gov / NWRSFiles / Legislation / HR1420 / hr1420enr.html > (January 23, 1998).

139 Wildlife Management Institute, *Wildlife,* 1.

140 See William R. Mangun, ed., *American Fish and Wildlife Policy: The Human Dimension* (Carbondale: Southern Illinois University Press, 1991); and Lynn Llewellyn, William R. Mangun, and Jean C. Mangun, "Human Dimensions in Fish and Wildlife Service Decisionmaking," *Transactions of the North American Wildlife and Natural Resources Conference* 63 (1998): 1–20.

141 Daniel A. Poole and James B. Trefethen, "Wildlife Management," in Howard P. Brokaw, ed., *Wildlife and America* (Washington, D.C.: CEQ, 1978), 339.

142 Conservation Foundation, *State of the Environment, 1982,* 276.

143 William R. Mangun, *U.S. Fish and Wildlife Service Important Resource Problem Source Document,* Important Resource Problem, no. 7 (Washington, D.C.: GPO, 1981).

144 Interagency Ecosystem Management Task Force, *The Ecosystem Approach: Healthy Ecosystems and Sustainable Economics,* vol. 3: *Case Studies* (Washington, D.C.: GPO, 1996), 33.

145 William R. Mangun, "Forecasting the Impacts of Global Forces on America's Wetlands," *Journal of Environmental Systems* 13, no. 1 (1983–84): 73.

146 William R. Mangun, "Introduction to Wildlife Policy Issues," in William R. Mangun, ed., *American Fish and Wildlife Policy,* 3–33.

147 Conservation Foundation, *State of the Environment, 1982,* 276–77.

148 Ibid., 277–78; also see William R. Mangun, Barbara A. Knuth, Jeffrey Keller, and Gary R. Goff, "Management Implications of Challenges to the Conservation of Biological Resources," in Daniel J. Decker, Maryann E. Krasny, Gary R. Goff, and Charles R. Smith, eds., *Challenges in the Conservation of Biological Resources: A Practitioner's Guide* (Boulder, Colo.: Westview Press, 1991), 333–34.

149 William R. Mangun and Andrew C. Brown, "Coalition Building in Wildlife Management: An Ecosystem-based Perspective," paper presented at Western Social Science Association annual meeting, Denver, Colo., April 16–19, 1998.

150 Thomas Kimball, "On Establishing a National Fish and Wildlife Policy," *International Wildlife* 10, no. 3 (May/June 1980): 24B.

151 William R. Mangun and Jean C. Mangun, "An Intergovernmental Dilemma in Policy Implementation," in William R. Mangun, ed., *Public Policy Issues in Wildlife Management* (Westport, Conn.: Greenwood Press, 1991), 3–16.

152 Thomas Fitch, *The Need for Comprehensive Wildlife Programs in the United States: A Summary* (Washington, D.C.: CEQ, 1980), 17–18.

153 Fish and Wildlife Conservation Act of 1980, Public Law 96-366, 96th Cong., 94 Stat. 1322 (September 29, 1980).

154 John B. Loomis and William R. Mangun, "Evaluating Tax Policy Proposals for Funding Nongame Wildlife Programs," *Evaluation Review* 11, no. 6 (1988): 715–38.

155 Fitch, *The Need for Comprehensive Wildlife Programs,* 18–19.

156 Ibid., 19; see William R. Mangun, "Environmental Impact Assessment as a Tool for Wildlife Policy Management," in Robert V. Bartlett, ed., *Policy through Impact Assessment* (Westport, Conn.: Greenwood Press, 1989), 51–61.

157 See William W. Weeks, *Beyond the Ark: Tools for an Ecosystem Approach to Conservation* (Washington, D.C.: Island Press, 1997).

158 CEQ, *Environmental Trends*, 148, 166.

159 Steven Lewis Yaffee, *The Wisdom of the Spotted Owl: Policy Lessons for a New Century* (Washington, D.C.: Island Press, 1994), 303.

160 CEQ, *Environmental Quality, 1994-1995* (Washington, D.C.: GPO, 1995), 429.

161 Robert E. Gordon, Jr., James K. Lacy, and James R. Streeter, "Conservation under the Endangered Species Act," *Environment International* 23 (1997): 359–419.

162 Conservation Foundation, *State of the Environment: An Assessment,* 181.

163 Ibid.

164 Michael J. Bean, "The Endangered Species Program," in Eno et al., eds., *Audubon Wildlife Report 1986,* 355–56.

165 Harvey Doerksen and Craig S. Leff, "Policy Goals for Endangered Species Recovery," *Society and Natural Resources* 11 (June 1998): 365–74.

166 Conservation Foundation, *State of the Environment: An Assessment,* 183–84.

167 Richard B. Primack, *Essentials of Conservation Biology* (Sunderlund, Mass.: Sinauer Associates, 1993), 411.

168 Jeff Wheelwright and Glenn Oakley, "Condors: Back from the Brink," *Smithsonian* 28 (1997): 48–56.

169 Susan George, William J. Snape, and Michael Senatore, *State Endangered Species Acts: Past, Present, and Future* (Washington, D.C.: Defenders of Wildlife, 1998); also, see Bean, "The Endangered Species Program," 356–58.

170 World Resources Institute and International Institute for Environment and Development, *World Resources 1986,* 87.

171 CEQ and Department of State, *Global 2000* (Washington, D.C.: GPO, 1980), 37.

172 CEQ, *Environmental Quality, 1980,* 6.

173 Ibid., 31.

174 Norman Myers, *The Sinking Ark* (New York: Pergamon, 1979), 4–5.

175 Paul R. Ehrlich and Anne H. Ehrlich, *Extinction* (New York: Random House, 1981), 1–10.

176 FWS, "Endangered Species General Statistics," May 31, 1997, available at the FWS website: <http://www.fws.gov/r9endspp/esastats.html>.

177 FWS and National Marine Fisheries Service, "Notice of Interagency Cooperative Policy for the Ecosystem Approach to the Endangered Species Act," *Federal Register,* July 1, 1994.

178 FWS, "Grizzly Bear Recovery in the Bitterroot Ecosystem," available at the FWS website: <http://www.r6.fws.gov/endspp/grizzly/bittereis/abstract.htm> (September 30, 1997).

179 Jeffrey P. Cohn, "Ferrets Return from Near-Extinction," *BioScience,* 41, no. 3 (1991): 132–35.

180 A. S. Leopold, S. A. Cain, C. M. Cotton, I. N. Gabrielson, and T. L. Kimball, "Predator Control in the United States," *Transactions of the North American Conference on Wildlife and Natural Resources* 29 (1967): 27–49.

181 Robert Prescott-Allen and Christine Prescott-Allen, *In Situ Conservation of Wild Plant Genetic Resources: A Status Review and Action Plan* (Gland, Switzerland:

International Union for the Conservation of Nature and Natural Resources, 1984), 13.

182   Michael J. Bean, "International Wildlife Conservation," in Eno et al. eds., *Audubon Wildlife Report 1986*, 567–68.

183   Robert Mendelsohn, "The Role of Ecotourism in Sustainable Development, Case Study 4," in Gary K. Meffe and C. Ronald Carroll, eds., *Principles of Conservation Biology* (Sunderland, Mass.: Sinauer Associates, 1994), 511.

184   IUCN, *World Conservation Strategy: Living Resource Conservation for Sustainable Development* (Gland, Switzerland: IUCN, 1980), sec. 4.

185   Susan R. Fletcher, *International Environment: Current Major Global Treaties*, CRS Report 96-884-ENR (Washington, D.C.: CRS, November 5, 1996).

186   Aldo Leopold, "The Land Ethic," in *A Sand County Almanac, with Essays on Conservation from Round River* (New York: Oxford University Press, 1949), 90.

187   Charles S. Elton, *The Ecology of Invasions by Plants and Animals* (London: Methuen, 1958), 143.

188   David Ehrenfeld, "The Conservation of Non-Resources," *American Scientist* 65 (1964): 655. Also see David Ehrenfeld, *The Arrogance of Humanism* (New York: Oxford University Press, 1981).

6   Nonrenewable Resource Management

1   U.S. Department of Agriculture (USDA), *A Geography of Hope: America's Private Land* (Washington, D.C.: Natural Resources Conservation Service, 1997), 36.

2   USDA, *Natural Resources Inventory Draft Report* (Washington, D.C.: GPO, 1984).

3   USDA, Natural Resources Conservation Service, *Natural Resources Inventory* (Washington, D.C.: GPO, 1992).

4   Frederick Steiner, "Soil Conservation Policy in the United States," *Environmental Management* 2, no. 2 (1987): 212.

5   GAO, *Agriculture's Soil Conservation Programs Miss Full Potential in the Fight against Soil Erosion* (Washington, D.C.: GPO, 1983), 1.

6   United States Soil Conservation Service (SCS), *Statutory Authorities for the Activities of the Soil Conservation Service* (Washington, D.C.: USDA, 1981), 1–20.

7   United States General Accounting Office (GAO), "Soil and Wetlands Conservation: Soil Conservation Service Making Good Progress but Cultural Issues Need Attention" (Washington, D.C.: GAO/RCED-94-241, September 27, 1994), 1.

8   Robert J. Gray, "Proving Out: On Implementing the Conservation Title of the 1985 Farm Bill," *Journal of Soil and Water Conservation* 41, no. 1 (1986): 31–32.

9   USDA, Natural Resource Conservation Service, *1992 National Resources Inventory* (Washington, D.C.: GPO, 1995).

10   John R. Block, "Conserving Soil For America's Future," *Journal of Soil and Water Conservation* 41, no. 1 (1986): 30.

11   USDA, Natural Resources Conservation Service, *America's Private Land: A Geography of Hope* (Washington, D.C.: GPO, 1997), 36.

12   Linda K. Lee and J. Jeffery Goebel, "Defining Erosion Potential on Cropland: A Comparison of the Land Capability Class-Subclass System with RKLS/T Categories," *Journal of Soil and Water Conservation* 41, no. 1 (1986): 41–44.

13   Information on NSDAF can be accessed at the NSDAF website: <http://www.statlab.iastate.edu/soils/nsdaf/main.html> (September 6, 1998).

14   Information on NASIS can be found at <http://www.itc.nrcs.usda.gov/nasis/index.html> (September 6, 1998).

15   CEQ, *Environmental Quality, 1978* (Washington, D.C.: GPO, 1978), 311.

16   Thomas E. Dahl, *Wetlands Losses in the United States 1780s to 1980s* (Washington, D.C.: U.S. Department of the Interior, FWS, 1990), 5.

17   CEQ, *Environmental Quality, 1983* (Washington, D.C.: GPO, 1983), 44.

18   USDA, Natural Resources Conservation Service, *America's Private Land,* 8.

19   CEQ, *Environmental Quality, 1994-1995* (Washington, D.C.: GPO, 1995), 281.

20   David Pimentel, "Soil Erosion," *Environment* 39, no. 10 (1997): 4-5.

21   CEQ, *Environmental Quality, 1980* (Washington, D.C.: GPO, 1980), 312.

22   NWF, "1982 Environmental Quality Index," *National Wildlife* 20, no. 2 (February/March 1982): 34.

23   "Scientists Fear Repeat of Dust Bowl Days," Associated Press (October 26, 1985).

24   Conservation Foundation, *State of the Environment, 1982* (Washington, D.C.: Conservation Foundation, 1982), 239-241.

25   Ibid., 243-45.

26   Environment and Natural Resources Policy Division, "The 1996 Farm Bill: Comparison of Selected Provisions with Previous Law," CRS Report 96-304 ENR (Washington, D.C.: CRS, April 4, 1996), 11-15.

27   Jerry R. Griswold, "Conservation Credit: Motivating Landowners to Implement Soil Conservation Practices through Property Tax Credit," *Journal of Soil and Water Conservation* 42, nos. 1, 2 (1987): 41.

28   CEQ and Department of State, *Global Future: Time to Act* (Washington, D.C.: GPO, 1981), 34.

29   Conservation Foundation, *State of the Environment, 1982,* 237.

30   CEQ, *Environmental Quality, 1979* (Washington, D.C.: GPO, 1979), 389-90.

31   Jeremy Rifkin, *Entropy: A New World View* (New York: Viking, 1980), 139.

32   Jon R. Luoma, "Havoc in the Hormones," *Audubon* 97 (July-August 1995): 67; see also Lee E. Limbird and Pamela Taylor, "Endocrine Disruptors Signal the Need for Receptor Models and Mechanism to Inform Policy," *Cell* 93 (March 17, 1998): 157-64.

33   Conservation Foundation, *State of the Environment, 1982,* 347.

34   Ibid., 239.

35   CEQ, *Environmental Quality, 1979* (Washington, D.C.: GPO, 1979), 395.

36   Robert H. Giles, "Wildlife and Integrated Pest Management," *Environmental Management* 4, no. 5 (1980): 373.

37   CEQ, *Environmental Quality, 1980,* 312.

38   CEQ and Department of State, *Global Future,* 28-29.

39   Ibid., 27.

40   Katherine Barton and Whit Fosburgh, "The U.S. Forest Service," in A. S. Eno, R. L. Silvestro, W. L. Chandler, eds., *Audubon Wildlife Report 1986* (New York: National Audubon Society, 1986), 89.

41   U.S. Forest Service, "Custer National Forest Plan," mimeography (Billings, Mont.: Custer National Forest, May 20, 1982), 2.

42 NWF, "Minerals Security Act," *National Wildlife* 19, no. 5 (August/September 1981): 24A.

43 NWF, "Our National EQ: The First National Wildlife Federation Index of Environmental Quality," *National Wildlife* 7, no. 5 (August/September 1969): 12.

44 U.S. Forest Service, "Custer National Forest Plan," 2–3.

45 CEQ, *Environmental Quality, 1979*, 428.

46 Quoted in CEQ, *Environmental Quality, 1980*, 331.

47 Laura Michaelis, "Economic, Ecological Climate Favors Mining Law Overhaul," *Congressional Quarterly Weekly Report* 51 (March 20, 1993): 662.

48 Raymond F. Dasmann, *Environmental Conservation* (New York: John Wiley and Sons, 1976), 362–63.

49 See Robert D. Cairns, "A Contribution to the Theory of Depletable Resource Scarcity and Its Measures," *Economic Inquiry* 28 (October 1990): 744–55.

50 Dennis Meadows and Donnella H. Meadows, lecture presented at the Workshop on Systems Dynamics of Sustainable Resource Attainment, Turrialba, Costa Rica, July 1–5, 1985, pp. 12–14.

51 Richard Osborne, "The Mining Industry: Developing Policies and Laws That Help," *Vital Speeches of the Day* 58 (January 15, 1992), 200–205 (speech presented at the National Minerals Policy Forum, November 4, 1991).

52 Stephen D. Smith, "Preliminary Statistical Summary," in *Minerals Yearbook, Vol. II—Area Reports: Domestic* (Reston, Va.: U.S. Geological Survey, 1997); see also Thomas W. Martin, David L. Edelstein, Garrett H. Hyde, "Mining and Quarrying Trends in the Metal and Industrial Mineral Industries," *1985 Minerals Yearbook Vol. 1* (Washington, D.C.: U.S. Bureau of Mines, 1987), 7.

53 John B. Wachtman, "National Materials Policy: Evolution and Prospects," *Journal of Resource Management and Technology* 14, no. 3 (1985): 191.

54 Peter Harben, "Strategic Minerals," *Earth* 1 (July 1992): 44.

55 Rifkin, *Entropy*, 114.

56 Sally K. Fairfax and Carolyn E. Yale, *Federal Lands: A Guide to Planning, Management, and State Revenue* (Washington, D.C.: Island Press, 1987), 113.

57 John D. Morgan, "U.S. and World Mineral Positions, 1985 to the Year 2000," *Mining Engineering* 38, no. 4 (1986): 245.

58 CEQ and Department of State, *Global Future*, 27.

59 Harben, "Strategic Minerals," 43.

60 CEQ, *Environmental Quality, 1985* (Washington, D.C.: GPO, 1985), 334–42.

61 U.S. Congress, Office of Technology Assessment, *Strategic Minerals: Technologies to Reduce U.S. Import Vulnerability* (Washington, D.C.: GPO, 1985).

62 Rifkin, *Entropy*, 115.

63 Public Land Law Review Commission (PLLRC), *One-Third of the Nation's Land* (Washington, D.C.: GPO, 1970), 121–22.

64 Dasmann, *Environmental Conservation*, 355.

65 Duane A. Thompson, "Mining in National Parks and Wilderness Areas: Policy, Rules, Activity," CRS Report 96-161 ENR (Washington, D.C.: CRS, February 12, 1996), 1.

66 PLLRC, *One-Third of the Nation's Land*, 124–27.

67 CEQ, *Environmental Quality, 1978*, 291.

68 Marc Humphries, "The 1872 Mining Law: Time for Reform?" CRS Issue Brief for Congress, 89-130 ENR (Washington, D.C.: CRS, January 14, 1999), 2.

69 Marc Humphries, "Mining Law Reform: The Impact of a Royalty," CRS Report 94-438 ENR (Washington, D.C.: CRS, May 12, 1994), 3.

70 George Cameron Coggins and Robert L. Glicksman, "Power, Procedure, and Policy in Public Lands and Resources Law," *Natural Resources and Environment* 10 (Summer 1995): 5.

71 Ibid., 7.

72 Montana Department of State Lands and U.S. Forest Service, *Draft Environmental Impact Statement: Anaconda Minerals Company Stillwater Project* (Stillwater County: Anaconda Mineral Company, 1982), v–vi.

7 Outdoor Recreation and Wilderness Management

1 U.S. Department of the Interior and USDA, *Recreational Fee Demonstration Program: Progress Report to Congress,* vol. 1 (Washington, D.C.: GPO, 1998), 1.

2 Marion Clawson, "Outdoor Recreation: Twenty-Five Years of History, Twenty-Five Years of Projection," *Leisure Sciences* 7, no. 1 (1985): 74.

3 President's Commission on Americans Outdoors (PCAO), *Americans Outdoors: The Legacy the Challenge* (Washington, D.C.: Island Press, 1987), 38; the 1995 figures are from U.S. Bureau of the Census, *1997 Statistical Abstract of the United States* (Washington, D.C.: GPO, 1997), 250–52.

4 PCAO, *American Outdoors,* 59.

5 Market Opinion Research, *Participation in Outdoor Recreation among American Adults and the Motivations Which Drive Participation* (Washington, D.C.: Market Opinion Research, 1986).

6 H. Ken Cordell, Barbara L. McDonald, J. Alden Briggs, R. Jeff Teasley, Robert Biesterfeldt, John Bergstrom, and Shela H. Mou, "Emerging Markets for Outdoor Recreation in the United States: Based on the National Survey on Recreation and the Environment," available at Outdoor Recreation Coalition of America website: <http://www.outdoorlink.com/infosource/nsre/index.html> (October 2, 1998).

7 Ibid., table 2.3.

8 U.S. Department of the Interior, *1996 National Survey of Fishing, Hunting, and Wildlife-Associated Recreation* (Washington, D.C.: FWS, 1997), 34.

9 U.S. Department of the Interior, *1982–1983 Nationwide Recreation Survey* (Washington, D.C.: National Park Service, 1986).

10 Outdoor Recreation Resources Review Commission (ORRRC), *Outdoor Recreation for America* (Washington, D.C.: GPO, 1962).

11 Clawson, "Outdoor Recreation: Twenty-Five Years of History," 78.

12 NWF, "Outdoor Recreation in Trouble, Study Says," *National Wildlife* 21, no. 3 (April/May 1983): 31.

13 PLLRC, *One-Third of the Nation's Land* (Washington, D.C.: GPO, 1970), 197.

14 U.S. Congress, The National Environmental Policy Act of 1969 (NEPA), Public Law 91-190, 91st Cong. (January 1, 1970).

15 George H. Stanley et al., *The Limits of Acceptable Change (LAC) System for Wilder-*

ness Planning (Ogden, Utah: U.S. Department of Agriculture, Forest Service Intermountain Forest and Range Experiment Station, 1985).

16 George H. Stanley and Robert E. Manning, "Carrying Capacity of Recreational Settings," in *A Literature Review* (Washington, D.C.: GPO, 1986), 49.

17 Robert E. Manning, David W. Lime, and Marilyn Hof, "Social Carrying Capacity of Natural Areas: Theory and Application in the U.S. National Parks," *Natural Areas Journal* 16, no. 2 (1996): 118.

18 PLLRC, *One-Third of the Nation's Land,* 205.

19 Timothy J. Farnham, Cameron P. Taylor, and Will Callaway, "A Shift in Values: Non-Commodity Resource Management and the Forest Service," *Policy Studies Journal* 23, no. 2 (1995): 293.

20 CEQ, *Environmental Quality, 1981* (Washington, D.C.: GPO, 1981), 152–53.

21 Clawson, "Outdoor Recreation: Twenty-Five Years of History," 91.

22 CEQ, *Off Road Vehicles on Public Lands* (Washington, D.C.: GPO, 1979), 36.

23 Todd Wilkinson, "Snowed Under," *National Parks* 69 (January/February 1995): 32–38.

24 Phillip O. Foss, *Federal Agencies and Outdoor Recreation,* Study Report no. 13 (Washington, D.C.: ORRRC, 1962), 1.

25 William R. Mangun and John B. Loomis, "An Economic Analysis of Funding Alternatives for Outdoor Recreation in the United States," *Policy Studies Review* 7, no. 2 (1987): 421–31.

26 "What's the Holdup?" *Environment* 40 (July/August 1998): 21.

27 Eric O'Brien, "The High Public Cost of Inaction," *Parks and Recreation* 32 (April 1997): 2; also see "Show Us the Money! Free the US Parks Fund!" *Earth Island Journal* 12 (Summer 1997): 12.

28 Conservation Foundation, *State of the Environment: An Assessment at Mid-Decade* (Washington, D.C.: Conservation Foundation, 1984).

29 SCS, *Two-Thirds of Our Land: A National Inventory* (Washington, D.C.: USDA, 1971), 3.

30 See James R. Kahn, *The Economic Approach to Environmental and Natural Resources* (Orlando, Fl.: Dryden Press, 1995); or V. Kerry Smith, *Estimating Economic Values for Nature* (Cheltenham, U.K.: Edward Elgar, 1996).

31 Paul Pritchard, "The Four Percent Solution," *National Parks* 61, nos. 11–12 (1987): 5.

32 Raymond F. Dasmann, *Environmental Conservation* (New York: John Wiley and Sons, 1976), 362–63.

33 Harold Eidsvik, "International Defense: IUCN—A Union That Ranks Protecting Wildlands with World Peace," *National Parks* 61, nos. 9–10 (1987): 13.

34 Robert Mendelsohn, "The Role of Ecotourism in Sustainable Development, Case Study 4," in Gary K. Meffe and C. Ronald Carroll, eds., *Principles of Conservation Biology* (Sunderland, Mass.: Sinauer Associates, 1994), 512.

35 Patricia Cahn and Robert Cahn, "Costa Rica: Coast of Riches," *National Parks* 61, nos. 9–10 (1987): 18–20.

36 Ross W. Gorte, "Wilderness: Overview and Statistics," CRS Report 94-976 ENR (Washington, D.C.: CRS, December 2, 1994), 3.

37 Ibid., 13.

38 Wilderness Society, "Toward the Twenty-First Century: A Wilderness Society Agenda for the National Wilderness Preservation System," *Wilderness* 48, no. 165 (1984): 34.

39 Roderick Nash, *Wilderness and the American Mind,* 3rd ed. (New Haven, Conn.: Yale University Press, 1982).

40 J. Douglas Wellman, *Wildland Recreation Policy* (New York: Wiley, 1987), 124.

41 Craig W. Allin, *The Politics of Wilderness Preservation* (Westport, Conn.: Greenwood, 1982), 276.

42 An Act to Establish a National Wilderness Preservation System, Public Law 88-577, 88th Cong. (September 3, 1964).

43 CEQ, *Environmental Trends* (Washington, D.C.: GPO, 1980), 22.

44 An Act to Establish a National Wilderness Preservation System.

45 Seymour Gold, "Recreation Planning for Energy Conservation," *Parks and Recreation* (September 1977): 61.

46 Katherine Barton and Whit Fosburgh, "The U.S. Forest Service," in A. S. Eno, R. L. DiSilvestro, and W. J. Chandler, eds., *Audubon Wildlife Report* (New York: National Audubon Society, 1986), 128.

47 Olen Paul Matthews, Amy Hank, and Katheryn Toffenetti, "Mining in the Wilderness," *Environment* 27, no. 3 (April 1985): 12.

48 The Wilderness Society, "Toward an Understanding of the Wilderness Act," undated mimeograph.

49 Randall G. Gloege, personal interview, Billings, Mont., December 12, 1986.

50 John Gibson, personal contact, March 1983.

51 Leonard Godwin, "Management Problems in Designated Wilderness Areas," *Journal of Soil and Water Conservation* 34, nos. 5-6 (1979): 141.

52 Barton and Fosburgh, "The U.S. Forest Service," 121.

53 The Wilderness Society, "Toward the Twenty-First Century," 36–38.

54 Wellman, *Wildland Recreation Policy,* 195–97.

55 George C. Coggins and Charles F. Wilkinson, *Federal Public Land and Resources Law* (Mineola, N.Y.: Foundation Press, 1981), 821.

56 Ibid.

57 Dasmann, *Environmental Conservation,* 252–53.

58 William Eddy, "Kenya: Rhythms of Survival," *National Parks* 61, nos. 9–10 (1987): 22.

59 Jeremy Harrison, Kenton Miller, and Jeffrey McNeely, "The World Coverage of Protected Areas, Development Goals and Environmental Needs," *Ambio* 2, no. 5 (1982): 238.

60 Ibid., 252–53.

61 Betsy Cody, "Major Federal Land Management Agencies: Management of Our Nation's Lands and Resources," CRS Report 95-599 ENR (Washington, D.C.: CRS, May 15, 1995), 1.

62 CEQ, *Environmental Trends,* 294–95.

63 Ibid.

64 CEQ, *Environmental Quality, 1980,* 295–96.

65 Daniel H. Henning, "The Public Land Law Review Commission: A Political and Western Analysis," *Idaho Law Review* (spring 1970), 77.

66 PLLRC, *One-Third of the Nation's Land,* 1.

67 Federal Land Policy and Management Act of 1976 (FLPMA), Public Law 94-579, 94th Cong. (October 21, 1976), sec. 102(a) 90, Stat. 2744.

68 "Land Sale of the Century," *Time* (August 31, 1982): 17–22.

69 FLPMA, sec. 103(c) 90, Stat. 2745–2746.

70 CEQ, *Environmental Quality, 1980,* 297.

71 Ibid., 297–98.

72 Daniel H. Henning, "Western Wilderness Values," *Western's World* (January/March 1973): 21–24.

73 BLM, *Public Land Statistics 1992* (Washington, D.C.: GPO, 1993).

74 USDA Forest Service, *Land Areas of the National Forest System As of September 30, 1993,* Report no. FS-383 (Washington, D.C.: February, 1994).

75 CEQ, *Environmental Quality, 1994–1995* (Washington, D.C.: GPO, 1995), 411.

76 U.S. Department of the Interior, "Managing America's Assets: Fact Sheet" (November 2, 1982), 1.

77 Phillip O. Foss, "Public Land Policy to the Year 2000: The Sagebrush Rebellion," paper presented at Western Social Science Association conference, Denver, Colo., April 14, 1982.

78 Ibid.

79 FLPMA, sec. 108(8) 90, Stat. 2745.

80 Paul J. Culhane, *Public Lands Politics: Interest Group Influence on the Forest Service and the Bureau of Land Management* (Baltimore, Md.: Johns Hopkins University Press, 1981).

81 BLM, *BLM and the Environment* (Washington, D.C.: BLM, 1980), 18–19.

82 Wayne A. Morissey, Jeffrey A. Zinn, and M. Lynne Corn, *Ecosystem Management: Federal Agency Activities,* CRS Report 94-339 ENR (Washington, D.C.: CRS, April 19, 1994), 13.

83 Ibid., 60.

84 Jean C. Mangun and William R. Mangun, "Reciprocal Relations Between Humans and Ecosystems: Toward a Social Exchange Perspective," in Dennis L. Soden, Berton L. Lamb, and John R. Tennert, eds., *Ecosystem Management: A Social Science Perspective* (Dubuque, Iowa: Kendall/Hunt, 1998), 59–69.

8 Environmental Pollution Control

1 EPA, *The U.S. EPA's 25th Anniversary Report: 1970–1995* (Washington, D.C.: Office of Policy and Program Evaluation, 1997), 7–10.

2 CEQ and Department of State, *Global Future: Time to Act* (Washington, D.C.: GPO, 1981), 126.

3 Clarence Davies, *The Politics of Pollution* (New York: Pegasus, 1970), 19.

4 Jeremy Rifkin, *Entropy: A New World View* (New York: Viking, 1980), 35.

5 CEQ, *Environmental Quality, 1970* (Washington, D.C.: GPO, 1970), 8.

6 Walter A. Rosenbaum, *The Politics of Environmental Concern* (New York: Praeger, 1977), 136.

7 John H. Baldwin, *Environmental Planning and Management* (Boulder, Colo.: Westview, 1985), 131.

8 Rosenbaum, *The Politics of Environmental Concern,* 136.

9 See William R. Mangun, "Environmental Program Evaluation in an Inter-

governmental Context," in Gerrit J. Knaap and Tschangho John Kim, eds., *Environmental Program Evaluation: A Primer* (Urbana: University of Illinois Press, 1998), 86–125.

10   See Denise Scheberle, *Federalism and Environmental Policy: Trust and the Politics of Implementation* (Washington, D.C.: Georgetown University Press, 1997).

11   See Evan J. Ringquist, *Environmental Protection at the State Level: Politics and Progress in Controlling Pollution* (Armonk, N.Y.: M. E. Sharpe, 1993).

12   Patricia McGee Crotty, "The New Federalism Game: Primary Implementation of Environmental Policy," *Publius: The Journal of Federalism* 17 (spring 1987): 53–54.

13   Baldwin, *Environmental Planning and Management,* 131.

14   U.S. Department of Commerce, Bureau of the Census, *1996 Statistical Abstract of the United States* (Washington, D.C.: Bureau of the Census, 1996), 238.

15   U.S. Department of Commerce, Bureau of Economic Analysis *Survey of Current Business* (Washington, D.C.: Bureau of Economic Analysis, May 1995).

16   U.S. Department of Commerce, Bureau of the Census, *Statistical Abstract of the United States—1987* (Washington, D.C.: Bureau of the Census, 1986), 193.

17   Robert Repetto and Dale Rothman, "Has Environmental Protection Really Reduced Productivity Growth?" *Challenge* 40, no. 1 (January/February, 1997): 46–47.

18   Adam B. Jaffe, Steven R. Peterson, Paul R. Portney, and Robert N. Stavins, "Environmental Regulation and the Competitiveness of U.S. Manufacturing: What Does the Evidence Tell Us?" *Journal of Economic Literature* 33 (March 1995): 132–63.

19   See Robert W. Hahn and Robert N. Stavins, "Economic Incentives for Environmental Protection: Integrating Theory and Practice," *AEA Papers and Proceedings* 82 (May 1992): 464–68; and Thomas H. Tietenberg, *Economics and Environmental Policy* (Brookfield, Vt.: Edward Elgar Publishing, 1994).

20   John H. Cumberland, "Ecology, Economic Incentives, and Public Policy in the Design of a Transdisciplinary Pollution Control Instrument," in Jeroen C. J. M. van den Bergh, Jan van der Straaten, and Sandra Koskoff, *Toward Sustainable Development: Concepts, Methods, and Policy* (Washington, D.C.: Island Press, 1994), 265–78.

21   Richard B. Stewart, "Economics, Environment, and the Limits of Environmental Control," *Harvard Environmental Law Review* 9, no. 1 (1985): 1.

22   William Ramsay, "On Assessing Risk," *Resources* (October 1981): 10.

23   Linda-Jo Schierow, "The Role of Risk Analysis and Risk Management in Environmental Protection," CRS Report 94-36 ENR (Washington, D.C.: CRS, October 2, 1997), 1.

24   EPA, *Environmental Progress and Challenges: An EPA Perspective* (Washington, D.C.: EPA, 1984), 72–73.

25   EPA, *Characterization of Municipal Solid Waste in the United States: 1995 Update* (Washington, D.C.: Office of Solid Waste and Emergency Response, 1996), 26.

26   EPA, *Environmental Progress and Challenges,* 72.

27   James E. McCarthy, "Solid Waste Issues in the 105th Congress" CRS Report 97-6 ENR (Washington, D.C.: CRS, May 14, 1998), 3.

28 Robert K. Ham, "Overview and Implications of U.S. Sanitary Landfill Practice," *Air and Waste* 43 (February 1993): 187.

29 Gwendolyn Holmes, Ben Ramnarine Singh, and Louis Theodore, *Handbook of Environmental Management and Technology* (New York: John Wiley and Sons, 1993), 251.

30 "A Problem That Cannot Be Buried," *Time* (October 14, 1985): 76.

31 EPA, *The Waste Minimization National Plan* (Washington, D.C.: Office of Solid Waste and Emergency Response, November 16, 1994), 1.

32 EPA, *Superfund: A Six-Year Perspective* (Washington, D.C.: Office of Solid Waste and Emergency Response, 1986), 2.

33 Steven Cohen, "Defusing the Toxic Time Bomb: Federal Hazardous Waste Programs," in Norman J. Vig and Michael E. Kraft, eds., *Environmental Policy in the 1980s: Reagan's New Agenda* (Washington, D.C.: CQ Press, 1984), 273–74.

34 Riley E. Dunlap, "Public Opinion and Environmental Quality," in James P. Lester, ed., *Environmental Politics and Policy: Theories and Evidence,* 2nd ed. (Durham, N.C.: Duke University Press, 1995), 63–114.

35 James P. Lester and Ann O. Bowman, eds., *The Politics of Hazardous Waste Management* (Durham, N.C.: Duke University Press, 1983); Charles E. Davis and James P. Lester, eds., *Dimensions of Hazardous Waste Politics and Policy* (Westport, Conn.: Greenwood, 1988); Charles E. Davis, *The Politics of Hazardous Waste* (Englewood Cliffs, N.J.: Prentice-Hall, 1993).

36 CEQ, *Environmental Quality, 1994–1995,* 485.

37 Steven Cohen, "Defusing the Toxic Time Bomb," 289–300.

38 Michael R. Overcash, "Environmental Legacies," *Environmental Science and Technology* 21, no. 2 (1987): 115.

39 IRPTC, *International Register of Potentially Toxic Chemicals* (Geneva: UNEP, 1985), 1–3.

40 J.H. Jennrich, "Environmental Control: Keeping the Good Earth Clean," *Nation's Business* (December 1979): 74.

41 Richard A. Dension, "Environmental Life-Cycle Comparisons of Recycling, Landfilling, and Incineration: A Review of Recent Studies," *Annual Review of Energy and Environment* 21 (1996): 193.

42 CEQ, *Environmental Quality, 1981,* 91–92.

43 U.S. House of Representatives, *EPA's Proposed Land Ban Regulations, Hearing before the Subcommittee on Commerce, Transportation, and Tourism of the Committee on Energy and Commerce, February 19, 1986* (Washington, D.C.: GPO, 1986), 93.

44 Peter A. Berle, "NIMBYs and LULUs," *Audubon* 88, no. 3 (1986): 4.

45 CEQ, *Environmental Quality, 1981,* 93.

46 Kirsten V. Oldenburg and Joel S. Hirschhorn, "Waste Reduction," *Environment* 29, no. 2 (1987): 17–18.

47 David J. Sarokin, Warren R. Muir, Catherine G. Miller, and Sebastian R. Sperber, *Cutting Chemical Wastes* (New York: INFORM, 1985); Environmental Defense Fund, *Approaches to Source Reduction* (Berkeley, Calif.: EDF, 1986); EPA, *Report to Congress: Minimization of Hazardous Wastes,* EPA/530-SW-86-033 (Washington, D.C.: EPA Office of Solid Waste and Emergency Response, 1986).

48 Oldenburg and Hirschhorn, "Waste Reduction," 20, 39–43.

49 CEQ, *Environmental Quality, 1970,* 106.

50 Ibid., 89.

51 Curtis D. Klaassen and John Doull, "Evaluation of Safety: Toxicologic Evaluation," in Curtis D. Klaassen, Mary O. Amdur, and John Doull, eds., *Casarett and Doull's Toxicology: The Basic Science of Poisons,* 2nd ed. (New York: Macmillan, 1980), 12.

52 CEQ, *Environmental Quality, 1981,* 101.

53 William R. Mangun, "A Comparative Analysis of Hazardous Waste Management Policy in Western Europe," in Davis and Lester, eds., *Dimensions of Hazardous Waste Politics and Policy,* 205–22.

54 Bruce Piasecki and Jerry Gravander, "The Missing Links: Restructuring Hazardous Waste Controls in America," *Technology Review* 88, no. 7 (1985): 49.

55 Alan C. Williams, "A Study of Hazardous Waste Minimization in Europe: Public and Private Strategies to Reduce Production of Hazardous Waste," *Environmental Affairs* 14 (1987): 166.

56 Terry Lash, "Radioactive Wastes," *Amicus* (Fall 1979): 26–27.

57 Mark Holt, "Civilian Nuclear Waste Disposal," CRS Report 92-59 ENR (Washington, D.C.: CRS, July 24, 1998), 9.

58 CEQ and Department of State, *Global 2000* (Washington, D.C.: GPO, 1980), 37.

59 Holt, "Civilian Nuclear Waste Disposal."

60 George E. Brown Jr., "U.S. Nuclear Waste Policy: Flawed but Feasible," *Environment* 29, no. 8 (1987): 7, 25. Also see Brandon B. Johnson, "Public Concerns and the Public Role in Siting Nuclear and Chemical Waste Facilities," *Environmental Management* 11, no. 5 (1987): 571–86.

61 See Gerald Jacob, *Site Unseen: The Politics of Siting a Nuclear Repository* (Pittsburgh, Pa.: University of Pittsburgh Press, 1990).

62 Holt, "Civilian Nuclear Waste Disposal," 3.

63 Luther J. Carter, "Siting the Nuclear Repository: Last Stand at Yucca Mountain," *Environment* 29, no. 8 (1987): 8–10.

64 CEQ and Department of State, *Global Future,* 4.

65 EPA, *Final Report to Congress on Benefits and Costs of the Clean Air Act, 1970 to 1990,* EPA 410-R-97-002 (Washington, D.C.: Office of Air and Radiation, 1997), ES-8.

66 EPA, *National Air Quality and Emissions Trends Report, 1995,* EPA 454/R-96-005 (Research Triangle Park, N.C.: Office of Air Quality Planning and Standards, October 1996), 2.

67 EPA, *The U.S. EPA's 25th Anniversary Report,* 2–8.

68 Baldwin, *Environmental Planning and Management,* 3–4.

69 Roper Organization, "Natural Resource Conservation: Where Environmentalism Is Headed in the 1990s," *Times-Mirror Magazine's National Environmental Forum Survey* (June 1992).

70 CEQ, *Environmental Quality, 1981,* 271.

71 CEQ, *Environmental Quality, 1970,* 66–71.

72 Rosenbaum, *The Politics of Environmental Concern,* 143.

73 Lennart J. Lundqvist, "Who Is Winning in the Race for Clean Air," *Ambio* (April 1979): 144–51.

74 "Code of Federal Regulations 61.01," *Environment Report* 16, no. 26 (October 1985): 1084.

75 Richard A. Liroff, "Targeting Air Toxics," *Environmental Professional* 9 (1987): 93–94.

76 "Total Exposure Assessment Methodology Study Measures Personal Exposure to Pollutants," *International Journal of Air Pollution Control and Hazardous Waste Management* 37, no. 12 (1987): 1466.

77 J. Held and Ronald Harkov, "Present Applications of the TLV Approach to Setting State Guidelines and Standards: Consistencies and Problems," paper presented at the seventy-ninth annual meeting of the Air Pollution Control Association, Minneapolis, Minn., 1986, cited in Liroff, *Environmental Professional* 9 (1987): 93–94.

78 EPA, *25th Anniversary Report*, 7.

79 Susan L. Cutter, "Airborne Toxic Releases: Are Communities Prepared?" *Environment* 29, no. 6 (1987): 12–14.

80 Susan L. Cutter and William D. Solecki, "The National Pattern of Airborne Toxic Releases," *Professional Geographer* 41 (1989): 152.

81 Brian Schimmoller, "NAAQS Attacks," *Power Engineering* 102 (May 1998): 10.

82 World Bank, *World Development Report 1992: Development and the Environment* (Washington, D.C.: World Bank, 1992) cited in Jane Vise Hall, "Air Quality Policy in Developing Countries," *Contemporary Economic Policy* 13 (April 1995): 77.

83 CEQ and Department of State, *Global Future*, 130–44.

84 CEQ and Department of State, *Global 2000*, 36.

85 See J. T. Houghton, G. J. Jenkins, and J. J. Ephraums, eds., *Climate Change: The IPCC Scientific Assessment* (Cambridge: Cambridge University Press, 1990); and National Research Council Board on Atmosphere Sciences and Climate, *Changing Climate* (Washington, D.C.: National Academy Press, 1983), 22–34.

86 Wayne A. Morrissey and John R. Justus, *Global Climate Change* CRS Report 89-5 ENR (Washington, D.C.: CRS, October 27, 1997).

87 Intergovernmental Panel on Climate Change (IPCC), *Climate Change: The IPCC Scientific Assessment* (Geneva: World Meteorological Organization/United Nations Environment Programme, 1990–1991); and IPCC, *Second Assessment Report* (Geneva: World Meteorological Organization/United Nations Environment Programme, 1995).

88 See Robert C. Balling Jr., *The Heated Debate: Greenhouse Predictions Versus Climate Reality* (San Francisco, Calif.: Pacific Research Institute, 1992); and William F. O'Keefe, "It's Time to Reconsider Global Climate Change Policy," *USA Today* 125 (March 1997): 81–83.

89 David Helvarg, "The Greenhouse Spin," *Nation* 263 (December 16, 1996): 24.

90 Conservation Foundation, *State of the Environment, 1982* (Washington, D.C.: Conservation Foundation, 1982), 44–77.

91 *Environmental Science and Technology* 22, no. 3 (1988): 238.

92 "Interagency Cooperative Scientific Program to Investigate Antarctic Ozone Hole," *International Air Pollution Control and Hazardous Waste Management* 37, no. 9 (1987): 1078–79.

93 Irving M. Mintzer and Alan S. Miller, "The Ozone Layer: Its Protection Depends on International Cooperation," *Environmental Science and Technology 21*, no. 12 (1987): 1167–68.

94 David E. Gushee, *Stratospheric Ozone Depletion: Regulatory Issues* CRS Report 89-21 ENR (Washington, D.C.: CRS, March 18, 1996).

95 CEQ and Department of State, *Global Future,* 142–43.

96 William H. Layden, "Food Safety: A Patchwork System," *GAO Journal* (spring/summer 1992): 50.

97 Sophie Boukhari, "Poison of the Earth," *UNESCO Courier* 51 (July/August 1998): 15–16.

98 Ibid., 83.

99 Rachel Carson, *Silent Spring* (1962; Greenwich, Conn.: Fawcett, 1970).

100 Ibid., 261.

101 Congressional Quarterly, *Environment and Health* (Washington, D.C.: CQ Press, 1981), 82.

102 Jon R. Luoma, "Havoc in the Hormones," *Audubon* 97 (July/August 1995): 60–67.

103 EPA, *Emerging Global Environmental Issues,* EPA 160-K-97-001 (Washington, D.C.: Office of International Activities, January 1997), 5.

104 CEQ, *Environmental Quality, 1979* (Washington, D.C.: GPO, 1979), 445.

105 Congressional Quarterly, *Environment and Health,* 82.

106 CEQ, *Environmental Quality, 1981,* 92.

107 Conservation Foundation, *State of the Environment: A View Toward the Nineties* (Washington, D.C.: Conservation Foundation, 1987), 144.

108 Ibid., 82–93.

109 David Pimentel and Marcia Pimentel, "The Risks of Pesticides," *Natural History* (March 1979): 30.

110 Kamal Tolba, "Pesticide Dangers," *UN Chronicle* (July 1979): 103–4.

111 Congressional Quarterly, *Environment and Health,* 82–83.

112 Ibid., 215.

113 Linda-Jo Schierow, *Pesticide Policy Issues,* CRS Report 95-16 ENR (Washington, D.C.: CRS, December 4, 1996), 2–3.

114 Christopher J. Bosso, *Pesticides and Politics* (Pittsburgh, Pa.: University of Pittsburgh Press, 1987).

115 Ibid., 242–43.

116 Thomas Schoenbaum, *Environmental Policy Law* (Mineola, N.Y.: Foundation Press, 1982), 81.

117 CEQ and Department of State, *Global Future,* 35–37.

118 UNEP, *The State of the Environment, 1985* (Nairobi, Kenya: UNEP, 1985), 10.

119 CEQ, *Environmental Quality, 1979,* 395.

120 CEQ and Department of State, *Global Future,* 35–37.

121 Anthony DeCosta, "The Two Faces of Pesticide Exports," *Organic Gardening* (March 1979): 1.

122 Boukhari, "Poison of the Earth," 15; also, see J. Jeyaratnam, "Acute Pesticide Poisoning: A Major Global Health Problem," *World Health Statistical Quarterly* 43 (1990): 139–44.

123 See Robert Repetto, *Paying the Price: Pesticide Subsidies in Developing Countries* (New York: World Resources Institute, 1985).

124 Ibid., 1–2.

125 CEQ, *Environmental Quality, 1979,* 533–34.

126 Ibid., 535–36.

127 Ibid., 534–36.

128 James R. Udall, "Silence under Siege," *Sierra* 82 (March/April 1997): 28.

129 Schoenbaum, *Environmental Policy Law,* 1049.

130 *Congressional Record,* S-18401-08 (November 18, 1970), 143–44.

131 CEQ, *Environmental Quality, 1979,* 559.

132 Robert V. Bartlett, "The Budgetary Process and Environmental Policy," in Norman J. Vig and Michael E. Kraft, eds., *Environmental Policy in the 1980s: Reagan's New Agenda* (Washington, D.C.: CQ Press, 1984), 125.

133 "A Cite for Sore Ears," *Science News* 142 (July 25, 1992): 61.

134 CEQ, *Environmental Quality, 1979,* 559–60.

135 Ibid., 555–62.

136 "Airport Hazards," *Newsweek* (September 18, 1979): 90.

137 France Bequette, "Defeating Decibels," *Unesco Courier* 47 (June 1994): 24.

138 Ibid., 25.

139 CEQ, *Environmental Quality, 1979,* 568–69.

140 UNEP, *Review of Major Achievements in the Implementation of the Action Plan for the Human Environment* (Nairobi, Kenya: UNEP, January 1982), 37.

141 CEQ and Department of State, *Global Future,* 237.

142 Congressional Quarterly, *Environment and Health,* 41.

143 Ibid., 42.

144 Albuquerque Urban Observatory, *Factors Pertaining to Water Quality in the Alburquerque Metropolitan Area* (Albuquerque, N. Mex.: Albuquerque Urban Observatory, 1970), vi.

145 Congressional Quarterly, *Environment and Health,* 42–43.

146 EPA, *The Quality of Our Nation's Water: 1996,* EPA 841-S-97-001 (Washington, D.C.: EPA, Office of Water, April, 1998), 14–19.

147 CEQ, *Environmental Quality, 1994–1995,* 225.

148 Carl F. Myers, "A New Nonpoint-Source Water Pollution Control Challenge," *Journal of Soil and Water Conservation* 42, no. 4 (1987): 222.

149 NWF, *1985 Environmental Quality Index,* 4–5.

150 Helen M. Ingram and Dean E. Mann, "Preserving the Clean Water Act," in Vig and Kraft, eds., *Environmental Policy in the 1980s,* 261.

151 EPA, *Report to Congress: Nonpoint Source Pollution in the U.S.,* EPA 841R-84-100 (Washington, D.C.: Office of Water Program Operations, 1984).

152 EPA, *A Groundwater Protection Strategy for the Environmental Protection Agency* (Washington, D.C.: GPO, 1984).

153 EPA, *National Water Quality Inventory: 1990 Report to Congress,* EPA 503/9-92-006 (Washington, D.C.: EPA, Office of Water, 1992).

154 Richard M. Dowd, "Analysis of the New SDWA," *Environmental Science and Technology* 20, no. 11 (1986): 1101.

155 CEQ, *Environmental Quality, 1994–1995,* 225.

156 Congressional Quarterly, *Environment and Health,* 43.

157 U.S. Senate, *Groundwater Contamination and Protection. Hearing before Subcommittee on Toxic Substances and Environmental Oversight. June 17, 20, 1986* (Washington, D.C.: GPO, 1985), 9; and U.S. Geological Survey, *Estimated Use of Water in the United States in 1990,* Circular 1081 (Washington, D.C.: GPO, 1992).

158 Ingram and Mann, "Preserving the Clean Water Act," 262–63.

159 EPA, *Watershed Protection: A Statewide Protection Approach,* EPA 841-R-95-004 (Washington, D.C.: EPA, Office of Water, August 1995), 1–3.

160 EPA, *Why Watersheds?* EPA 800-F-96-001 (Washington, D.C.: EPA, Office of Water, February 1996), 3.

161 Rosenbaum, *The Politics of Environmental Concern,* 157.

162 Ibid., 165.

163 Bruce Ackerman and William Hassler, *Clean Coal/Dirty Air* (New Haven, Conn.: Yale University Press, 1981), 2.

164 U.S. House of Representatives, *Clean Air Standards, Hearing before Subcommittee on Health and the Environment. February 19, 1987* (Washington, D.C.: GPO, 1987), 102–5.

165 Donna Engelau, "EPA Plan Would Give Cities Up to Eight Years to Attain Ozone, CO Standards," *International Journal of Air Pollution Control and Hazardous Waste Management* 38, no. 1 (1988): 72–73.

166 CEQ, *Environmental Quality, 1981,* 51.

167 Lawrence J. Jensen, "The Challenge of A New Generation of Wastewater Treatment," *EPA Journal* 12, no. 9 (1986): 2.

168 CEQ, *Environmental Quality, 1994–1995,* 236.

169 William R. Mangun, "A Comparative Evaluation of the Major Pollution Control Programs in the United States and West Germany," *Environmental Management* 3 (1979): 387–401.

170 Frank J. Humbrik, Michael D. Smolen, and Steven A. Dressing, "Pollution from Nonpoint Sources: Where We Are and Where We Should Be," *Environmental Science and Technology* 21, no. 8 (1987): 737.

171 EPA, *25th Anniversary Report,* 27.

172 CEQ, *Environmental Quality, 1980,* 85–87.

173 Conservation Foundation, *State of the Environment, 1984,* 116.

174 CEQ and State Department, *Global 2000,* 35.

175 Andrew C. Gross, "Water Quality Management Worldwide," *Environmental Management* 10, no. 1 (1986): 38–39.

176 CEQ and State Department, *Global 2000,* 123–24.

177 UNEP, *Annual Report of the Executive Director,* 1984 (Nairobi, Kenya: UNEP, 1983), 59.

9 Urban and Regional Environmental Policy

1 H. Hengeveld and C. DeVocht, eds., *The Role of Water in Urban Ecology* (Amsterdam: Elsevier Scientific, 1982), 12.

2 John H. Baldwin, *Environmental Planning and Management* (Boulder, Colo.: Westview Press, 1985), 5.

3   EPA, *Community-Based Environmental Protection: A Resource Book for Protecting Ecosystems and Communities,* EPA 230-B-96-003 (Washington, D.C.: EPA Office of Policy, Planning, and Evaluation, September 1997), 1–7.

4   Jeremy Rifkin, *Entropy: A New World View* (New York: Viking, 1980), 148–49.

5   Joel Garreau, "Edge Cities," *American Demographics* 16 (February 1994): 24–33; and Joel Garreau, *Edge City: Life on the New Frontier* (New York: Doubleday, 1991).

6   CEQ, *Environmental Trends* (Washington, D.C.: GPO, 1980), 45.

7   "End of the Population Explosion?," *Discover* 18 (July 1997): 14–16.

8   CEQ and Department of State, *Global 2000: Report to the President* (Washington, D.C.: GPO, 1980), 12.

9   George Moffett, "World's Cities Running Out of Room," *Christian Science Monitor* 88 (March 25, 1996): 4.

10 Ibid.

11 Eugene Linden, "Megacities," *Time* 141 (January 11, 1993): 28.

12 Deil Wright, *Understanding Intergovernmental Relations,* 3rd ed. (Belmont, Calif.: Wadsworth, 1988), 83–85.

13 Walter Scheiber, "Regionalism: Its Implications for the Urban Manager," *Public Administration Review* (January/February 1971): 43.

14 Michael P. Conzen, "American Cities in Profound Transition: The New City Geography of the 1980s," *Journal of Geography* (May–June 1983): 100.

15 Scheiber, "Regionalism," 45–46.

16 Baldwin, *Environmental Planning and Management,* 278; see also EPA, *People, Places, and Partnerships: A Progress Report on Community-Based Environmental Protection,* EPA-100-R-97-003 (Washington, D.C.: EPA, Office of the Administrator, July 1997).

17 Melvin G. Marcus and Thomas R. Detwyler, "Urbanization and Environment in Perspective," in Thomas R. Detwyler and Melvin G. Marcus, eds., *Urbanization and Environment* (Belmont, Calif.: Wadsworth, 1972), 10.

18 Dennis C. McElrath, "Public Response in Environmental Problems," in Lynton K. Caldwell, ed., *Political Dynamics of Environmental Control* (Bloomington: Indiana University, Institute of Public Administration, 1967), 3.

19 Ibid., 1–8.

20 EPA, *Performance Partnerships: Building Stronger Relationships between EPA and the States,* EPA-100-R-97-001 (Washington, D.C.: EPA, Office of the Administrator, January 1997), 5.

21 Gail Tapscott, "Community Right-to-Know: A New Environmental Agenda," *The Environmental Forum* 3, no. 6 (1984): 8–15.

22 Clifford Grobstein, "The Role of the National Academy of Sciences in Public Policy and Regulatory Decision Making," in J. Daniel Nyart and Milton M. Carrow, eds., *Law and Science in Collaboration* (Lexington, Mass.: Lexington Books, 1983).

23 Elder Witt, "The Tedious Chore of Preparing for Chemical Disaster is in the Lap of Local Government," *Governing* 1, no. 7 (1988): 24–28.

24 EPA, *Title III Fact Sheet: Emergency Planning and Community Right-to-Know* (Washington, D.C.: GPO, 1986).

25 Witt, "The Tedious Chore of Preparing for Chemical Disaster," 28.

26 J. Gordon Arbuckle and Paul A. J. Wilson, "New Potentials for Criminal Liability of Environmental Managers under the Emergency Planning and Community Right-to-Know Act," *Environmental Management Report,* no. 4 (1987): 114–15.

27 EPA, *Harnessing the Power of the Internet: EPA Responds to the Rising Public Demand for Environmental Information,* EPA-100-R-98-04 (Washington, D.C.: Office of Reinvention, August 1998), 4.

28 Organization for Economic Cooperation and Development (OECD), "Making the Cities More Livable," *OECD Observer,* no. 144 (January 1987): 18.

29 CEQ, *Environmental Quality, 1971* (Washington, D.C.: GPO, 1971), 189.

30 Ibid., 190.

31 Stewart Marquis, "Ecosystems, Societies, and Cities," in Roy L. Meek and John A. Straayer, eds., *The Politics of Neglect: The Environmental Crisis* (Boston: Houghton Mifflin, 1971), 122–28.

32 Ibid., 130–31.

33 CEQ, *Environmental Quality, 1971,* 204.

34 Ibid., 207.

35 Floyd A. Malveaux and Sheryl A. Fletcher-Vincent, "Environmental Risk Factors of Childhood Asthma in Urban Centers," *Environmental Health Perspectives Supplements* 103 (September 1995): 59.

36 Claudia Spix and H. Ross Anderson, et al., "Short-Term Effects of Air Pollution on Hospital Admissions of Respiratory Diseases in Europe: A Quantitative Summary of APHEA Study Results," *Archives of Environmental Health* 53 (January/February 1998): 54.

37 CEQ, *Environmental Quality, 1971,* 191–96.

38 Ljiljana Skender and Visja Karacic, "Assessment of Urban Population Exposure to Trichloroethylene and Tetrachloroethylene by Means of Biological Monitoring," *Archives of Environmental Health* 49 (November/December, 1994): 445–52.

39 John Adamson, personal communication, May 1983.

40 CEQ, *Environmental Quality, 1971,* 196–97.

41 Ibid., 197–200.

42 Kathryn R. Mahaffey et al., "National Estimates of Blood Lead Levels: United States, 1976–1980," *New England Journal of Medicine* 307, no. 10 (1982): 573.

43 Ibid., 200–201.

44 Charles E. Little, "Linking Countryside and City: The Uses of Greenways," *Journal of Soil and Water Conservation* 42, no. 3 (1987): 167.

45 Paul R. Ehrlich and Anne H. Ehrlich, *Population, Resources, and Environment* (San Francisco: Freeman, 1970), 141–43.

46 Conservation Foundation, *State of the Environment, 1982* (Washington, D.C.: Conservation Foundation, 1982), 358.

47 President's Council on Recreation and Natural Beauty, *From Sea to Shining Sea: A Report on the American Environment Our Natural Heritage* (Washington, D.C.: GPO, 1968), 104.

48 CEQ, *Environmental Quality, 1970,* 171.

49 Alfred Kahn, "Tyranny of Small Secessions," *Kyklos* 19 (1966): 23–45.

50 Thomas Rudel, "The Quiet Revolution in Municipal Land Use Control: Competing Expectations," *Journal of Environmental Management* (October 1980): 125.

51 CEQ, *Environmental Quality, 1981,* 48–49.

52 Glenn V. Fuguitt, Tim B. Heaton, and Daniel T. Lichter, "Monitoring the Metropolitanization Process," *Demography* 25, no. 1 (1988): 125–26.

53 Conzen, "American Cities in Profound Transition," 94–95.

54 Ibid., 95.

55 Judith Kunofsky and Larry Orman, "The Greenbelts and the Well-Planned City," *Sierra* (November/December 1985): 45–46.

56 CEQ, *Environmental Quality, 1970,* 171–72.

57 Ian M. McHarg, *Design with Nature* (Garden City, N.Y.: Doubleday/Natural History, 1971), 80–85.

58 Robert S. Lorch, *State and Local Politics: The Great Entanglement,* 2nd ed. (Englewood Cliffs, N.J.: Prentice-Hall, 1986), 226.

59 Max Ways, "Urban Sprawl: A Smothering of People," in Bruce Wallace, ed., *People, Their Needs, Environment, Ecology* (Englewood Cliffs, N.J.: Prentice-Hall, 1972), 178–79.

60 Darlene Himmelspack, "Council OKs River Flood-Control Project," *Oceanside* [Calif.] *Blade Tribune* (February 28, 1988): B1.

61 McHarg, *Design with Nature,* 81–87.

62 CEQ, *Environmental Quality, 1970,* 173.

63 UNEP, *The State of the World Environment, 1972–82* (Nairobi, Kenya: UNEP, 1982), 53–54.

64 Ibid., 54.

65 Ibid., 55.

66 EPA, *Community-Based Environmental Protection: A Resource Book for Protecting Ecosystems and Communities,* EPA 230-B-96-003 (Washington, D.C.: Office of Policy, Planning, and Evaluation, September 1997).

67 EPA, *People, Places, and Partnerships,* 14–15.

68 Ibid., 19–21.

69 Baldwin, *Environmental Planning and Management,* 278.

70 Thomas R. Dye, *Politics in States and Communities,* 6th ed. (Englewood Cliffs, N.J.: Prentice-Hall, 1988), 444.

71 CEQ, *Environmental Quality, 1971,* 188.

72 Ibid., 184–85.

73 Congressional Quarterly, *Man's Control of the Environment* (Washington, D.C.: CQ Press, 1970), 35.

74 William R. Mangun, "The Role of Restraint and Noninterference in the Evolution of Dutch Environmental Policy," *Environmental Review* 6, no. 1 (1982): 38–53.

75 CEQ, *Environmental Quality, 1971,* 61.

76 Baldwin, *Environmental Planning and Management,* 46.

77 McHarg, *Design with Nature,* 43.

78 Conservation Foundation, *State of the Environment, 1982,* 16–19.

79 Ibid., 19.

80 Baldwin, *Environmental Planning and Management,* 295–96.

81 UNEP, *Annual Report of the Executive Director, 1984* (Nairobi, Kenya: UNEP, 1985), 77.

10 International Environmental Administration

1   Hilary F. French, *Partnership for the Planet: An Environmental Agenda for the United Nations* (Washington, D.C.: Worldwatch Paper 126, July 1995), 10.

2   Lynton K. Caldwell, *International Environmental Policy* (Durham, N.C.: Duke University Press, 1996), 3rd ed. 2–5.

3   Ibid., 7–8.

4   Ibid., 8–9.

5   Conservation Foundation, *State of the Environment: An Assessment* (Washington, D.C.: Conservation Foundation, 1984), 7.

6   CEQ, *Environmental Quality, 1985* (Washington, D.C.: GPO, 1985), 322.

7   Nigel Sitwell, "Our Trees Are Dying," *Science Digest* (September 1984): 39–40.

8   John McCormick, *Acid Earth: The Global Threat of Acid Pollution,* Earthscan paperback series (Washington, D.C.: International Institute for Environment and Development, 1985), 1–187.

9   Mostafa K. Tolba, Osama A. El-Kholy, E. El-Hinnawi, M. W. Holdgate, D. F. McMichael, and R. E. Munn, eds., *The World Environment 1972–1992: Two Decades of Challenge* (London: Chapman and Hall, 1992), 18–21.

10 McCormick, *Acid Earth.*

11 Anne LaBastille, "Acid Rain: How Great a Menace," *National Geographic* (November 1981): 680–82.

12 CEQ and Department of State, *Global Future: Time to Act* (Washington, D.C.: GPO, 1981), 140–41.

13 "Acid Rain: A Problem for the Eighties," *IUCN Bulletin* (April/May/June 1982): 40.

14 Marc A. Levy, "European Acid Rain: The Power of Tote-Board Diplomacy," in Peter M. Haas, Robert O. Keohane, and Marc A. Levy, eds., *Institutions for the Earth: Sources of Effective International Environmental Protection* (Cambridge, Mass.: MIT Press, 1993), 75–132.

15 Peter Sand, "Air Pollution in Europe—International Policy Responses," *Environment* 29, no. 10 (1987): 16–17.

16 LaBastille, "Acid Rain," 680.

17 UNEP, *The State of the Environment, 1983* (Nairobi, Kenya: UNEP, 1983), 21.

18 Christer Agren, "The Acid Test," *New Statesman and Society* 6 (September 10, 1993): 30.

19 Sand, "Air Pollution in Europe," 18.

20 Ibid.

21 See Haas et al., eds., *Institutions for the Earth.*

22 Kenneth A. Dahlberg et al., *Environment and the Global Arena: Actors, Values, Policies, and Futures* (Durham, N.C.: Duke University Press, 1985), 110–13.

23 James L. Regens and Robert W. Rycroft, *The Acid Rain Controversy* (Pittsburgh, Penn.: University of Pittsburgh Press, 1988), 123.

24 Lynton K. Caldwell, *International Environment Policy*, 3rd ed. (Durham, N.C.: Duke University Press, 1996), 210.

25 Richard J. Tobin, "Revising the Clean Air Act," in Norman J. Vig and Michael E. Kraft, eds., *Environmental Policy in the 1980s: Reagan's New Agenda* (Washington, D.C.: CQ Press, 1984), 244–45.

26 Joseph F. Schuler, Jr., "Climate Change at the Stack: Posturing toward Kyoto," *Public Utilities Fortnightly* 135 (August 1997): 20–23.

27 International Joint Commission (IJC), *1980 Annual Report* (Washington, D.C.: IJC, 1980), 5–6. See also Lynton K. Caldwell, ed., *Perspectives on Ecosystem Management for the Great Lakes* (Albany: State University of New York Press, 1988).

28 Ibid., 5–7.

29 See Lynton K. Caldwell, "Emerging Boundary Environmental Challenges and Institutional Issues: Canada and the United States," *Natural Resources Journal* 33 (Winter 1993): 10–31.

30 Personal communication, June 21, 1982.

31 Emmanuel Somers, "Transboundary Pollution and Environmental Health," *Environment* 29, no. 5 (1987): 8.

32 IJC, *An Inventory of Chemical Substances Identified in the Great Lakes Ecosystem, Report to the Great Lakes Water Quality Board, International Joint Commission* (Windsor, Ontario: IJC, 1983).

33 Charles Caccia, "Acid Rain: The Canadian Perspective," *School of Natural Resources News* 30 (1984): 5–7.

34 Phillip Shabecoff, "Reagan to Back Study Calling for Acid Rain Curb," *New York Times,* large print edition (March 17, 1986), 9–10.

35 LaBastille, "Acid Rain," 680.

36 CEQ, *Environmental Quality, 1981* (Washington, D.C.: GPO, 1981), 193.

37 Lynton K. Caldwell, "The World Environment," in Vig and Kraft, eds., *Environmental Policy in the 1980s,* 330.

38 "Canadian Ads Assail Acid Rain," Associated Press (May 21, 1986).

39 Leslie R. Alm, "Scientists and the Acid Rain Policy in Canada and the United States," *Science, Technology and Human Values* 22 (summer 1997): 349–69.

40 Klaus Berk Mullier, *Environmental Education about Rain Forests* (Mt. Pleasant: Central University of Michigan, Department of Wildlife Management, 1984), 1–10.

41 World Wildlife Fund, *Tropical Forest Campaign* (Gland, Switzerland: World Conservation Centre, 1982), 1–90.

42 Ibid.

43 Forest Resources Assessment 1990 Project, Food and Agriculture Organization of the United Nations (FAO), "Second Interim Report on the State of Tropical Forests," paper presented at the 10th World Forestry Congress, Paris, France, September 1991.

44 CEQ, *Environmental Quality, 1985,* 381.

45 World Wildlife Fund, *Tropical Forest Campaign* (Gland, Switzerland: World Conservation Centre, 1982), 1–90.

46 Geraldo Badowski, lecture presented at the Workshop on Sustainable Resource Systems, Turrialba, Costa Rica, July 1–5, 1985.

47 CEQ and Department of State, *Global Future,* 66–67.

48 Thomas A. Lewis, "Will Species Die Out As the Earth Heats Up?" *International Wildlife* 17, no. 6 (1987): 18–20.

49 Jose A. Lutzenberger, "Who Is Destroying the Amazon Rain Forest?" *Agapan* (December 1983): 8.

50 CEQ and Department of State, *Global Future,* 66.

51 CEQ and Department of State, *Global 2000: Report to the President* (Washington, D.C.: GPO, 1980), 25–26.

52 Ans Kolk, *Forests in International Environmental Politics: International Organizations, NGOs, and the Brazilian Amazon* (Utrecht: International Books, 1996), 72.

53 Ronald B. Nigh and James D. Nations, "Tropical Forests," *Bulletin of the Atomic Scientists* (March 1980): 10–13.

54 Nicholas Guppy, "Tropical Forest Deforestation: A Global View," *Foreign Affairs* (spring 1984): 955.

55 "The Dwindling Forests of Thailand," *Thailand Business* (August 1981): 25.

56 IUCN, *World Conservation Strategy: Living Resource Conservation for Sustainable Development* (Gland, Switzerland: International Union for the Conservation of Nature, 1980), sec. 11.

57 Guppy, "Tropical Forest Deforestation," 955.

58 Office of Technology Assessment, *Technologies to Sustain Tropical Forest Resources* (Washington, D.C.: U.S. Congress, Office of Technology Assessment, 1984), 12.

59 CEQ and Department of State, *Global Future,* 69–70.

60 Natural Resources Defense Council, *International Activities Report: November 1979–January 1981* (Washington, D.C.: Natural Resources Defense Council, 1981), 3.

61 Peter M. Haas, Marc A. Levy, and Edward A. Parson, "Appraising the Earth Summit: How Should We Judge UNCED's Success," *Environment* 34 (October 1992), 14.

62 Lynton K. Caldwell, "Garrison Diversion: The Constraints on Conflict Resolution," paper presented at the Northwest Political Science Association annual meeting, Portland, Or. October 21–26, 1983, pp. 23–24.

63 Prince Philip Mountbatten, "President's Message," *World Wildlife Fund Twentieth Anniversary Review* (Gland, Switzerland: WWF International, 1980), 2.

64 Howard T. Odum, *Environment, Power, and Society* (New York: Wiley Interscience, 1971), 11.

65 Lynton K. Caldwell, "A World Policy for the Environment," *UNESCO Courier* (January 1973): 4–5.

66 Ibid.

67 Caldwell, *International Environmental Policy,* 3rd ed., 2.

68 Harold Sprout and Margaret Sprout, *Toward a Politics of Planet Earth* (New York: Van Nostrand Reinhold, 1971), 14.

69 Dennis Pirages, *Global Ecopolitics: The New Context for International Relations* (North Scituate, Mass.: Duxbury, 1978), 31.

70 Dahlberg et al., *Environment and the Global Arena,* xiv–xvii.

71 Pirages, *Global Ecopolitics,* 5.

72 Ibid., 5–10, and Jeremy Rifkin, *Entropy: A New World View* (New York: Viking, 1980).

73 Caldwell, *International Environmental Policy,* 3rd ed., 32–62.

74 Pirages, *Global Ecopolitics,* 5–6.

75 Ibid., 5.

76 Dahlberg et al., *Environment and the Global Arena,* 68–85.

77 Mostafa Tolba, "UNEP Looks at a Decade of Cancer," *IUCN Bulletin* (April–June 1982): 29.

78 Thomas N. Gladwin, "Environment, Development, and Multinational Enterprise," in Charles S. Pearson, ed., *Multinational Corporations, Environment, and the Third World: Business Matters* (Durham, N.C.: Duke University Press, 1987), 8.

79 Caldwell, "A World Policy for the Environment," 32.

80 Lynton K. Caldwell, "Is the World Environment Threatened?" *Indiana Alumni* (May/June 1982): 17.

81 Norman Myers, "What Happened to Utopia?" *International Wildlife* 17, no. 4 (1987): 36–37.

82 Thomas W. Wilson Jr., "International Environmental Management: Some Preliminary Thoughts," in Albert E. Utton and Daniel H. Henning, eds., *Environmental Policy: Concepts and International Implications* (New York: Praeger, 1973), 104.

83 David A. Kay and Harold K. Jacobson, *Environmental Protection: The International Dimension* (Totowa, N.J.: Allenheld, Osmun, 1983), 329–30.

84 Wilson, "International Environmental Management," 104–5.

85 IUCN, *World Conservation Strategy,* sec. 19.

86 Caldwell, "Is the World Environment Threatened?" 19.

87 Caldwell, *International Environmental Policy,* 87.

88 Maurice Strong, "Strong Views on the Environment," *IUCN Bulletin* (April–June 1982): 27.

89 IUCN, UNEP, and WWF, *Caring for the Earth: A Strategy for Sustainable Living* (Gland, Switzerland: IUCN, 1991), 3.

90 Lynton K. Caldwell, "Organizational and Administrative Aspects of Environmental Problems at the Local, National, and International Levels," *Department of Economic and Social Affairs, Organization, and Administration of Environmental Programs* (New York: United Nations, 1974), 16–17.

91 Kay and Jacobson, *Environmental Protection,* 327–28.

92 Lynton K. Caldwell, *In Defense of Earth* (Bloomington: Indiana University Press, 1972), 110.

93 Caldwell, "Is the World Environment Threatened?" 20.

94 Ibid.

95 Editorial, *Bulletin of the Atomic Scientists* (August/September 1986): 2. "Chernobyl: the Emerging Story," discusses many aspects of Chernobyl as an international political and environmental problem.

96 Erik P. Hoffmann, "Nuclear Deception: Soviet Information Policy," *Bulletin of the Atomic Scientists* (August/September 1986): 36.

97 Ibid., 36.

98 Jim Harding, "An Earthshaking Analysis That Shows Why Chernobyl Sent the Nuclear Industry's Best Theories on Reactor Safety Up in Steam," *Earth Island Journal* (fall 1986): 22.

99 Ibid., 25.

100   Caldwell, "Is the World Environment Threatened?" 20.

101   Pirages, *Global Ecopolitics,* 16–40.

102   Ibid., 260–63.

103   Ibid.

104   Strong, "Strong Views on the Environment," 36.

105   Robert C. Repetto, ed., *The Global Possible* (New Haven, Conn.: Yale University Press, 1985), 12.

106   Julian Simon and Herman Kahn, eds., *The Resourceful Earth* (New York: Basil Blackwell, 1984), 44–47.

107   Center for International Environmental Information, *Environmental Training in Developing Countries* (New York: Center for International Education Information, 1981), 1.

108   Ibid., 5.

109   Brent Blackwelder, "The Role of the U.S. in Shaping the Global Environment," *Environmental Professional* 9 (1987): 102–4.

110   U.S. Agency for International Development, *Environmental and Natural Resource Management in Developing Countries* (Washington, D.C.: Department of State, 1979), vol. 1, report 1.

111   Caldwell, "Is the World Environment Threatened?" 20.

112   Lynton K. Caldwell, *International Environmental Policy,* 3rd. ed., 132.

113   Lawrence E. Susskind, "What Will It Take to Ensure Effective Global Environmental Management? A Reassessment of Regime-Building Accomplishments," in Bertram I. Spector, Gunnar Sjostedt, and I. William Zartman, eds., *Negotiating International Regimes: Lessons Learned from the United Nations Conference on Environment and Development (UNCED)* (London: Graham and Trotwood, 1994), 221–232.

114   UNEP, *The State of the Environment 1984: The Environment in the Dialogue between and among Developed and Developing Countries* (Nairobi, Kenya: UNEP, 1984), 36–37.

115   Michael McCally, "Five Years down the Road from Rio," *British Medical Journal* 315 (July 5, 1997): 3–5; and Richard Scrader, "The Fiasco at Rio," *Dissent* 39 (Fall 1992): 431–32.

116   Thomas J. Schoenbaum and Ronald H. Rosenberg, *Environmental Policy Law,* 3rd edition (Westbury, N.Y.: Foundation Press, 1996), 12–13.

117   "Programme for the Further Implementation of Agenda 21," *Environmental Policy and Law* 27 (September 1997): 423–29.

118   CEQ and Department of State, *Global 2000,* 1–5.

119   Dennis J. O'Donnell, "Resource Scarcity and Population Growth: International Planning and Environmental Imperatives," in Richard N. Barrett, ed., *International Dimensions of the Environmental Crisis* (Boulder, Colo.: Westview, 1981), 217.

120   Lester Brown, "Facing Food Scarcity," *Earth Island Journal* 12 (spring 1997): 38.

121   Sandra Postel, "Running Dry," *Unesco Courier* 5 (May 1993): 19.

122   Peter H. Gleick, "Water in Crisis: Paths to Sustainable Water Use," *Ecological Applications* 8 (August 1998): 571.

123   Rifkin, *Entropy,* 2–3.

124 Dennis Meadows and Donnella Meadows, lecture presented at the Workshop on Systems Dynamics of Sustainable Resource Attainment, Terrialba, Costa Rica, July 1–5, 1985, p. 1.

125 Donella H. Meadows, comments and discussion at the UNESCO Eden International Seminar on Education and Environment in Budapest, Hungary, November 10–14, 1980. See also Donella H. Meadows et al., *The Limits to Growth* (New York: Universe Books for Potomac Associates, 1972).

126 UNEP, "Economic Instruments for Environmental Management and Sustainable Development," Environmental Economics Series Paper no. 16 (Nairobi, Kenya: UNEP Environment and Economics Unit, 1995).

127 Caldwell, *International Environmental Policy,* 2nd ed. (1984), 280.

128 "A Strategy Is Launched," *IUCN Bulletin* (March 1980): 33–37.

129 Gilberto C. Gallopin and Paul Raskin, "Windows on the Future," *Environment* 40 (April 1998): 7–19.

130 Becky J. Brown, Mark E. Hanson, Diana M. Liverman, and Robert W. Meredith, "Global Sustainability: Toward Definition," *Environmental Management* 11, no. 6 (1987): 717.

Glossary

This glossary is based on professional environmental and natural resource literature, dictionaries, and glossaries. Special effort was made to incorporate concepts and terms from various fields and disciplines, especially ecology, associated with environmental affairs. The *World Conservation Strategy* was also consulted as a source for terminology and concepts. Many terms have been incorporated into the text and defined there rather than in the glossary. The glossary itself provides clear and concise definitions of selected environmental terms.

Acculturation. The processes and results of external change imposed on a human population with loss, or degrees of loss, of traditional social and cultural values and institutions.

Aesthetics. Considerations, values, and judgments pertaining to the quality of the human perceptual experience (including sight, sound, smell, touch, taste, and movement) evoked by phenomena or components of the environment.

Air, ambient. Surrounding outdoor air. That portion of the atmosphere, external to building, to which the general public has access.

Air quality standards. The level of air pollution prescribed by law or regulation that cannot be exceeded during a specified time in a defined area. Ambient air quality standards refer to the maximum allowable levels of specific polluting materials permitted under the law.

Ambient. The natural conditions (or environment) in a given place or time.

Ambient standard. Maximum allowable levels of specific polluting materials permitted by state, federal, or local laws.

Anthropocentric. A view conceiving of everything in the environment/universe in terms of human values, ends, or aims without recognition or considerations of other forms (plant or animal) of life or responsibilities thereof. Interpreting reality (environment) exclusively in terms of human values, interests, and experiences.

Aquatic life. Growing or living in or frequenting water (plant or animals), not terrestrial.

Aquifer. An underground, water-bearing bed or stratum of earth, gravel, or porous stone in which water may move for long distances. It allows groundwater to move to springs and wells.

Biomass. The total mass or amount of living organisms in a particular area or volume.

Biome. A complex community of all living organisms characterized by a distinctive type of vegetation, including the successional stages of the area. E.g., tundra biome, grassland biome, desert biome.

Biosphere reserves. Protected land, water, and/or coastal environments that, together, constitute a worldwide network of scientific information and include significant examples of natural biomass and/or unique, representative biological areas throughout the world.

Biota. All living organisms, both plant and animal, that exist within a given area or period.

Buffer zone. A designated land or water area along the edge of some land (often nature or other reserves) use, whose own use is regulated so as to absorb, or otherwise preclude, unwanted development or other intrusions into areas beyond the buffer.

Carcinogen. Any chemical substance or form of energy (e.g., radiation) capable of producing cancer.

Carnivore. An organism that acquires life-sustaining nutrients by utilizing animals as food.

Carrying capacity. The maximum number of a wildlife species that can be supported indefinitely by a given ecosystem or area without deterioration. The limit as to the number of any one species that can be maintained in a particular environment during the most critical period (dry, winter, etc.) of the year.

CITES. Convention on International Trade in Endangered Species of wild plants and animals. Membership composed of various nations that have a regulatory network to control trade of endangered species on a worldwide basis.

Community. All the plants and animals in a particular habitat that are bound together by food chains and other interactions that are self-perpetuating.

Community, climax. A relatively stable, biotic community that appears to perpetuate itself in the absence of disturbance. The final, culminating stage of ecological succession for a given environment.

Conflict resolution. Process of resolving or reducing conflicts through negotiation, change, and/or mediation with two or more conflicting sides/interests being willing to accept the results.

Conservation. Management of the biosphere so that it may yield the greatest sustainable benefits to present generations while maintaining its potential to meet the needs and aspirations of future generations.

Conservation, living resources. Processes to (a) maintain essential ecological processes and life-support systems; (b) preserve genetic diversity (range of genetic material found in world's species); and (c) to ensure the sustainable utilization of species and ecosystems.

Cost-Benefit analysis. An analytical technique approach to solving problems of choice and worth (including development proposals) that identifies for each objective the alternative that yields the greatest benefit for a given cost, or the alternative that produces the required level of benefits at the lowest cost. It is recognized that certain benefits and costs (e.g., quality, intangibles, scenery, wildlife, etc.) cannot be quantified in monetary terms.

Culture. The complex whole of knowledge, achievements, technology, traditions, perceptions, customs, values, habits, and other capabilities of society and human inherited traditions and patterns. Culture influences societal/individual behavior and its environmental relationships.

Desertification. The gradual destruction or reduction of the capacity of drylands (characterized by low rainfall with high evaporations) for plant and animal production due to the inherent vulnerability of the land and to the pressure of human activities, e.g., overgrazing, deforestation, poor soil management, etc.

Development. The modification of the biosphere through the application of human, financial, living, and nonliving resources to satisfy human needs and to improve the quality of life.

Development, sustainable. Integrates development and conservation of living resources. Sustainable development comprehensively takes into consideration social, ecological, and economic factors, the living and nonliving resources base as well as short- and long-term advantages and disadvantages of alternative actions.

Diversity. The biological complexity of numbers of species of organisms of an ecosystem. In many instances, the ecosystem becomes more stable as diversity increases.

Ecological impact. The total effect on the ecology of an area caused by an environmental change, either natural or human.

Ecological niche. The role, status, and position of a species in the environment, its activities and relationships to the biotic and abiotic environment. Also refers to specific places where individual organisms can live (special niche).

Ecological succession. The gradual and progressive sequence of communities and organisms that replace each other in a given place. The changes, over time, in the structure and function of an ecosystem with the replacement of one kind of community or organisms with a different one. Primary succession occurs on sites that supported vegetation previously.

Ecology. The branch of biological science that studies the relationships of living organisms with each other and with their environment.

Economics. The study of how humans allocate scarce, productive resources in the production of different commodities over time and how these commodities are distributed for consumption across time periods and among members of a society.

Ecosystem. A complex system composed of a community of fauna and flora, taking into account the chemical and physical environment with which the system is interrelated.

Ecotone. A boundary and/or transition area (zone) between two or more communities that has some of the characteristics of each.

Effluent. Waste material, including liquid or gas, discharged into the environment, treated or untreated. Generally refers to water pollution.

Emission. A discharge of particulate gaseous or soluble waste material/pollution into the air from a polluting source.

Energy. The capacity to do work or transfer heat. Energy may take a number of forms, among them mechanical, chemical, and radiant, and can be transferred from one form to another.

Energy flow. The one-way passage (transfer) of energy through an ecosystem, including the way in which energy is converted and used at each trophic (food) level of an ecosystem.

Entropy. A measure of the degree of disorder within a system.

Environment. The aggregate of surrounding things (biotic and abiotic) and conditions that influence the life of an individual organism or population, including humans. The sum of all external things (living and nonliving), conditions, and influences that affect the development and, ultimately, the survival of an organism.

Environmental awareness. The growth and development of awareness, understanding, and consciousness toward the biophysical environment and its problems, including inter-actions and effects. Thinking "ecologically" or in terms of an ecological consciousness.

Environmental degradation. Any action that makes the environment less fit for human, plant, or animal life. Also associated with the lowering and reduction of environmental quality.

Environmental ethic. An ecological conscience or moral that reflects a commitment and responsibility toward the environment, including plants and animals as well as present and future generations of people. Oriented toward human societies living in harmony with the natural world on which they depend for survival and well-being.

Environmental impact assessment. An activity designed to identify, predict, interpret, and communicate information about the effects of an action and to ensure ecological and sociological information is included with physical and economic information as the basis for making decisions. An evaluation activity that involves the consideration of the interaction of physical, natural, social, and economic factors and a determination of probable effects and consequences of the plan or proposal upon these operating systems before the proposal or actions are undertaken. An evaluation and objective prediction of the environmental impacts of a proposed action, using a systematic, interdisciplinary approach. Preliminary assessment for determining the need for an environmental impact statement (EIS) relative to significant impacts.

Environmental impact statement. A document (report) prepared on the possible negative and positive effects and influences (impacts) of a proposed project and/or development that would have significant impacts on the environment and society. An EIS provides information to decisionmakers and the public on the proposed undertaking and list alternatives to the proposed action, including taking no action.

Environmental indicators. Characteristics and factors for determining present and future conditions of the environment.

Environmental law. Rules and controls of conduct in environmental/natural resources affairs that are prescribed by or accepted by the governing authority of a state and enforced by the government and the courts. Environmental law may derive from a constitution, legislative acts, administrative rules, or common law. The body of law regulating or incidentally affecting the relationships between humans and their environment.

Environmental management. Various international, state, and local measures and controls that are directed at environmental conservation, the rational and sustainable allocation and utilization of natural resources, the optimization of interrelations between society and the environment, and the improvement of human welfare for present and future generations.

Environmental monitoring. Periodic and/or continued measuring, evaluating, and determining environmental parameters and/or pollution levels in order to prevent negative and damaging effects to the environment. Also includes the forecasting of possible changes in ecosystems and/or the biosphere as a whole.

Environmental perception. Consciousness, understanding, and awareness of elements, inter-relationships, and problems of the environment through sensory knowledge and judgment.

Environmental protection. Measures and controls to prevent damage and degradation of the environment, including the sustainability of its living resources. To protect the environment from negative or destructive effects, influences, and consequences.

Environmental quality standards. Normative documents and guidelines for determining the degree of environmental conditions and requirements to avoid negative and damaging effects, influences, and consequences. The maximum amount of a pollutant that can be discharged from a given source.

Environmental science. The interdisciplinary study (programs and courses) of environmental problems within the framework of established physical and biological principles, i.e., oriented toward a scientific approach.

Environmental studies. University and college (postsecondary) programs and courses that deal with environmental problems/affairs on a broad, interdisciplinary basis/approach. Utilizes disciplines from the natural sciences, social sciences, applied sciences/technologies, and humanities, i.e., oriented toward a broad-based approach to environmental problems.

Erosion. The wearing away of the land surface/soil by water or wind. Erosion occurs naturally from weather or runoff, but is often intensified by land clearing or disruption.

Evolution. The biological theory or process whereby species of plants and animals change with the passage of time so that their descendants differ from their ancestors, i.e., development from earlier forms by hereditary transmissions of slight variations in successive generations.

Exotics. Plants, animals, or microorganisms that are introduced by humans into areas where they are not native. Exotics are often associated with negative ecological consequences for native species and their ecosystems.

Extinction. The process by which a species ceases to exist.

Fauna. All animal life associated with a given habitat, area, country, or period.

Floodplain. A lowland fringing a watercourse. It serves a valuable and natural function by containing large volumes of water in times of flood. Development on floodplains, therefore, is considered unwise.

Flora. All plant life associated with a given habitat, area, country, or period. Bacteria are considered flora.

Food chain. A sequence of transfers of food energy from organisms in one trophic (food) level to those in another.

Food web. The complex and interlocking series of food chains. A given organism may obtain nourishment from many types of organisms in a food web. The biomass and energy flow of the food web are in pyramid form (from a wide bottom to a narrow top).

Forest. Generally, an ecosystem characterized by a more or less dense and extensive tree cover. Specifically, a plant community composed mainly of trees and other woody vegetation that grow, more or less, closely together. Coniferous forests (evergreen) retain their leaves throughout the seasons while deciduous forests shed their leaves entirely at the end of the growing period (season).

Forester. A professional who has the responsibility for planning and executing activities that allow the full values of forest resources to be perpetually obtained for human benefit and that recognize the forest as a living biological community with interrelationships.

Forestry. Management of forestlands for the provision of the various goods and services that forests can continuously supply in such a manner that yields are sustainable and that the resource base (essential ecological processes and genetic diversity) is secured.

Global commons. Tracts of land (e.g., Antarctica) or water owned or used jointly by the members of the community of nations. Global commons include those parts of the earth's surface beyond national jurisdictions, including the atmosphere.

Greenbelts. Open spaces maintained between urbanized areas where construction of buildings and homes is prohibited or restricted. These areas contain plant life for aesthetic purposes and often serve as a buffer between pollution sources and population concentrations.

Habitat. The sum of environmental conditions in a specific place that is occupied by an organism, population, or community and where it naturally lives and grows.

Health. The natural/normal state of an organism, characterized by harmonious interrelationships with the environment and by the absence of adverse/harmful changes caused by diseases, pollution, and other negative factors. The ability to live and adapt to living conditions as well as for the environment to have a wholesome/well-being effect on an organism. Human health depends not only on biological factors, but also upon social, cultural, and psychological aspects and includes preventative measures to avoid future, negative factors.

Herbivore. An organism that acquires life-sustaining nutrients by feeding on vegetation.

Holistic approach. Thorough and comprehensive analysis of interrelations between the natural environment, social, cultural, technological, and other factors, i.e., that the environment can only be understood by viewing it as a general complex of its parts. The view/approach that an integrated whole has reality independent of and greater than its parts.

Human ecology. The study of the growth, distribution, and organization of human communities relative to their interrelationships with other humans and other species, and with their environment.

Indigenous. Refers to plant or animal species that are restricted to and characteristic of a certain area or location. A native species (one not introduced).

Integrated approach. Unified, combined, and coordinated approach toward environmental problems, which correlates relevant organizations, groups, individuals, and disciplines by bringing the "parts" together for a complete or whole approach. Also, to signify the various measures and processes used in integrating conservation (combinations and contributions) with development.

Integrationists. People who possess the knowledge, skills, understanding, and attitudes to act as links between the researchers/specialists who develop environmental information and the policymakers/generalists who make decisions on environmental matters. Utilizing a holistic approach, integrationists can play a vital role in environmental education and training.

International environmental law. The most important form of international environmental action is in the development of international environmental law. Strong international conventions of agreements provide a legally binding means of ensuring the conservation of living resources that cannot be conserved by national legislation (e.g., those living resources that are shared, occur—temporarily or permanently—in areas beyond national jurisdiction, or are affected by activities carried out in another state).

Inversion. An atmospheric condition where a layer of cool air is trapped by an upper layer of warm air, precluding the release of the bottom air. Inversions spread polluted air horizontally rather than vertically so that contaminating substances cannot be widely dispersed. An inversion of several days can cause an air pollution "episode" with serious problems.

Land capability. Suitability and feasibility of an area of land for use(s) on a sustained basis. Possibilities of degradation and depletion should be taken into account when assessing land capability (suitability).

Landscape planning. The aspect of the land use planning process that deals with physical, biological, aesthetic, cultural, and historical values and with the relationships and planning between these values, land uses, and the environment.

Limiting factor. A condition or factor whose absence, short supply, or excessive concentration exerts some restraining or negative influence upon a population that is incompatible with a given species' requirements or tolerance.

Natural area. A physical and biological unit, in as near a natural condition as possible, that exemplifies typical or unique vegetation and associated biological, geological, and/or aquatic features.

Natural disaster. Violent and sudden change in the environment due to destructive, natural phenomena, e.g., floods, earthquakes, fire, hurricanes, etc.

Natural resources. A feature or component of the natural environment that is of value in serving human needs, e.g., soil, water, plant life, wildlife, etc. Some natural resources have an economic value (e.g., timber) while others have a "noneconomic" value (e.g., scenic beauty).

Natural resources management. The integrated and harmonious management of natural resources through utilization, protection, manipulation (change), and conflict reduction measures and activities. The management of human use of natural resources on a sustained use basis for present and future generations of human, animal, and plant life.

Natural resources, nonrenewable. Resources not capable of perpetuating themselves, e.g., coal, oil, and other minerals. A nonliving resource of finite supply that cannot be replaced.

Omnivore. A biological system that sustains itself by feeding on both animal and vegetable tissue.

Open space. A relatively undeveloped green or wooded area usually provided within an urban development to minimize feelings of congested living.

Outdoor recreation. Leisure time activities that utilize an outdoor area or facility. A self-rewarding experience, occurring in outdoor settings during unobligated time, that results from a free personal choice and commitment by the individual.

Park. Any public area of land set aside for the aesthetic, educational, recreational, or cultural use of the urban resident or visitor.

Pathogen. Any virus, microorganism, or other substance capable of causing disease.

Pesticide. Any chemical substance used to kill plant, animal, and insect pests. Some pesticides can contaminate water, air, or soil and can bioaccumulate in humans, plants, animals, and the environment with negative effects.

Plan, comprehensive. A plan that indicates the principal acts by which all of the most important ends are to be attained in a comprehensive manner. A unified design for an area or community that relates developmental goals to social, economic, and environmental goals. Includes environmental assessment and compatible use allocations.

Planning, anticipatory. A planning process that attempts to foresee potential problems and to develop solutions to them before they become real, current problems.

Planning, environmental. Concerned with the consequences of human activities on the environment in terms of forecasting, anticipating, evaluating, and reconciling the demands for and impacts upon the environmental resources/amenities and ecology with reference to the present and future values and options at stake.

Policy, anticipatory. Policies that attempt to anticipate significant economic, social, and ecological events rather than simply react to them. Involves actions to ensure that conservation and other environmental requirements are taken fully into account at the earliest possible stage of any major decision likely to affect the environment. A policy/planning process that attempts to foresee potential problems and to develop solutions to them before they become real and current problems.

Policy, environmental. Official statements of principles, intentions, values, and objectives that are based on legislation and the governing authority of a state and that serve as a guide for the operations of governmental and private activities in environmental affairs.

Pollutant. Any extraneous material or form of energy whose rate of transfer between two components/factors of the environment is changed so that the well-being of organisms/ecosystems is negatively affected. Any introduced gas, liquid, or solid that makes a resource unfit for a specific purpose or that adversely affects human, plant, or animal life.

Pollutant, point source. A stationary location where a pollutant is discharged, usually from industry. A located source of a pollutant discharge. A nonpoint pollutant would imply an unlocated source for that given pollutant.

Pollution. The presence of matter or energy whose nature, location, or quantity produces undesirable environmental effects. The contamination or alteration of the quality of some portion or aspect of the environment and its living organisms by the addition of harmful impurities.

Pollution, thermal. Discharge of heated water from industrial activities that can affect the life processes of aquatic plants and animals. Even small deviations from normal water temperatures can affect aquatic life. Thermal pollution usually can be controlled by cooling towers.

Pollution control. Systems of measurement, criteria, standards, laws, and regulations that are directed at the sources and causes of various forms of pollution and its effects in terms of control and prevention. Control measures involve both quantity (degree) and quality (value) considerations.

Population control. All methods utilized for conception/birth control in order to control population growth, including natural or deliberate changes in economic, political, and social conditions. All factors that regulate the size of a population.

Population density. Number of organisms in a particular population in a given area at a given time.

Predation. An interaction in which one organism (predator) kills and eats another organism (prey).

Public health. Social, medical, and environmental measures and controls undertaken by the state in order to prevent diseases and improve the health of the general public, including sanitary hygiene areas concerned with planning and implementing a healthy environment for the public. Public health involves various preventative measures, e.g., ensuring sanitary water supplies to prevent diseases from water that might become a source of disease-causing organisms (pathogens).

Public inputs. Various suggestions, information, questions, views, opinions, values, and critiques that are expressed by individual, group, or collective members of the general public in efforts to influence a given government policy, plan, and/or decision in development and/or environmental affairs. Public inputs may be made through both formal and informal public participation processes and may be solicited and unsolicited.

Public interest. An abstract and symbolic concept that refers to the ends, values, benefits, or costs for the general or common interests of all the public. The public interest is often subject to various justifications and interpretations but implies the overall interest of the general public of the whole society over short- and long-term considerations, as contrasted to the private interests of given individuals, groups, and organizations that make up part of society over short-term/immediate considerations.

Public participation. The involvement, informing, and consultation of the public in planning, decisionmaking, and management activities in environmental affairs. The public actively sharing in the decisions that government makes in environmental affairs by having individual and group views taken into account through various participation measures that involve the public. Public participation requires adequate encouragement and opportunities.

Quality control. The control and protection of the natural environment under quality consideration. A system of measurements and controls directed at the protection and enhancement of environmental quality factors and considerations.

Quality of life. A subjective concept that characterizes the measure of the degree to which a given society offers effective opportunity to a combination of physical, social, and cultural components in the total environment. A broad and all-encompassing concept that refers to the quality characteristics of all aspects of one's environment and life.

Recycling. Converting wastes (normally solid) into new products by using the resources contained in the discarded materials to produce new materials.

Reserves. Natural or near-natural areas of land and/or genetic resources (threatened or endangered species) of interest, including representative or unique ecological communities. Economic and human activity is usually controlled, compatible, or prohibited in terms of the natural state of the reserve and its category, e.g., strict nature reserve, managed nature reserve, wildlife, sanctuary, etc.

Sanctuary. An area, usually in natural condition, that is reserved (set aside) by a governmental or private agency for the protection of particular species of animals during part or all of the year.

Sanitation. Control of physical factors in the human environment that can harm the health or survival of human and plant/animal life, e.g., drainage and disposal of sewage, or threats to pure water supply.

Sediment. Soil particles (sand, silt, clay, and minerals) washed from land into water systems, including reservoirs, as consequences of human and natural activities.

Shared resources. Ecosystems and species shared by two or more states, including species that move between one national jurisdiction and another. Ecosystems and species that depend on or are affected by events in another state, e.g., international river basins, fisheries, migratory species, etc.

Smog. The irritating, visible haze resulting from the sun's effect on certain pollutants in the air (photochemical), and particularly those pollutants from transportation (exhausts) and industry. Also, a mixture of fog, smoke, and gaseous waste.

Social costs. The net loss considerations of long-range societal values and conditions as the result of a given human action and/or development. Social costs can include both economic and noneconomic costs that are passed on to groups of individuals, communities, or society as a whole, e.g., air pollution, loss of health, disruption of lifestyles, etc.

Social indicator. A measure of human welfare in terms of the opportunity or accommodation for a public or private good or service.

Socioeconomic. Involving and combining both social and economic factors, indicators, and considerations. Implies measures of human welfare through the correlation of economic and social aspects.

Solar energy. Power (energy) collected from sunlight. Alternate energy source used most often for heating purposes but occasionally to generate electricity.

Spaceship earth. A concept/philosophy for understanding the earth as a spaceship with a limited life-supporting system or as a finite, complex ecosystem in which survival requires wise management of limited resources and harmonious human and environmental relationships.

Species. Natural population or group of populations of plants or animals that transmit specific characteristics from parent to offspring. They are reproductively isolated from other populations (species).

Species diversity. The number of different species occurring in a given location or under some condition. The ratio between the number of species in a biotic community and number of individuals in a given species. Diversity is generally correlated with ecological stability.

Standard. A measure for relating an allocation of resources to existing or potential needs as determined by stated goals, objectives, and policies. The maximum allowable levels of specific polluting materials that are permitted.

Stewardship. The wise use and management of the environment and its resources in terms of the recognition of living relationships and responsibilities for the environment and for future generations of all forms of life. Stewardship implies that humankind respect, oversee, and conserve the environment for present and future considerations for all life through individual and collective efforts and responsibilities.

Stress. A state of particular strain or tension for an organism under the influence of unfavorable factors, impacts, or situations of the environment. Involves protective physiological reactions.

Subspecies. A division of a species based most often on geographical distribution and/or taxonomic characteristics. Subspecies have interbreeding potential.

Sustainable utilization. To keep up, maintain, and perpetuate living resources through wise utilization and management. Ensures that various (present and potential) uses, modifications, and developments of the environment and its living resources can continue for present and future generations. Safeguards the ecological processes and genetic diversity essential for the maintenance of the living resources concerned.

Symbiosis. An association of two or more organisms of different species in which one or more may benefit and none are harmed.

System analysis. Study and analysis of complex problems in terms of the relations between the subsystems that form its component parts with the objective of predicting the behavior of the whole from that of its parts. The application of scientific and social science disciplines to a system to determine its relative worth and/or the relationships between elements or components of a system.

Technology assessment. Methods and processes of projecting, evaluating, and predicting possible short- and long-range physical, social, economic, and ecological benefits and costs—including negative side effects and impacts—of introducing a particular technology in general or for a given area of the environment.

Terrestrial. Of the land, not the water.

Territory. An area over which an animal or group of animals establishes jurisdiction. Activity associated with an organism claiming an area and defending it against members of its own (or similar) species. Area within the home range of an organism that is actively defended against other organisms.

Topography. The relief of an area of land, e.g., mountainous, flat, hills, meadows, swamp, etc. The physical shape of the ground surface.

Tolerance. The relative capacity of an organism(s) to endure an unfavorable environmental factor or change. The safe level/amount of a chemical on any food consumed by an organism.

Toxicity. Refers to the degree of danger posed by a toxic substance to human, animal, and/or plant life.

Toxic substance. Substances that are dangerous and harmful to human, plant, and/or animal life. A chemical or mixture that may present an unreasonable risk of injury to health or to the environment.

Trade-offs. Occur when incremental increase or gains in one desirable situation or output necessitate incremental decreases or losses in another equally desirable situation or output, i.e., giving up (or trading) something to gain something else that is more preferred.

Tropical forests. Forest communities that are maintained by the rainy/moist climates of tropical regions. Contain the greatest abundance and diversity of plant and animal species and are identified as a conservation priority by the World Conservation Strategy due to their rapid rate of exploitation and disappearance.

Tundra. High altitude, arctic latitude, or mountainous areas/ecosystems that are too cold to support trees and are above the treeline. They are characterized by meadows, low-growing and fragile plant life—including flowering plants, shrubs, grasses, and mosses—during the summer season and by a permanently frozen subsoil in the case of arctic tundra (permafrost).

Ubiquitous. Refers to a plant or animal species that is capable of thriving under varying environmental conditions.

Waste, nonbiodegradable. Inorganic or mineral waste materials, (industrial and domestic) that cannot be broken down (decomposed) into their basic elements by the action of microorganisms and, consequently, remain in the environment for a long or indefinite period of time, e.g., synthetic materials (plastics), metal, etc.

Water conservation. Various measures and methods to prevent the exhaustion, pollution, and destruction of fresh water sources and their aquatic life and features. The protection, improvement, and utilization (wise use) of water to ensure its highest social/economic/environmental benefits for present and future generations.

Water quality standards. Characteristics and degrees/levels of water quality for comparison in terms of different sources and uses. A management plan that considers (a) water uses; (b) setting water quality criteria levels to protect those uses; (c) implementing and enforcing the water treatment plans; (d) protecting existing high-quality waters; and (e) establishing regulations designating standards for bodies of water.

Watershed. An area of land from which all precipitation drains to a specific water course or outlet. The boundary line of a watershed is the natural ridge that divides one drainage area from another. The area drained by a stream.

Wetland. An area that is regularly wet or flooded, and where the water table (the upper level of the groundwater) stands at or above the land surface for at least part of the year.

World Conservation Strategy. A global conservation document that provides strategies and principles for the integration of conservation and development to ensure that modifications to the earth do indeed secure the survival and well-being of all people. The strategies deal with working with nature and with conservation as the mainstream of human progress to protect and sustain the life-support systems of the planet.

Index

About the Authors

William R. Mangun is Professor of Political Science in the Department of Political Science at East Carolina University, Greenville, North Carolina. He received his doctoral degree in political science from Indiana University with an emphasis in environmental and urban policy and administration. Previously, he received degrees from the University of Oklahoma, Syracuse University, and the University of Akron in public administration, Russian, and history. Dr. Mangun has an extensive list of publications in the area of environmental policy and natural resource administration. He has produced several other environmental policy books, including *American Fish and Wildlife Policy: The Human Dimension, Public Policy Issues in Wildlife Management, The Public Administration of Environmental Policy: A Comparative Analysis of the United States and West Germany,* and *U.S. Fish and Wildlife Service Important Resource Document.* Dr. Mangun also edited the Wildlife Conservation and Public Policy Symposium issue of *Policy Studies Journal.* He has presented papers at professional meetings and authored or coauthored articles, book chapters, and monographs on such topics as air and water pollution control, hazardous waste management, wetlands, wildlife-associated recreation, financing wildlife management, outdoor recreation, human dimensions of wildlife, endangered species, ecosystem management, and the value of wildlife. His writings deal with American and European policies, drawing upon his experiences while living in Germany and Turkey. Professor Mangun has taught environmental policy and management courses at Indiana University, Mississippi State University, Salisbury State College, University of Akron, American University, George Washington University, George Mason University, and Indiana State University, in addition to East Carolina University. Dr. Mangun also has considerable administrative experience in natural resource management through his employment in the U.S. Department of the

Interior. He served as the National Important Resource Problem Coordinator, National Resource Management Coordinator, and Project Manager for Policy Analysis and National Surveys in the U.S. Fish and Wildlife Service in Washington, D.C. This experience gave Dr. Mangun the unique opportunity to become familiar with complex natural resource issues that confront environmental managers throughout the United States. For a brief period of time Dr. Mangun also worked in the Office of the Administrator of the U.S. Environmental Protection Agency as an Environment Protection Specialist in Washington, D.C. Previously he worked for the Open Lands Project in Chicago, Illinois, identifying alternatives for preservation of open space.

Daniel H. Henning is Professor Emeritus of Political Science and Environmental Affairs at Montana State University–Billings, Montana. He received a B.S. in social science and biology from Bowling Green State University, an M.S. in public administration and conservation from the University of Michigan, and a Ph.D. in public administration, political science, and forestry from Syracuse University. He has taught at the University of New Mexico and Chinese University of Hong Kong. Most recently, he served as Senior Fulbright Research Scholar, Southeast Asia Region, where he studied tropical forest administration and training in Thailand, Indonesia, and Malaysia, with emphasis on national parks and nature tourism. He has authored and coedited books and articles on environmental and natural resources affairs and has presented numerous papers at national and international conferences. Dr. Henning holds a number of positions in national and international organizations, including the International Union for Conservation of Nature and Natural Resources, International Council on Environmental Law, and American Society for Public Administration, and he has served as an environmental administration and training consultant for various agencies, including the United Nations. He has also served as a ranger with the National Park Service and USDA Forest Service. Dr. Henning has been the recipient of various fellowships and awards, including those from the National Academy of Sciences, Smithsonian Institution, National Science Foundation, and foreign governments, which provided research travel grants to pursue his environmental work in Germany, Norway, Czechoslovakia, India, Thailand, Malaysia, Indonesia, and other countries. He is the holder of the distinguished scholar professor award from Eastern Montana College and the Vice Chancellor's Faculty Award of Merit from the Chinese University of Hong Kong.

Library of Congress Cataloging-in-Publication Data

Managing the environmental crisis : incorporating competing values in natural
resource administration / William Mangun and Daniel H. Henning ; with a new
foreword by Lynton Keith Caldwell. — 2nd ed., rev. and updated. Rev. ed. of:
Managing the environmental crisis / Daniel H. Henning and William R. Mangun.
1989.

Includes index.

ISBN 0-8223-2379-6 (cloth : alk. paper). — ISBN 0-8223-2413-X (pbk. : alk. paper)
1. Environmental policy—Decision making.   2. Conservation of natural resources—
Government policy—Decision making.   I. Mangun, William Russell.   II. Henning,
Daniel H.   III. Henning, Daniel H. Managing the environmental crisis.
HC79.E5H46   1999   363.7—dc21   99-26267CIP